CAMBRIDGE LIBRARY COLLECTION

Books of enduring scholarly value

Life Sciences

Until the nineteenth century, the various subjects now known as the life sciences were regarded either as arcane studies which had little impact on ordinary daily life, or as a genteel hobby for the leisured classes. The increasing academic rigour and systematisation brought to the study of botany, zoology and other disciplines, and their adoption in university curricula, are reflected in the books reissued in this series.

Elements of the Philosophy of Plants

This treatise on scientific botany brings together the works of two leading European scientists from the beginning of the nineteenth century, the Swiss botanist Augustin Pyramus de Candolle (1778–1841) and the German botanist and physicist Kurt Polycarp Joachim Sprengel (1766–1833). First published in German in 1820, it was almost immediately translated (anonymously) into English and published in Edinburgh by Blackwood in 1821. This collaborative volume includes three chapters from de Candolle's *Théorie élémentaire de la botanique* published in Paris in 1819, while the remaining texts and the preface were written by Sprengel; at the time, it provided significant advances on previous botanical theories such as the work of German botanist Carl Ludwig Willdenow (1765–1812). A fascinating document on the evolution of botanical science, the book contains a practical section detailing the characteristics of over forty plants, as well as eight illustrations.

Cambridge University Press has long been a pioneer in the reissuing of out-of-print titles from its own backlist, producing digital reprints of books that are still sought after by scholars and students but could not be reprinted economically using traditional technology. The Cambridge Library Collection extends this activity to a wider range of books which are still of importance to researchers and professionals, either for the source material they contain, or as landmarks in the history of their academic discipline.

Drawing from the world-renowned collections in the Cambridge University Library, and guided by the advice of experts in each subject area, Cambridge University Press is using state-of-the-art scanning machines in its own Printing House to capture the content of each book selected for inclusion. The files are processed to give a consistently clear, crisp image, and the books finished to the high quality standard for which the Press is recognised around the world. The latest print-on-demand technology ensures that the books will remain available indefinitely, and that orders for single or multiple copies can quickly be supplied.

The Cambridge Library Collection will bring back to life books of enduring scholarly value (including out-of-copyright works originally issued by other publishers) across a wide range of disciplines in the humanities and social sciences and in science and technology.

Elements of the Philosophy of Plants

Containing the Principles of Scientific Botany

Augustin Pyramus de Candolle
Kurt Sprengel

CAMBRIDGE UNIVERSITY PRESS

Cambridge, New York, Melbourne, Madrid, Cape Town,
Singapore, São Paolo, Delhi, Tokyo, Mexico City

Published in the United States of America by Cambridge University Press, New York

www.cambridge.org
Information on this title: www.cambridge.org/9781108037464

© in this compilation Cambridge University Press 2011

This edition first published 1821
This digitally printed version 2011

ISBN 978-1-108-03746-4 Paperback

ELEMENTS

OF THE

PHILOSOPHY OF PLANTS:

CONTAINING

THE PRINCIPLES OF SCIENTIFIC BOTANY;

NOMENCLATURE, THEORY OF CLASSIFICATION, PHYTOGRAPHY;
ANATOMY, CHEMISTRY, PHYSIOLOGY, GEOGRAPHY,
AND DISEASES OF PLANTS:

WITH A HISTORY OF THE SCIENCE, AND
PRACTICAL ILLUSTRATIONS.

BY

A. P. DECANDOLLE AND K. SPRENGEL.

TRANSLATED FROM THE GERMAN.

EDINBURGH:

PRINTED FOR WILLIAM BLACKWOOD, EDINBURGH;
AND T. CADELL, STRAND, LONDON.

1821.

P. Neill, Printer, Edinburgh.

TO

ROBERT JAMESON, Esq. F. R. S. E.

REGIUS PROFESSOR OF NATURAL HISTORY IN THE UNIVERSITY OF EDINBURGH,
KEEPER OF THE MUSEUM, PRESIDENT OF THE WERNERIAN SOCIETY,
&c. &c. &c.

WHOSE ZEALOUS AND ENLIGHTENED LABOURS

HAVE SO GREATLY CONTRIBUTED TO THE ADVANCEMENT

OF NATURAL HISTORY IN THIS COUNTRY ;

THIS WORK

IS MOST RESPECTFULLY DEDICATED

BY THE

TRANSLATOR.

TRANSLATOR S PREFACE.

Every person acquainted with the recent history of Botany, must have been struck with the remarkable fact, that there does not at present exist in this country any work which embraces all the speculations and views which that science has now opened up; or in which its higher branches are treated in a manner suited to their importance. Much has lately been done for elucidating the Anatomy, Chemistry, Physiology, and Diseases of Plants; and the travels of those enlightened and indefatigable men who have recently traversed the globe in all directions, with the view of illustrating its Natural History, have thrown the most instructive and pleasing light on the laws which regulate the distribution of vegetables over the face of the earth. But the facts and speculations which have originated in these labours, have hitherto remained scattered throughout the various occasional works in which they first appeared: in many cases they continue buried in the obscurity of foreign languages; and the student who wishes to become acquainted with every thing of importance that has lately been done for the elucidation of this field of inquiry, is forced to explore his way amidst intricacies and thickets, not less formidable than those which were traversed by the men who first gleaned from the great book of Nature the manifold riches of the vegetable kingdom.

The work of WILLDENOW, which has hitherto been the chief introduction to the study of the Physiology of Plants, has now lost almost all its value.—It never was remarkable either for the philosophical spirit which it displayed, or for the powers of arrangement by which its materials were disposed; and by the recent progress of the more advanced branches of the science, it has been rendered more capable of misleading and perplexing the student, than of serving even as an introductory treatise to more correct speculations.

Several other works upon the same subject have lately been given to the public; and those which have proceeded from the pen of the distinguished President of the Linnean Society, are confessedly remarkable, both for the correctness of the facts and views which they contain, and for the scientific elegance which pervades all their descriptions. But these seem to be the only qualities which the author was desirous to secure for his works. To the merit of being complete elucidations of all the higher principles and discoveries of the science, they make no pretensions; and, certainly, there is no other work hitherto published in this country, which can be considered as surpassing these in that higher species of merit to which allusion is now made.

At the same time, it is evidently a matter of much importance, that a complete and well executed system of Physiological Botany should exist. The vegetable world comprehends many of the most interesting and beautiful productions of Nature; and beside those who are professedly engaged in the study of these productions, there are so many persons who take an interest in some one or other of the forms which the vegetable world assumes, and which we have at all moments before us and around us,—that there is perhaps no part of science in which the want of just and comprehensive views must be more generally or constantly felt.

To execute such a work, however, requires talents which are not always found united in the same person. Its successful accomplishment requires, in the first place, that he who devotes himself to it should be acquainted with a vast body of knowledge, which can now only be acquired by the study of innumerable works, which are often difficult to be procured, and laborious in the perusal,—with a multitude of facts of the most interesting kind indeed, but which are either buried in the obscurity of volumes that have long been laid aside, or which lie dispersed among the occasional productions of those Travellers and Natural Historians,

who have lately done so much for the improvement
of this part of science. We expect of any person who
should undertake a work of this description, that he
should be gifted, in the second place, with that sound
judgment in matters of speculation, which may prevent him from being led aside, by the very beauty of
the facts which he has to disclose, into theories that
are rather amusing or adventurous than well founded ; that he should cherish, in short, that truly philosophical spirit which spurns equally the contracted
views of vulgar minds, and the fanciful reveries of
men of genius and enthusiasm. We expect of him,
in the last place, that he should not only think sound-
ly and philosophically, but that he should also have
the kindred talent of disposing his reasonings in the
most luminous order, and that his work should imitate the great system of nature, of which it professes
to illustrate a part, by having all things well arranged,
and in due proportion.

The Translator of the work which is now offered
to the public, is not expressing his own conviction
merely, but that of men much better qualified than
himself to form a correct opinion upon this subject,
when he ventures to claim for this Treatise all the excellencies which the qualities now enumerated are necessary to secure. The extent of reading which has
been gone through for obtaining the materials of the
work, may be seen from the list of authors prefixed
to its more important chapters ; and perhaps there is

no work on the Vegetable World that displays more
varied or instructive information. With respect to
the second qualification already mentioned, it may also be maintained, that the philosophical views exhibited in the work are no less sound than they are frequently ingenious and original; and that in no one
part of the performance is it possible to discover a
trace of that visionary mode of considering facts and
appearances, which has been unjustly represented as
belonging to all German writers. Nor is the merit
of the work less distinguished with respect to the
powers of arrangement which it displays;—condensation and perspicuity, indeed, are among its most
striking excellencies,—so that it bears, in all respects,
evidence of having proceeded from men, who not only
knew their subject extensively and well, and who
thought justly on all its parts, but of men who were
in possession of the best means of conveying their information with effect, or who had made the higher
laws of composition their study.

The Translator cannot feel any hesitation in speaking warmly of a work, which appears to him to be
marked by such excellencies. He ventures, indeed,
to believe, that its essential merits will speedily be recognised; and he has no doubt that its influence will be
considerable, both in enlarging the views of those who
are prosecuting Botany as a science, and in spreading
just notions respecting the structure and distribution

of vegetable bodies, among all the liberal and enlightened classes of the community.

The reader must have already perceived, that the
work is the joint production of two authors, the first
three parts being extracted from the " Théorie Elementaire de la Botanique," of DE CANDOLLE, published at Paris 1819, and all the rest being furnished by
SPRENGEL, who superintended the publication of
the whole. The separate merits of the style of these
two authors may not perhaps be discernible in the
translation ; but in the original these merits are
strongly marked ; and, as DE CANDOLLE is distinguished by the subtlety, the flexibility and metaphy-
sical cast of his expression, SPRENGEL seems to possess a style, occasionally abrupt indeed, but always
luminous, condensed, and bearing evident marks of a
mind of no common powers.

Respecting the merits of the translation, it is not to
be expected that any thing should here be said :—
fidelity and perspicuity are all that have been aimed
at, and with the attainment of these the Translator
would be satisfied. But, in a work involving so many
technical terms, and so much recondite learning,—in
which the views exhibited are sometimes such, as
even well informed botanists were not formerly acquainted with,—and in which, along with many facts
borrowed from microscopical observations, there are
occasionally reasonings as subtle as any that are to
be found in our most ingenious systems of metaphy-

sics,—in an attempt to present such a work in a new dress to the public, the Translator cannot but be diffident. lest inaccuracies may sometimes have gained admittance both into his definitions of terms, and into his statements of reasonings. He ventures, at the same time, to hope, that these inaccuracies may be but of little moment,—that they are not such as in any instance to affect materially the meaning of any passage,—and that, as they are not likely, therefore, to mislead the inexperienced inquirer, they can easily be corrected by those whose information is more profound and accurate.

SPRENGEL'S PREFACE.

In my " Introduction to the Knowledge of Plants,"
it was my object to promote a knowledge of the vegetable world among the public at large,—and I may
venture to say, that the result has surpassed my ex-
pectations. But the progress which the *higher and
scientific* knowledge of plants has lately made, seemed to demand an introductory Treatise for the use of
students, which, embracing the discoveries that have
recently been made throughout the whole extent of
the science, might supply the place of the introductory work of WILLDENOW, which has now become
completely useless. With the help of the latest edition of DE CANDOLLE's " Théorie Elementaire de
la Botanique," published at Paris 1819, I flatter my-
self that I have been able to present to the public
such a work. But only the first three parts of this
treatise, namely, the Nomenclature, the Theory of
Classification, and Descriptive Botany, are to be considered as extracts from the book of my excellent
friend. All the rest is my own ; and the reader will
perceive that I have used my utmost exertions to fulfil well the task I had undertaken.

CONTENTS.

PART II.

TAXONOMY, OR THE THEORY OF CLASSIFICATION.

PART III.

PHYTOGRAPHY, OR DESCRIPTIVE BOTANY.

PART IV.

PHYTONOMY, OR ON THE STRUCTURE AND NATURE OF PLANTS.

PRACTICAL PART.

CONTENTS. **xxi**

EXPLANATION OF THE PLATES.

PLATE I.

Fig. 1. The chalaza of the Citron, opposite to the umbilicus, (Sect. 120.)

2. Cotyledons with the radicle. Seeds destitute of albuminous substance The embryon erect, (121, 383.)

3. Section of Cardamon seed, having the umbilicus turned upwards. The mealy albuminous substance is dotted. The *vitellus* is marked by lines. Within this lies the embryon, having its upper (in this case its lower) end bent into a hooked shape, (121, 385, 386.)

4. Seeds of *Cardiospermum Halicacabum* with the heart-shaped *strophiolus*, (120.)

5. Thick fleshy cotyledons wound within one another, (121.)

6. Achenium of *Centaurea ruthenica*, with its bristly *pappus* and lateral *umbilicus*, (109, 120.)

7. Silicle of *Thlaspi bursa*, (114.)

8. Galbulus of *Thuia orientalis*, (116.)

9. Strobilus of *Alnus incana*, (116.)

10. Section of the seed of *Strelitzia reginæ*. The albuminous substance is finely dotted: the embryon is in the centre, unevolved. A red tomentum forms the strophiolus, (strophiolus stuppeus, 25, 120, 121.)

11. Section of a grain of Wheat, in the upper end of which is the superficial unevolved embryon lying on the *scutellum*, by means of which it is connected

with the albuminous substance which is denoted by the lined portion, (121, 386.)

Fig. 12. Pencil-shaped *pappus* of *Leyssera capillifolia.*

13. Seed of Epilobium angustissimum, with its hairy *coma*, (109.)

14, 15. *Achenium* of *Laserpitium latifolium*, with four membranaceous wings, and the intervening cavities, (109.)

16. Legume of *Vicia consentina*, (114.)

17. Tailed *utriculus* (arillus Linnæi) of *Geranium Bohemicum*, (109.)

18. *Lomentum* of *Hedysarum coronarium*, (114.)

19. Winged seed of Tritonia flava *Ker*, (109.)

20. Section of the seed of *Mirabilis Jalappa*. The embryon in the circumference marked by lines. The albuminous substance in the centre, (28, 121.)

21. Capsule of *Maurandia antirrhiniflora* Willd. It has double sides : the exterior side is composed of five regular valves ; the interior displays two completely dissimilar, irregular loculi, the larger of which shews the rudiments of four loculi in the four projections of the side. The placentation is central. There are properly, therefore, two united capsules, the larger of which shews the intended quadrilocular structure of the capsule of this family in its first rudiments, (110, 111, 185, 189, 192.)

22. Achenium of *Asterocephalus Caucasius*. Membranaceous pappus. Five bristles proceeding from within, (§ 109, page 354.)

23. Apple with five loculi, (115.)

24. Seed of *Anona squamosa* L. The albuminous substance is formed into plaited wrinkles, (albumen ruminatum, 26.) The small embryon lies at the upper part, in a cavity of the umbilicus.

25, 26. *Craterium pyriforme* Ditm. One of the Gastromyci, (page 462.)

27, 28. The sporidia of the former, with the tufty texture, (119, page 72.)

Fig. 29. *Agaricus amanita*, with the volva, the annulus, and the pileus, (88.)

30. *Hymenium* of *Geoglossum viride* Pers. with the thecæ sporophoræ, and intervening sap-tubes, (119, page 461.)

31. *Botrytis polyspora* Link. The bushy, branchy flocci are externally set with sporæ. One of the *Nematomyci*, (119, page 463.)

32. *Fusidcum gryseum* Link. Fusiform sporæ, (119, page 463.)

33. The radicle of *Erysimum hieracifolium* lying on the back of the cotyledons (Cotyledones incumbentes), (38, p. 405.)

34. The radicle of *Sinapis nigra* lying between the cotyledons (Cotyledones accumbentes), (38.)

35. Berry of *Basella rubra*. Snail-shaped embryon, (41.)

36. Sarcobasis of *Ochna squarrosa*, (85, 105.)

37. Bilocular capsule of *Justicia paniculata*, with the hooks on the dissepimentum, (120.)

38. Silique of *Leucoia*, with the seeds on both sides, (114.)

PLATE II.

Fig. 1. Hymenium of *Peziza cerea* Pers. The sporidia contain eight sporæ, (119, p. 459.)

2. Apparent seeds of *Solerina saccata* Ach (118, p. 456.)

3. Leconera straminea *Ach.* (118, p. 456.)

4. Grimmia controversa *Hedw.* with its calyptre cleft laterally, (88.)

5. Perichætial leaves of *Neckera disticha* Hedw. (88.)

6. Club-shaped, reticulated, apparent antheræ of Gymnostomum pyriforme, with the pistilla and sap-tubes.

7. Exterior *peristomium* of *Hypnum alopecurum*.

Fig. 8. Interior *peristomium* of the same plant, (117.)

    9. Capsule of *Aspidium spinulosum* Sw. with its notched ring and rough seeds, (117.)

  10. Indusium of the same Fern, fringed on the margin and studded with glands, (88.)

  11. Flowers of *Aponogeton distachys* Thunb. Appendages for the petals, (90.) Heptandria.

  12. Flowers of *Chironia frutescens*. Stamina perigyna, (35.) Corolla infera ; germen superum, (34.)

  13. Twisted antheræ of the same plant, (41.)

  14. Phylica ericoides; folia sparsa, revoluta, sessilia, lanceolate. Flores fasciculati, (29, 38, 40, 44, 84.)

  15. Calyx corollina of the same plant, ciliated (51, 90.), the antheræ under *nectarilymata*, (102.) These scales may also be called petals. The filaments are united with the calyx corollina, (191, 192.)

  16. Urceolate corolla of Arbutus unedo, (31.)

  17. Bilocular antheræ of the same plant with spurs. The filaments ciliated, (334.)

  18. *Oxalis purpurea* Jacqu. The pistilla are shorter than the interior, but longer than the exterior filaments, (380.)

  19. *Oxalis macrostylis* Jacqu. The pistilla are longer than the exterior and interior filaments. The exterior filaments are always shorter than the interior, (203, 205, 380.)

  20. Five pistilla, having their extremities set with stigmata, and their lower parts with glands. Perhaps the collectores, (341.) Connate ovaria of Oxalis macrostylis, (107, 189.)

  21. Flowers of *Agathosma pubescens* Willd. Beside the five principal petals are five subordinate fringed bodies,—abortive filaments. Of the five filaments, three are higher than the two others, (175, 185, 203, 205.)

  22. Similar abortive filaments.

  23. Veiled stigma of *Lobelia discolor*, (106.) The an-

theræ connate. The fruit inferior to the calyx,
and connected with it, (34.)
24. Pollen of *Amaryllis reginæ*, (335.)

PLATE III.

Fig. 1. Male and female flowers of Phyllanthus epiphyllan-
thus *W.* out of their buds. The apparent flower-
stalk is an abortive leaf. The male flowers have
abortive pistilla, (174.) The three filaments are
united in one pillar. On their base are nectareous
glands, which are wanting in the female flowers,
(138, 187.) Of the six flower-stalks, the interior
alternate with the exterior, and the former may
therefore be considered as petals, (196.)

2. *Piqueria trinervia* Cav. The common calyx, anthodium or periclinium, consists of four leaves, and
contains four florets, (87.)

3. A single floret, the rim of which has five lobes. The
stigma cleft. *Syngenesia æqualis*, (137.)

4. Racemus of Ottonia anisum. Neue Entdeck., i. s.
255, (84.)

5. A single floret of the same. At the base of the stalk,
a fringed scale or bractea. No corolla. Four bilocular antheræ. Spherical germen, with a four
cornered stigma, (86.)

6. Crassula spathulata. The ovaria superior to the calyx and corolla. These alternate with the filaments,
the latter parts with the petals, and these last with
the teeth of the calyx, (34, 196.)

7. *Schmidtia subtilis* Trattin. Two glume-valves enclose two hypogynous filaments with moveable antheræ, a proportionately very large germen, and two
simple linear stigmata, (35, 40, 198^a.)

8. Capsules of *Targionia hypophylla*, surrounded by a
notched ring, and furnished with *catenulæ*, as they
had never before been observed. Its affinity with

Jungermannia and Marchantia is thus established, (109, 278.)

Fig. 9. Correa alba *Sm.* with eight perigynous, unequal filaments, and nectareous glands between them, (35, (101.)

10. The germen of the same plant enlarged by its greater maturity. The nectaries are divided, and embrace the four two-seeded cocci (86, 109.), which being united form the quadrilocular fruit. That they were originally separate is evident from the inferior linear processes, which might be mistaken for collectores, (178, 189, 341.)

11. Flower of *Acacia lophanta*, Willd. Monadelphia, (131, 187.)

12. Urceolate corolla of *Erica aggregata*, with its four-lobed margin. Neue Entdeck., 1. 270, (31.)

13. The sexual parts of the same plant after the corolla has been taken away. The four-leaved calyx inferior to the germen, (34.) Eight bent filaments, inclosed in the corolla, and between them the nectareous glands. Eight bilocular, unarmed, or awnless anthers, (11, 56, 80.) The pistillum with the four-lobed stigma, (188.)

14. The germen in particular · there appears a fivefold division, as an evidence of the original numerical proportion, (178, 199, 389.) The unripe germen also is multilocular, and it is only when it is fully ripe, that the individual capsules become so united, that their dissepimenta become simple, and extend from the central column, (178, 189, 192.)

15. Two lipped, ringent corolla of *Salvia Boosiana*, Jacq. (96.)

16. Two filaments, with processes, which seem to be two other abortive filaments, (178.)

17. Gynobasis, along with the laterally projecting nectary, and the four caryopses standing on it. The pistillum between these, with the divided stigma,

the consequence of union, (85, 101, 105, 109, 188, 198^a, 202.)

18, 19. Antheræ of *Erica tenella* Andr. (antheræ cristatæ), with cristate appendages. Neue Entdeck. 1. 271, (31.)

20. *Mycoporum tenuifolium* Forst. The corolla with its quinque-partite ciliated margin. Four filaments of unequal length at the entrance of the tube of the corolla. The original numerical proportion is indicated by the segments of the corolla. Of the five filaments one is abortive, (198^b, 199.)

21. Quadrilocular drupe of the same plant, (112.)

22. Flower of *Duchesnea fragiformis* Smith. The calyx has tripartite subordinate leaves. The hypogenous filaments, in indeterminate number, are inserted on the corollar integument of the calyx; Icosandria, (35, 131.) The fruit arises from the enlarged receptacle, having its surface studded with granular bodies.

23. Flower of *Barleria flava* Jacq. The calyx consists of four unequal segments, of which the two exterior are much larger than the other two, and the uppermost is so much dentated, that it is impossible not to perceive the original tendency towards a fivefold division. The tubular corolla has a quinque-partite margin, two of the lobes of which are reflex, and the other three are emarginated. Of the five filaments which there ought properly to be, and which sometimes do appear, only two commonly come to perfection, (178, 181, 389.)

24. Flower of *Veltheimia viridifolia* W A calyx corollina, with the filaments inserted in it : these are hypogynous. The germen is superior. The pistil bent downwards, (35, 43, 90.)

25. Indications of the spiral vessels on the superficial cellular texture of *Sphagnum obtusifolium*, (278.)

PLATE IV.

Fig. 1. Punctured tubes and spiral vessels of the root of *Cissampelos Pareira* L., (276, 277.)

   2. Section of the buds of the common Alder, (304.)

   3, 4. Simple buds of the Tulip-tree, with the three following series, (304.)

   5. Buds of the common Ash, (ib.)

   6. Buds of *Mespilus glandulosa*, (ib.)

   7. Buds of *Salisburia adiantifolia* Sm., (ib.)

   8. Buds of the Guelder-Rose bush, (243.)

   9. Leaf of *Gymnostomum ovatum*, the nerve of which passes into two processes (nervus lamellatus), (49.)

  10. Columna genitalium of *Bletia Tancarvilleæ* R. Br. At its upper part the column is shaped into the clinandrium of Richard, on both sides of which the staminodia stand. The clinandrium has under it the rostellum, which again has the stigma strongly shaded under it, with which it is closely connected. The anther is consequently epigynous.

  11. Four connected masses of pollen of the same plant, (107.)

  12. The four loculi of the clinandrium, in which the pollenous masses are contained, (107.) I have not found the partition any where so distinct as it is here figured.

  13. Flower of *Malva umbellata*, Monadelphia. The veins in the petals are spiral vessels, (324.)

  14. Section of the same flower; the tube of filaments united with the corolla: expanded below, the filaments and corolla being of the same nature, (191.)

  15. Pollen of the same plant, (107, 335.)

  16. Flower of *Pogostemon plectranthoides* Desfont. The corolla is almost reversed (resupinata): for the filaments are bent downwards, whilst in the related genera they stand erect. The undivided lip is pro-

perly the superior one, that with three lobes the inferior. The long ciliated filaments are of the didynamia structure : the pistillum is divided, (43, 187.)

17, 18. Flower of *Büttnera cordata*. Quinque-partite, open calyx. Five lunulate petals, furnished at their upper part with two auriculæ, and between them with a long slightly ciliated bristle. Five filaments with bilocular antheræ, between which and the pistillum a five-lobed fringe stands, the lobes of which are emarginated, and carry the nectaries on their outer-side, (101.)

19. Spiral vessels and sap-tubes of *Marantha lutea* Jacqu. (276, 277.)

PLATE V

Fig. 1. Spiral vessels and sap-tubes of *Alpinia nutans* Rosc., (276, 277.)

2. Slits of the epidermis of the leaves of Hyacinthus, with the cellular texture lying under them, (310, 311.)

3. Vasa scalaria of *Lycopodium clavatum*, with the inner bark, (282. 297.)

4. Tubes with interstitial spaces, and the sap-vessels and radiating vessels of the wood of the Fir, (273, 293.)

5. *Sporophleum gramineum* Nees. Miser. A doubtful Inomycus.

6. *Camptosporium glaucum* Link. Micros. On the inner side of an old Oak bark; (Ehrenb. Sylv. Mycol. p. 11.; Nees, Radix Plant Mycetoid. p. 5.)

7. *Eurotium epixylon* Link., (Berlin. Mag. 3. s. 31.) One of the Gastromyci.

8. *Diatoma flabellulata* Jürg. A Bacillaria, according to Nitsch. Transition from the vegetable to the animal kingdom.

9. Bacillaria vexillum.

10. Strictures of *Nodularia fluviatilis* Lyngb. Hydroph.
p. 99. t. 29., with the granular germs ranged in
the shape of a chain.
11. Anthocorynium of *Surubea Guianensis* Aubl. Boyl.
Meyer. Fl. Essequeb. p. 120, (86.)
12. The flower-bearing leaf-stalk of *Turnera cuneiformis*
Juss., with two cup-shaped glands.
13. Flower of *Panicum leiocarpon*, (Neue Entdeck. 1.
s. 243.

PLATE VI.

Fig. 1. Tubers of *Trevirania coccinea* Willd., (288.)
2. Tubers of *Tritonia squalida* Ker, (ib.)
3. Tubers of *Ixia leucantha*, (ib.)
4, 5. Flowers of *Euphorbia Characias*. Double antheræ on the filaments, which are furnished with
a joint; and the four pistilla, (138, 331, 333.)
6. *Buginvillæa spectabilis* Juss., with red coloured
bracteæ, (177.)
7. The seven filaments of the same plant united at the
base, (191.)
8. Star-shaped scales on the leaves of Croton Eluteria,
(27.)
9. Panduræform leaf of *Solanum Belfortianum* Dunal,
(29.)
10. Star-shaped hairs of the same plant, (25.)
11. *Cassia flexuosa*, with its bent stem, pinnated leaves,
and ciliated stipulæ.
12. *Sauvagesia Adima* Aubl., (Neue Entdeck. 1. 294,
with its root-shaped ascending stem, and bristly
pinnated stipulæ, (42, 54, 109.)
13, 14. Flower of the same plant. Between the calyx
and corolla stand bodies of an intermediate nature between filaments and nectaries, as in Parnassia. Three lobed capsule, (Neue Entdeck. 1.
295. 296.)

PLATE VII.

Fig. 1. Bunchy panicle of *Hirtella glandulosa* (Neue Entdeck. 1. s. 303.) Reflex bracteæ. Bush of petiolated glands.

2. Ovate-oblong, somewhat pointed, leaf of the same plant, full of veins.

3. The flower magnified. The calyx quinque-partite, reflex, internally set with silky hairs. The corolla fallen off. Five long, linear filaments. The pistillum ciliated below. The dry berry set with rough hairs.

4. The flower opened before the evolution. The filaments convoluted.

5. *Alyssum nebrodense* Lin. (Neue Entdeck. 1. s. 286.) shrubby stem, with rose-shaped aggregated leaves.

6. Elliptical silicle, crowned with the pistillum.

7. Inverted ovate leaf of the same plant, set with radiated hairs.

PLATE VIII.

Fig. 1, 2. *Cyphia serrata*, (Neue Entdeck. 1. s. 274.)

3. The flower in particular.

4. The filaments not attached to one another.

5. *Hydrocotyle plantaginca*, (Neue Entdeck. 1. s. 284.)

6. The fruit.

7. Section of the same.

8, 9. Notched pappus of *Pteronia glabrata* Thunb

CORRIGENDA.

Page 9, line 24. *for* lanceolate-shaped *read* lanceolate
—— 10. — 3. *for* points *read* extremity
—— 62. — 14. *instead of* to *read* for
—— 124. line 17. *for* of involucrum *read* of the calyx
—— 239. — 25, *for* also *read* both

PRINCIPLES

OF

SCIENTIFIC BOTANY.

―――――

INTRODUCTION.

1.

BOTANY comprehends the knowledge of Plants. These, like all other natural bodies, may be considered under a two-fold aspect; either in relation to their external properties, or with a view to their internal structure, their nature, and the causes of the phenomena which they present.

2.

Botany, accordingly, divides itself into two principal departments, which, in our days, can no longer be separated, since they mutually support and illustrate each other.

I. *The Natural History of Plants*, which has been exclusively denominated Botany. This comprehends the knowledge of the external marks of plants, and the means of distinguishing them; their Description, Determination, and Classification. This department includes three subdivisions.

1. The *Nomenclature*, (*Glossologie* DE CANDOLLE, falsely called *Terminology*). Under this subdivision, we include the knowledge of the expressions, by which the different organs,

A

of which plants are made up, and their properties, are designated.

2. *Taxonomy*, or the Theory of the Classification of the Vegetable Kingdom.

3. *Phytography*, or the Art of describing Plants in a conformable manner. As applications of this art, we must consider *Descriptive Botany*, or the technical representation of all the essentially different forms of the Vegetable Kingdom ; and *Botanical Synonymes*, or the knowledge of the different names under which plants are mentioned by writers. This latter kind of knowledge has a necessary connection with an insight into the fortune and progress of the science, as well as with mere book-learning in this department. *The history and literature* of Botany are hence essential requisites.

II. If we turn our attention to the *internal* structure, the nature, and the principles of the phenomena of plants, these lead us to the *Natural Science*, or Physics, of Plants; which has also been called the *Physiology* of Plants, *Phytonomy*, or *Phytology*. This department again comprehends three principal subdivisions.

1. The doctrine of the structure of plants; or what has been called their Anatomy ;—the *Organography* of De Candolle.

2. The doctrine of the composition of the constituent parts of plants; the Chemistry of Plants, or *Phytochemy*.

3. The proper explanation of the manner in which plants originate, grow, form their parts, and propagate themselves. This is properly the philosophical part of Botany, or the true *Phytonomy*.

3.

To these two essential parts, we may add the following departments of knowledge, as more or less connected with them, or derived from them.

1. The before-mentioned History and Literature of Botany. These in no other science are so necessary as in this, where we can only hope to attain a perfect acquaintance with

the objects of our study, by being acquainted with their different names and representatives among writers, and where every new step must be accomplished by a comparison of all earlier observations.

2. The Geography of Plants, or the examination of the causes which determine plants to choose certain regions and stations. This knowledge respecting the distribution of plants is naturally enough included under the second great department, or that which relates to the *Physiology* of plants. It has of late begun to be studied with particular zeal, and is fitted to afford the most important assistance in the *Classification* of plants.

3. The knowledge of the Anomalies to which the forms of the Vegetable Kingdom are subject; to which belongs also the doctrine of the Malformation and Diseases of plants. If we consider these variations in their most comprehensive relations, their study is one of the most difficult, but it is also one of the most instructive parts of Botany. The doctrine of the diseases in particular, is called the Pathology of Plants.

4. The applications of these different parts of knowledge to the arts and business of life are excluded from this Treatise; yet it cannot be denied, that these applications often reflect an important light upon the science itself. The knowledge of the officinal plants has been called Medical Botany; the knowledge of the plants which are employed in agriculture and husbandry, is called Œconomical Botany; that of the plants which are useful in arts and trades, is called Technical Botany; and the knowledge of forest trees, is the Botany of Forests.

4.

All these different branches of the science, or parts of Botany, are connected together in the closest manner. They cannot well be treated or learned separately, without disadvantage to the science; and one chief cause of the interruptions which are given to the progress of Botany, lies in the separation of these individual branches.

Especially it ought to be most zealously inculcated, that the applications of this science should only be taught or learned after a previous study of its general principles. That part of Botany which relates to forest trees, is as little capable, as the medical department, of dispensing with the helps which are afforded by the Nomenclature, the Classification, and the Physiology of Plants.

PART I.

NOMENCLATURE.

===

CHAP. I.

GENERAL PRINCIPLES.

===

5.

As all the departments of human knowledge and art, and even all trades and occupations, have a multitude of peculiar expressions, by which they designate certain things, properties, operations, and appearances; and as, in Botany, amidst the immense number of its different forms, every thing depends upon our having a clear idea of these differences, the necessity of a general agreement in the choice of expressions for these different terms and properties is obvious; since no man will ever convey to another a distinct idea of any object, if he either uses such expressions as that other person does not understand, or if he employs them in a sense different from that in which they are to be understood.

6.

The necessity of a general agreement imposes, no doubt, a certain restraint, to which every person must submit; and there has been no want of writers, both in our own and in former times, who have entertained the idea of releasing themselves from this restraint, which to them was so oppressive; and who have, for this purpose, either indulged themselves in a certain negligence of expression, or have dared to choose a language for themselves, which it was necessary to have explained in a preliminary vocabulary. This practice is highly reprehensible; because it proceeds partly from ignorance and disregard of the laws of the already recognised Nomenclature, and partly from conceit, arrogance, fondness for novelty, and national prejudice; and because it creates unnecessary difficulties in the science, and affords it no essential advantage whatever.

At the same time, our reprehension is not meant to fall upon those who designate forms that are really different and peculiar, by new and suitably selected expressions; because the farther the knowledge of plants is extended, the greater number of altogether new forms, or of such as were hitherto misunderstood, do we discover; and these could only be designated in a very defective manner, if we should confine ourselves to the expressions that are already used for them.

7.

Nomenclature has its difficulties; but these would be unnecessarily increased, by compounding terms to too great an extent, or by applying them to the most subtle subdivisions of our ideas. Without making any breach on the solidity of the structure, it may be simplified and relieved, by, on every occasion, consulting nature, and giving life to our demonstrations by examples; by considering many terms as in general useful, without constantly repeating them in the description of every part; and by supposing the knowledge of the learned languages to have been already acquired.

8.

The Nomenclature is of Latin derivation, because this language is understood by the learned of all nations, and of all times. This cannot be objected to, since the descriptions of the plants are also given in the language of each particular country. These descriptions, however, are not generally intelligible; and as long as there is no agreement in the choice of terms, they must also be defective in respect of certainty.

It is not to be expected that the Latin Nomenclature of Botanists should always display the purity of the golden, or even of the silver age, of Roman literature : because it is impossible to select such expressions only as are found in the Roman writers of that period, as the designations of objects which were altogether unknown to those writers; yet it is to be expected, that he who writes Latin, should neither mistake nor disregard the laws of grammar and of composition, nor the spirit of the language in which he expresses himself.

9

Where the Latin language cannot be employed, where the necessary compositions are either foreign or adverse to the spirit of that language, we betake ourselves to the richer and more pliant dialect of the Greeks. Only here also, the terms must be chosen according to the laws of the grammar and composition of that language. We must be on our guard not to employ what have been called hybrid expressions, or words compounded from both the learned languages, (as, for example, *muscologia, algologia, ovöides*); or to exchange customary and intelligible Latin terms for unusual, often strange and falsely compounded Greek expressions.

10.

The first principle of Botanical Nomenclature is, That each distinct form, and every different organ, be designated by a peculiar expression. By following out this principle, all wavering of ideas, all uncertainty of knowledge, is avoided. In conformity with this law, we call the leaves of the branches and stem, *folia ;* the leafy appendages of the leaves, *stipulæ ;* the leaves in the neighbourhood of the flower, *bracteæ ;* the parts of the cover of the flower and of the calyx, *sepala ;* and, lastly, the divisions of the corolla, *petala.* In the same manner must the elliptical form of the leaves, the varied rounding or tapering of their summits, be designated by peculiar expressions.

We may easily, however, go too far in this process, by assigning different terms to every small additional variety of an already well known form, to every insignificant appendage or part of an organ. Much circumspection is necessary, not to mistake the happy middle course.

11.

The properties of forms are expressed by adjectives or participles. These have a different meaning according to their different terminations.

1. Adjectives in *atus* designate the presence of a certain organ: *radicatus*, that which has a root; *foliatus*, which has a leaf. But this rule must not be understood as universal; because *lanceolatus*, *ovatus*, and many others, signify only a resemblance.

2. Those which end in *osus* express an abundance of particular organs; *nervosus*, *foliosus*, *cicatricosus*.

3. Adjectives in *inus* and *aceus* denote the nature of the organ. Thus, *foliaceus* means having the nature of a leaf: *radicinus* denotes that which has the consistence of a root. A *corolla calycina* is one which partakes of the nature of a calyx. But there are many exceptions also to this mode of expression, however much a general agreement is to be wished.

4. When an organ is wanting, we put the Greek *a* privative before a Greek word, or the *e* before a Latin word; (*aphyllus*, *apetalus*, *enervis*, *exstipulatus*, *eglandulosus*). Some hybrid expressions, however, have crept in, which we must endure: for instance, *avenius;* and also *acaulis*, where the better term would be *acaulus*. In some instances, *subnullus* is placed immediately after *nullus: Margo nullus*, *subnullus*. We often express the absence of certain organs or forms, although in an indefinite manner, by *nudus* or *simplex :* thus, the Corolla is said to be *nudus*, when it has no calyx; the branches, *nudi*, when they have no leaves: the stalk, *simplex*, when it is destitute of branches. Finally, it is usual, when an organ is wanting, to designate positively the opposite property : thus, we say *inermis*, and *muticus*, in opposition to *spinosus*, *aculeatus*, *aristatus*, and *cuspidatus*.

When a part or a property is not distinctly evolved, or is not seen, we say that it is *obsoletus inconspicuus*.

5. Smaller variations of form are expressed by diminutive terminations; by the termination *oïdes* in Greek adjectives, or by the preposition *sub*. This latter syllable particularly is put before the word, where the property or organ is not found every where in the same degree. We thus use the phrase *folium subdentatum*, where the teeth in many positions are not observable, or pass into the smooth margin. The expressions *hirsutiusculus, acutiusculus, obtusiusculus*, are also frequently used; as also are not unfrequently *rhizoïdes, calycoïdes*.

6. When the property is present in a higher degree, we commonly use the superlative; by which means, all particles expressive of abundance, as, *valdè, maximè, insigniter*, are avoided. We thus say, *integerrimus, spinosissimus, aculeatissimus, glaberrimus*.

7. The intermediate condition between two varying forms is frequently expressed by the compounding of two adjectives; thus, *oblongo-lanceolatus, repando-dentatus, palmato-lobatus*. But we fail in these compositions, when we put together words which exclude each other, which are co-ordinate, and are therefore self-evident. *Elliptico-lanceolatus* involves a contradiction; because the lanceolate-shaped leaf is pointed, and the elliptical is symmetrically rounded at both ends. *Pubescenti-hirtellus* involves a contradiction, and, at the same time, expresses subordinate ideas of the same class; because *hirtus* denotes long stiff hairs: *hirtellus* is therefore a nonentity; but *pubescenti-hirtellus* is self-evident, and *pubescens* alone were sufficient.

8. When we would express the reversed form of an organ, we usually put the syllable *ob* before the adjective: thus, *obovatus, obcordatus*, are very common. *Oblanceolatus* can scarcely be permitted, because the positive term *spathulatus* is more definite. When an organ has the external shape of another organ, without actually fulfilling the same purpose, or even without having the same structure, we usually designate it as *spurium*: we thus say, *antheræ spuriæ* of Tradescan-

tia; *margo thallodes spurius* of Lecidea petræa; *peridium spurium* of Æcidium.

12.

Terms taken from common life remain without any explanation, when they are used with no variation of their usual meaning: otherwise, they must be more exactly defined. Thus, *pilosus,* in technical language, means something different from the idea suggested by it in the usual written language. In this latter use, it means the hairy condition in general; but, in the former, it denotes particularly the presence of soft hairs, somewhat bent.

CHAP. II.

CHARACTERISTIC EXPRESSIONS FOR FORMS AND QUALITIES.

13.

WE enumerate under this title, those terms which relate to the properties of the organs, or parts of plants, and we seek, especially, to make those terms general, in order to avoid all repetition. It ought to be remarked, that we are only furnishing an introduction to the comprehension of the works of other writers, and that we do not, therefore, take upon us, to answer for the correctness and necessity of every term.

I. *Measure of the Parts.*

14.

We call the measure of the parts *absolute,* when it has a reference, not to any comparison with other parts of the same plant, but to some other commonly assumed scale. The measure is *relative,* when we compare the extension of one part with the size of others, belonging to the same plant.

15.

The absolute measure may be best taken from the parts of the human body, because these are universally present, and are the same among all nations.   If these parts be compared with the fixed civil measure, we readily perceive that they cannot, by any means, be determined with geometrical exactness.   But this exactness is the very thing which in Botany we ought not to aim at, because organised bodies neither form geometrical lines and figures, nor can be calculated according to a geometrical measure.   Where the measure of the human body is not sufficient, we can only have recourse to a civil scale: but, in this case, we can only form an approximation to the truth, and ought not to consider the individual extension as general.

16.

The smallest measure in Botany is the breadth of a hair, (*capillus*, hence *capillaris*).   If this be compared with the civil measure, it forms about the twelfth part of a geometrical line.   But we must not confound the term *capillaris* with *capillaceus*, since this latter term signifies, according to the rule formerly given, (§ II. 5.), having the nature or consistence of a hair.

After the hair, follows the breadth of the white crescent on the nail, which corresponds more or less with a geometrical line.   This measure is named *linea*, whence *linearis ;* which term, when applied to surfaces, signifies, that they preserve the same breadth throughout.   It must not be confounded with *lineatus*, which may be translated, *marked with lines, or streaks.*

To this succeeds the length of the nail of the finger ; (*unguis*, whence *unguicularis*).   This is about six lines, or half an inch.

Next comes the breadth, or even the length of the outermost division of the thumb, (*pollex*, whence *pollicaris*). This forms exactly an inch ; hence it is called *uncia, uncialis.*

The length of the middle finger follows, or; which is the same thing, the breadth of the hand reckoned from the thumb. From *digitus* and *palmus*, we form *digitalis* and *palmaris*: this last, however, must be carefully distinguished from *palmatus*, which means, divided like the hand. The measure of which we are now speaking, is about three inches.

The small span (*spithama*, whence *spithameus*) is the distance between the thumb and middle-finger, when they form a span. This measure, in full grown men, is about seven inches.

The great span (*dodrans*, whence *dodrantalis*) is the distance between the thumb and little finger, when they form a span. It amounts to about nine inches.

The foot (*pes*, whence *pedalis*) is the length of the sole of the foot, when it is of a large size, or the distance of the elbow-joint from the wrist. It is usually taken for a geometrical foot. But *pedalis* must not be confounded with *pedatus*, which denotes the foot-shaped position of the leaves.

We next take the distance from the elbow-joint to the point of the middle finger. This is called *cubitus*, whence *cubitalis*; its length is about a foot and a half.

Next comes the length of the arm, (*brachium*, whence *brachialis*: as also *ulna, ulnaris*). This is reckoned about two feet.

Lastly, there is the fathom, (*orgya, orgyalis*), or the distance of the points of the fingers, when the arms of a man are extended. This measure is taken from the length of a man of large size: it is hence reckoned about six feet.

Whatever exceeds this measure, is reckoned in so many fathoms or feet.

17.

It is an unaccommodating practice which has been adopted by the French Natural Historians, of constantly employing the decimal measure, which is entirely unknown out of France, and which is particularly unsuitable to this purpose, on account of the very great exactness with which its subdivisions are given. We only state here, therefore, that the *metre* amounts to three feet and something more than eleven lines; that the *decimetre* is three inches and something more than eight lines; the *centimetre* four lines and a half; and the *millimetre* exactly $\frac{445}{1000}$, or about half a line. This measure is not adopted by any other nation.

18.

The relative measure is taken by comparison with other parts of the same plant. We say *æqualis* or *æquans*, *major*, *minor*, *longior*, *brevior*, *duplo*, *triplo major* or *minor*. We say also *superans*, *excidens*, *æquans*.

19.

We also frequently assume the relative size and extension from comparison, but without always naming the compared parts. We thus name Calyx *maximus*, a calyx which, in proportion to all the other parts of the plants, is very large. We also usually say, *rami*, or *pedunculi elongati*, *petioli brevissimi* or *abbreviati*, *stipulæ minimæ*, *folia angustissima*, *calyx ampliatus*, *planta pusilla* or *pumila*, *arbor gigantea*: by which expressions every one knows what he is to understand.

20.

The relative measure of the parts of one and the same organ, determines their regularity or equality.

The parts of an organ are *equal* (*æqualis*), when they have throughout the same measure and the same form. Inequality (*inæqualitas*), expresses the reverse. Like, (*conformis, similis*), relates only to the correspondence of form; unlike, (*dissimilis*, and in Greek compounds *hetero*), signifies the opposite quality. Variable, (*varius, variabilis, mutabilis*), relates to the disposition of an organ readily to change its form.

An organ is called *regular* (*regularis*) the parts of which shew a certain correspondence, but not a complete similarity, of parts in size and form; as when, for instance, larger parts are interchanged with those that are smaller. The structure of an organ, again, is *irregular*, when no correspondence of parts, either in shape or size, is observable.

When plants of the same kind have different shapes, they are said to be *dispares*. *Dimorphus* is the word applied to one and the same part, which has different forms in the same plant. An organ is called *difformis*, when its shape cannot be reconciled with the usual form of the part.

II.　*The Colours of the Parts.*

21.

As colour produces an impression upon the senses which is universal, it cannot be precisely defined, nor can it be employed as a character of the invariable Species. It is chiefly of use, therefore, in description. Yet we must put a great value even on this, when we are treating of the lower organic bodies, where other characters fail, and where those derived from colour are very stedfast.

22.

There are two colours, which have been usually considered as essential, and which have, hence, been universally taken as characters. These are the *hoary*, (*canus, incanus*) ; the weaker variety of which is called *canescens*, and the greenish-blue (*glaucus*), the weaker gradation of which is named *glaucescens*.

Besides, it is usual, when the ordinary green colour of the integument of the leaves fails, to express this simply by the word coloured, (*coloratus*) ; and when there is a difference of colours, as, for instance, on the two sides of the leaves, we employ the word *discolor*, as in Cornus *alba*. The opposite is, therefore, *unicolor*, that is, when the colours are of the same kind.

A difference of colouring is also expressed generally by the words spotted (*maculatus*), variegated (*variegatus*), and by the words *notatus* and *sphacelatus*. The former expresses a dark spot ; the latter the swarthy discolorations, which are yearly the consequences of the decay of the parts. We use the word *zonatus* to express differently-coloured curved lines upon a surface : *marginatus* signifies that the margin is dif-

ferently coloured, or of another substance, (§ 28.) Round
spots, surrounded by a circle, are said to be *ocellati*. We also use the words *halonatus, halone cinctus*.

23.

Among the many colours which appear in the vegetable
kingdom, the metallic are of least frequent occurrence.
Their various gradations have very properly been referred to
certain fundamental colours; and the transitions from one
colour to another have been pointed out.

When there is scarcely any colour, and the parts are almost transparent, we use the words *pellucidus, diaphanus,
hyalinus, aqueus* or *vitreus*. The opposite of this is opake
(*opacus*.)

The addition of the words *sordidè, intensè, saterrimè, pal-
lidè, dilutè*, to colours, expresses the different gradations of the
dirty, the intense, and the pale colours.

1. The white colour (*albus*, in Greek compounds *leuco-*,)
has for its ground-tone the snow-white (*niveus*), which we
particularly observe, in its most beautiful state, in *Camellia
Japonica*. If the white colour is still very pure, but not so
clear as in the case already noticed, it is said to be *candidus*.
In Greek compounds this is expressed by the word *argo-*.
The white lilies afford the best example. If this last white
be united with a certain lustre, the term *eburneus* is used,
although but seldom.

If the lustre of the white colour is still more distinct, but
the colour itself not quite pure, we employ the term *silvery*,
(*argenteus*, and in Greek compounds *argyro-*.)

A faded white passing somewhat into a bluish, is called
milk-white (*lacteus*, and in Greek compounds *galacto-*.)

Indeterminate varieties of the white colour constitute the
whitish (*albidus* and *albescens*).

If the white colour passes into the grey, it becomes hoary
(*canus, incanus*), especially when the surface is spreckled with
a greater or less number of distinguishable hairs.

2. The grey colour (*griseus*) has its ground-tone in the
ash-grey (*cinereus*, and in Greek compounds *tephro-* and *spo-*

do-). Its varieties are the lead-grey (*plumbeus*), which approaches near to the bluish; and the smoke-grey (*nebulosus*), which goes more into a brown. The mouse-grey also (*murinus*) is a soiled variety of the brownish-grey.

3. The black colour (*niger*) has as its ground-tone the full or coal-black (*ater*). In Greek compounds we express this by the words *mela-* or *melano-*.

If the black passes somewhat into a greenish, it is called raven-black (*pullus*); if it passes into a brownish, it is called pitch-black (*piceus*).

4. The brown colour (*brunneus*) has as its ground-tone the chesnut-brown (*badius*). If it falls more into the red, it is reddish-brown (*fuscus*, in Greek compounds *phæo-*). If it goes more into a yellow, it is liver-brown; and when a mixture of reddish is added, it becomes rust-coloured (*ferrugineus*).

If the brown colour is pure and somewhat splendent, it is said to be *spadiceus*. If it is quite dirty, and passing into black, it is said to be smoky (*fuligineus*). If along with this smoky state, it is also mixed with undefined shadings of colour, it is said, generally, to be lurid (*luridus*).

5. The yellow colour (*luteus*, in Greek compounds *xantho-*,) has as its ground-tone the golden-yellow (*aureus*, in Greek compounds *chryso-*). This is the same with the citron-yellow (*citrinus*).

A paler shade is denominated, generally, *flavus*. More exactly defined varieties are the wax-yellow (*cerinus*), the sulphur-yellow (*sulphureus*), and last the straw-yellow (*stramineus*), which approaches pretty much to the white colour, and is hence denominated by a Greek compound *ochroleucus*. The dark-yellow approaches either to the red or to the brown. It is generally called tawny (*fulvus*). Dark-yellow, with reddish-brown, is called *helvolus*. *Aurantiacus*, is dark-yellow, passing somewhat into a reddish. Saffron-yellow (*croceus*) is the distinctly intermediate colour between red and yellow.

If the yellow is soiled and pale, it is called *Isabella* colour, and ochre-yellow (*ochreus*). If along with its dirty appearance it passes into a brownish, it is fawn-coloured (*cervinus*).

6. The red colour (*ruber*, in Greek compounds *erythro-*), the ground-tone of which is carmine red (*puniceus*), has many gradations.

A pale clear red constitutes the rose-red (*roseus*) A still paler tint, and somewhat soiled, is called *incarnatus* and *carneus*, according to its different gradations. The former is still a pure, but pale red : the latter, or the flesh colour, is mixed with a yellowish tint. If the pale red falls still more into the yellow, it is called the *tile* colour (*lateritius*).

A pure red, which is clear and passes into yellow, is called *flame-coloured (flammeus)*, and also *vermilion-coloured* (*miniatus*). The highest degree of it is the scarlet-red (*coccineus*).

If the red falls into a brownish, it is called *clove-brown* (*xerampelinus*), which is nearly related to the brown-red.

If the red passes into a dusky black, it is called *blood-red* (*sanguineus*) ; and a complete similar mixture of pure black and red produces the black-red (*atro-purpureus*).

If the clear red has a slight shade of blue, it is called *purple* (*purpureus*). If the mixture of blue and red is almost equal, it is called *violet-colour* (*violaceus*) ; and the palest shade of this is lilac (*lilacinus*).

7. The blue colour (*cœruleus*, and in Greek compounds *cyaneo-*) has as its ground-tone the Berlin-blue (*cyaneus*), the most complete state of which is denominated *sky-blue* (*azureus*).

The *lavender*-blue is a pale blue (*cæsius*) : it is mixed with a little grey.

If the blue passes into the reddish, it approaches the *violet* colour. It is expressed by the words *purpureo-cœruleus*.

8. Lastly, The *green* colour (*viridis*, and in Greek compounds *chloro-*) has as its ground-tone the emerald-green (*smaragdinus*).

Its varieties are chiefly the *celandine*-green ; when mixed with blue and ash grey (*berillus*).

B

Glaucus is a mixture of blue and green. When it is a still clearer green, it is called *æruginosus*. *Prasinus* is a slight variety of it, with which a little ash-grey is mixed.

A dusky green, mixed with brown, forms the olive-green (*olivaceus*)

III. *The Surface of the Parts.*

24.

The surface of the parts has sometimes no covering, and no prominent substances. It is then called *even* (*lævis*). A higher degree of this evenness is denominated *shining* (*nitidus*. The highest degree, which exhibits the surface as a mirror, is called *lucid* or *splendent* (*lucidus*, *splendens*) ; and when there is at the same time a gloss upon the part, it is said to be *vernicosus*. Opake (*opacus*) is the opposite of lucid, as it is also of transparent, (23.)

The want of hairs, or of substances resembling hairs, constitutes the surface smooth (*glaber*.) The term *nudus* is also used, to express the want of hairs or of a covering.

Surfaces that are uniformly even are called *æquabiles*: those that are not uniform, but which have prominences and hollows, although they are at the same time smooth, are said to be *inæquabiles*.

25.

With respect to the hairs in particular, they are in general called *pubes*, and *pubescens* denotes a gentle and almost indistinguishable covering of hairs. Soft (*mollis*, *mollissimus*) is nearly the same.

If the hairs are soft, somewhat long and bent, the covering is said to be *pilosus*; but it is denominated *villosus* or shaggy, when the soft hairs stand parallel and erect.

When the hairs, again, have other hairs attached to them, the idea of plumes is generated (*plumosus*). In *Hieracium undulatum*, Ait. the whole surface is covered with these. In *Dampiera*, R. Br. the hairs of the whole are plumose

Pencil-like (*penicillatus*) is that sort of hair which has its extremity completely set with small attached hairs. (Tab. I. Fig. 12.)

When the soft hairs lie thick upon the surface, so that they give it a silky or satiny lustre, it is said to be silky (*sericeus*).

If the soft hairs are complicated, but yet so that the single hairs can be distinguished, the covering is said to be woolly (*lanatus, lanuginosus*).

Shag (*tomentum*), and shaggy (*tomentosus*), denote that the hairs are so thickly matted together, that the individual hairs cannot be distinguished. The shag is commonly white (*tomentum album, candidum, niveum*); frequently it is hoary (*t. canum, incanum*); less frequently it is rust-coloured (*ferrugineum*); and still more rarely it is of a golden colour (*t. aureum*).

Tufts (*flocci*) are short, thick, soft, irregularly hispid hairs, as they appear upon the leaves of a great many kinds of *Verbascum*, and upon the *corolla* of *Scœvola*.

Stiff, very short hairs, make the surface hispid (*hispidus*). When the stiff hairs are somewhat longer, the surface is said to be rough-haired (*hirsutus*). When the stiff hairs are very long, the surface is said to be rigid (*hirtus*).

When the stiff hairs stand singly, and resemble bristles, the surface is said to be bristly (*setosus*). When the stiff hairs spring from a small prominence or knoll, the surface is said to be strigose (*strigosus*).

When hairs, considerably stiff, are crowded together in a heap, the surface is said to be bearded (*barbatus*).

Stiff matted hairs form what is called *stuppa* or tow. The filaments of *Dianella* and *Hypandra*, R. Br., as also the abortive buds of *Acacia undulata*, W. are *stuppeæ*.

Small, star-shaped hairs constitute the *pubes stellata*. (Tab. VI. Fig. 10. ; Tab. VII. Fig. 7.)

A *coma* is formed when long soft hairs arise from the base of an organ, especially from the seed. *Semina comosa Epilobii*, (Tab. I. Fig. 13.) See another meaning of the term (85)

26.

Besides hairs, there are many other inequalities and roughnesses of the surface.

Thus, the surface is said to be dotted (*punctatus*) when fine small hollows or glands are observable on it. But those transparent points, which do not form any peculiar hollows on the surface, commonly receive the same appellation.

When the surface is beset with such small inequalities, as can only be distinguished by the touch, and not by the naked eye, it is said to be scabrous (*scaber*). *Scaberrimus* corresponds with the next definition. In the lichens, a surface is said to be *leprosus* when it is scabrous, somewhat rent, scaly, but also uniform.

Rough (*asper*) is the attribute of a surface when its inequalities can be distinctly seen. Commonly they pass into short, stiff hairs, and *asperrimus* is hence the same with *hispidus*.

Short herbaceous spines make the surface muricated (*muricatus*).

Stiff points, lastly, form the prickly (*echinatus*) surface.

Small, solid, visible inequalities make the surface granulated (*granulatus*) ; and when the inequalities are larger, the surface becomes warty (*verrucosus* and *papillosus*). When the warts are evidently filled with air or water, the surface is called pustular (*papulosus*). When they are very hard and white, the surface is said to be callous (*callosus*). A coarse granular substance is called grumose (*grumosa*): *Xyloma, stromata Sphæriarum variarum*.

If the warts are considerably larger, the surface is denominated *bunched* (*torosus, torulosus*).

When the inequalities proceed from successive risings and depressions, the surface is said to be wrinkled (*rugosus*). A considerable size in these risings makes the surface blistery (*bullatus*). If the depressions are nearly parallel or streaked, as we see them in the seed of the *Annoneæ*, they are called runcinate (*runcinatus*), (Tab. I. Fig. 24.) When the depressions resemble small hollows, the surface receives different names, according to the size and regularity of the hollows. It is called porous (*porosus*) when the hollows are indistinguishably small, so that they approach to the *dotted* surface. When they are larger, the surface is called *scrobiculate* (*scrobiculatus*). When, in the last place, the hollows are contiguous, and have a regular structure, the surface is said to be honeycombed *(favosus, alveolatus)*.

When the depressions are linear, they are called *furr ow* (*sulci*), and the surface is said to be furrowed (*sulcatus*).

When the depressions are intersected by raised lines of a net form, the surface is said to be reticulated (*reticulatus*), (Tab. VIII. Fig. 6, 7.) ; and when these lines run together into distinct regular squares, the surface is called *tessellated* (*tessellatus*). The surface is streaked (*lineatus*), when parallel straight lines run through it, and are frequently elevated, and frequently also of various colours.

A part is said to be striated, when fine straight lines, which project but a little, are seen running longitudinally on it.

Ringed (*annulatus*) is commonly applied to a roundish or tubular body, when it shews small circular prominences and depressions ; *Conferva muralis.* If these rings go quite through the body, like true partitions, the tube is called *septata ; Puccinia, Didymosporium Nees.*

Rifted (*rimosus*), is applied to a surface, when it has small deep and regular clefts. When these form small fields, these latter spaces are called *areolæ*, and the surface is said to be *areolata* (*Lecidea fumosa, atro-alba*).

A furrowed surface, where the spaces between the furrows are raised in folds, is said to be folded (*plicatus*).

Undulated (*undulatus*) is a surface which rises and sinks successively in gentle lines. When this bending and sinking are very irregular, the surface assumes a crispid appearance, (*crispus*).

27.

There are still some other apparently foreign substances, which form a covering on plants. To these belong

Hoariness (*pruina*), which makes the surface hoary (*pruinosus*). It is a fine, commonly a bluish substance, which is exhaled from the juices of the plant itself, and which can commonly be wiped off.

Meal *(farina)* makes the surface mealy *(farinosus)*, and has a similar origin.

Substances still finer than those form Dust *(pulvis)*, and make the surface dusty *(pulveraceus)*.

Scales *(squamæ)* are dry membranous substances, which rise mostly from the surface ; they render it scaly *(lepidotus)*, (Tab. VI. Fig. 8.) If they are flat, the surface is called *squamulosa*. If they lie upon one another, and are thickly placed, we say, as in the lichens, that the surface is globose *(globosus)*.

Chaff *(paleæ)* is formed by large, commonly dry and point-ed, skins, and renders the surface chaffy *(paleaceus)*.

Gluten, which is likewise an exhaled juice, renders the sur-face more or less glutinous *(glutinosus,* or *viscosus)*.

IV.   *Universal Forms.*

We here treat of such forms only as belong to all the parts, and to which all others have a reference.

The completion of a certain form is expressed by the term *effiguratus.* Thus, it is used with respect to the *peristomæ* of the mosses, and with respect to the lichens, where, for ex-ample, in *Lecanora fulgens,* it expresses the lobed crust of the margin, in opposition to the granular and uniform crust.

28.

The *base* of an organ or part refers always to the point, by which it is inserted, and through which it derives its nourishment.

The *apex* is the point or region which is opposed to the base.

The *axis* is an imaginary or actual line, which proceeds from the base to the apex.

The sides *(latera)* are the parts which lie on both sides of the axis.

The surface (*pagina*) is the extension of a part, and is commonly divided into the upper and under surface.

The margin (*margo*) is the boundary which unites the two surfaces. In a more common sense, when any thing is carried from the sides to this part, it is called the *circumference* or *ray* (*ambitus, radius*). That which lies in the circumference is called *radialis* or *periphericus*, (Tab. I. Fig. 20.) When the margin is perceptibly distinguished by its substance or colour from the central part or disc, the organ is said to be marginated (*marginatus*), as the leaves of Saxifraga aizoon, Bryum marginatum, and the fruits of many of the lichens, (22.)

Limitate (*limitatus*), is, when the boundary is sharply marked; as, for instance, in *Lecidea geographica, parasema*, and others.

The opposite of *limitatus* is expressed by *effusus*, (*Stroma Sphæriæ serpentis.*)

The *central* part of the surface, in opposition to the circumference, is denominated the *disc*, (*discus*). But this word is used equally with respect to leaves and flowers.

The *limb* (*limbus*, or *lamina*), is used with respect to the extended surfaces of flowers; the former term, when the corolla consists of but one petal, the latter, of more than one.

The *keel* (*carina*), is the prominent line or ridge on the under side of a horizontal surface.

Angle (*angulus*), is the point where two lines or surfaces meet one another. It is hence applicable as well to the circumference of the leaves as to the stem and stalk.

The *sinus*, means the curvilineal indentation between two projecting angles. It is chiefly used with respect to the margins of the leaves.

Umbilicus, is a depressed surface, surrounded by an elevated margin.

Appendage (*apophysis*), is a smaller body attached to a greater, which, in the capsules of the Mosses, is commonly called the *crop*, (*struma*).

29.

In order that we may be able to distinguish surfaces exactly, we must attend to the following expressions.

Linear (*linearis*, § 16.), is, as was formerly remarked, a surface which preserves the same small dimensions throughout, and is every where about the breadth of a geometrical line.

Orbicular (*orbicularis*), is a surface which has pretty nearly the form of a circle. If it deviates more or less from the circular form, it is called *roundish*, (*subrotundus*). The term *rounded* (*rotundatus*) is also used, when one part of the circumference approaches to the round form.

Ovate (*ovatus*), is when a surface is rounded at the base, but tapers towards the apex, and when its length is a little greater than its breadth.

Elliptical, or oval, (*ellipticus*, *ovalis*), is when the length of a surface exceeds its breadth twice or thrice, and it is equally rounded at both ends. (Tab. VII. Fig. 6.)

Oblong (*oblongus*), is a surface, the length of which exceeds the breadth more than three times, and which has its ends variously defined. (Tab. VIII. Fig. 2. 5.)

Lanceolate (*lanceolatus*), is a surface which tapers gradually towards its apex, and is of considerable length.

Spathulate (*spathulatus*), when a surface is rounded at the apex, and tapers towards the base.

Wedge-shaped (*cuneatus*, or *cuneiformis*), denotes properly a surface which terminates in a right line at the apex, and tapers towards the base. It is usual also, without regarding the apex, to designate this tapering alone, under the name of the *wedge-shaped*.

Tongue-shaped (*lingulatus*), or band-shaped (*ligulatus*), denotes a surface of some length, with blunt ends, and parallel sides.

Sword-shaped (*ensiformis*), is an oblong surface, one margin of which is hollowed, and the opposite one is elevated.

Sickle-shaped (*falcatus*), is a curved surface, one part of which remains straight.

Obovate (*obovatus*), is the ovate form reversed, with round-ed apex, tapering base, and having the length but a little more than the breadth. This form passes into the fan-shaped (*flabelliformis*), when the apex is much extended, and con-vex.

Triangular (*triangularis*), is a surface, the sides of which are nearly right-lined, and which meet in three angles.

Rhomboidal (*rhomboideus*), is a square moved forward, as it were, at the top. Trapezoidal (*trapezoideus*), is a surface, the sides of which meet in four unequal angles.

Panduræ-form (*panduræformis*), is an oblong surface hav-ing both its sides cut into a *sinus*. (Tab. VI. Fig. 9.)

Heart-shaped (*cordatus*), is when a surface is hollowed at the base, when this hollowing is rounded off at the sides, and the apex tapers. The opposite is *obcordatus*, when the apex of a surface has a heart-shaped hollowing and rounding-off, and the base is tapered.

Kidney-shaped (*reniformis*), is a surface of which the apex is very broad and flat, the sides much rounded, and the base emarginated.

Half-moon-shaped (*semilunatus*, *lunulatus*), is properly a higher degree of the kidney-shaped, when the sides are drawn to some length, and the apex is commonly sloped.

Arrow-shaped (*sagittatus*), when the sides terminate in two straight-lined pointed lobes at the base; and when these lobes are sloped outwards, the surface is said to be spear-shaped, (*hastatus*).

30.

When we attend not merely to the surfaces, but to the substance of the organs, we make use of the following ex-pressions.

Fleshy (*carnosus*), when a part is considerably thick and soft, and has a fleshy consistence.

Membranaceous (*membranaceus*), denotes the thin, but commonly also the coloured character of the part.

Scariose (*scariosus*), when the membranes are destitute of sap, and commonly discoloured.

Chartaceous (*chartaceus*), is a firmer variety of the former, as in the fruit-cover of the Notolæna, Venten.

Inflated (*inflatus*), when a part has thin extended sides, and is internally hollow. Calyx *inflatus* Silenes Hermanniæ ; Faux *inflata* Dracocephali. An inferior degree of this is ventricose, (*ventricosus*).

Crustaceous (*crustaceus, fragilis*), when a part is dry, and is composed of small pieces or scales ; as in the Peridiæ of Physarum, and the fruits of Leucopogon, R. Br.

Dry (*exsuccus*), is used chiefly with regard to fruits, and is opposed to juicy, (*pulposus*, or *succulentus*).

Bony (*osseus*), or stony (*lapideus*), denote the highest degree of hardness ; as in the fruit of Scleria, Lithospermum, Styphelia, and Ventenatia, Cav.

Cartilaginous (*cartilagineus*), is an inferior degree of hardness, but which still allows the parts to be separated with difficulty ; as the margins of many leaves, and the albuminous parts of seeds.

Cork-like (*suberosus*), and spongy (*spongiosus*), explain themselves.

Leathery (*coriaceus*), expresses the union of a certain degree of hardness with elasticity, as in many leaves, and in the corolla of many Styraceæ.

Horny (*corneus*), is something harder than the cartilaginous. It is applicable chiefly to seeds.

If we attend to the substantial configuration of the Organs, we employ the following expressions.

A part is said to be round (*teres, cylindricus*), of which the section is more or less circular. From the gradations of this form, there arise the ideas of half-round (*semiteres*), and roundish (*teretiusculus*). If the cylindrical body is very fine, it is called *capillaceous* (*capillaceus*, 16.), or *filiform*, (*filiformis*) ; in which last case, the thickness may be somewhat greater, and the section does not always exhibit a circle.

Compressed (*compressus*), when the body has two flat surfaces, which meet in projecting angles. When these angles are acute, and the surfaces parallel, the body is said to be *two-edged* (*anceps*).

Depressed (*depressus*), when the surface or centre of a body is sunk : this, therefore, corresponds with the navel-form.

Gibbous (*gibbus*), when a section of the body displays elevations of its surface in particular parts. The simple swelling up of a body is expressed by *tumidus*.

When a section of the body displays angles, and we do not pay any particular attention either to their number or regularity, the body is said, generally, to be angular (*angulatus*). When the angles are acute, and the surfaces flat, the part is said to be three-sided (*triqueter*), four-sided (*tetraqueter*). When the angles are obtuse, and the surfaces even, the part is called three-edged (*trigonus*), four-edged (*tetragonus*), five-edged (*pentagonus*), and so forth. When the sides are sunk, we are accustomed, without regard to the acuteness or obtuseness of the angles, to say, *triangularis, quadrangularis, quinquangularis,* and so on.

Deltoid (*deltoideus*), when a section of the body is three-sided, and when the body itself is not much longer than broad.

Prismatic (*prismaticus*), when the body is somewhat long, and, along with flat surfaces, has a considerable number of angles.

Spherical (*globosus, globulosus, sphæricus, sphæroideus*), when the body approaches more or less to the shape of a sphere. Hemispherical (*hemisphæricus*), when the body has the form of a divided sphere.

An indefinite elevation is expressed by *convexus*; and the opposite, the flat condition of a body, by *planus*, (44.)

Pear-shaped (*pyriformis*), when the body has a thick, but expanded summit, and tapers gradually towards the base. To this is nearly related the. turbinated shape (*turbinatus*). In this case, the summit is commonly truncated, and the base gradually tapers. This also is the form which is sometimes called the *reversed conical* (*obconicus*).

Conical (*conicus*), on the contrary, is when the section of the body is round, its apex tapering, and its base truncated.

Pyramidal (*pyramidatus, pyramidalis*), when the section of the body is angular, its apex tapering, and its base truncated. A variety of this is the mitre-shaped (*mitræformis*), resembling a night-cap; which form is very common in the calyptræ of Mosses, and is opposed to the obliquely-cut form, (calyptra *dimidiata*).

31.

We are also accustomed to derive the appellations of bodies from their resemblance to various substances. Hence the rule, That objects, either natural or artificial, which are seldom met with, or, at least, which are not universally known, ought not to be employed in comparing bodies. The most usual expressions scarcely need any more particular explanation.

Mammillated (*mammillatus*), is commonly a hemispherical body, with a small wart upon its top.

Urceolate (*urceolatus*), when the body is expanded at both ends, and contracted in the middle. (Tab. II. Fig. 16. Tab. III. Fig. 12.)

Cotyloid (*cotyloideus*), when the body is depressed in the centre, and elevated at the margin. (Tab. V. Fig. 12.) A slighter degree of this figure is called saucer-shaped (*scutelliformis*), a higher degree cup-shaped (*scyphiformis*), or cupola-shaped (*cupulatus*). When the surface is scarcely concave, it is called plate-shaped (*patelliformis*). This passes into the disc-shaped (*disciformis*).

Lenticular (*lenticularis*), is a body depressed both on the upper and lower surface, and of which the circumference is round. This is called *shield-shaped*, when the upper surface is raised, and the lower somewhat hollow. The higher degree of the shield-shaped passes into the meniscoid (*meniscoideus*).

The spiral form (*trochlearis*), resembles a screw, as in the capsules of Helicteres. (See *Spiralis*, § 37.)

Spindle-shaped (*fusiformis*), is the conical form reversed, with the lower end drawn out into a fine tapering form. But we must distinguish this from *fusinus*, which denotes a cylinder, which tapers at both ends. (Tab. 1. Fig. 31.)

Capitate (*capitatus*), when the body has a round thick end, or is of a roundish stalk shape.

Cap-shaped (*pileatus*, *pileiformis*). This form is derived from the capitate, when the top is expanded, or has a margin which hangs down. A still greater expansion of the cap constitutes the umbrella form (*umbraculiformis*). Necklace-shaped (*moniliformis*), denotes the connection of round or oval bodies by threads, as in a necklace,—Acrosporium, Antennaria, Nees : Legumen Parkinsoniæ.

Club-shaped (*clavatus*) is when the body has a thick apex, and a gradually tapering base.

Awl-shaped (*subulatus*), when a round body tapers conically towards the point, and becomes extremely fine.

Scobiform (*scobiformis*), as the fine long seeds of Leptospermum and the Orchidiæ.

Scymitar-shaped (*acinaciformis*), when a body, having commonly a section with three sides, has also a long projecting edge, and a thick back.

Axe-shaped (*dolabriformis*), when the body is compressed, rounded, obtuse, and becoming gibbous towards the apex.

Crested (*cristatus*), when the body has erect, rough points. Comb-shaped (*pectinatus*), on the other hand, is when a body has its parts deeply cut, parallel to one another, and lying in the same plane, (Tab. VI. Fig. 12.)

Pillow-shaped (*pulvinatus*), when several individuals or parts are so pressed together, that they form an eminence, a small hill, or pillow.

Wing-shaped or winged (*alatus*), when the body has membranous appendages on both sides, which go out from it in the manner of wings. Semina alata Gladioli, (Tab. I. Fig. 19.)

Arched (*fornicatus*), when the body is concave on the under side, but on the upper side is raised and arched. See the upper lips of many of the Labiatæ.

32.

With respect to the cavities which are found in bodies, many expressions have at different times been employed to denote them ; here we notice only the following.

Concave (*concavus*), when one surface of the body is depressed. A higher degree of this depression is denominated hooded (*cuculatus*).

When the hollowing is performed in a straight line, and a section of it is hemispherical, it is said to be channelled (*canaliculatus*).

When the hollowing is continued throughout a considerable length, and it shews an angular section, it is called *boat-shaped* or *keel-shaped* (*naviculatus*, and *carinatus*).

When a body is hollow throughout its whole length, it is called *tube-shaped* (*tubulosus, fistulosus*). When the lower and narrow part of the tube is gradually stretched into a very wide circumference, the funnel-shape is produced (*infundibuliformis*).

Bell-shaped (*campanulatus, campaniformis*), when the interior hollow channel is closed at one end, and is somewhat narrower in that part than at the other, where it is open and expanded.

When a narrow tube passes suddenly into a somewhat depressed margin, but raised in the middle, it is said to be salver-shaped.

If the tube is so short, that it is scarcely attended to, but the rim is very flat and expanded, the body is said to be wheel-shaped.

If the tube is bent, pointed at one end, and expanded at the other, it is said to be proboscis-shaped (*proboscideus*). To this belongs the shut and self opening cavity of some organs (*pars clausa et dehiscens*). If the opening is very small and round, it is said to be *pertusus*, as the fruit of Endocarpon tephroides. A small hollowing in the base is said also to be *exculptus*, as in the seeds of the Anchusa.

V.   *Insertion, or Relative Position.*

33.

The position of organs, or of their parts, is a very variable expression, which may be referred to several kinds of charac-

ters.   1. We may consider the situation of parts, in reference
to the things which surround them. We thus call them
floating, *submersed*, *buried*. Of these we shall speak on another occasion.   2. We may consider the situation of a part,
in relation to the organ which lies under it.   This is called
the *insertion* (*insertio*), which is either *mediate* or *immediate*.
The mediate insertion supposes an intervening substance, by
which means a kind of articulation is produced.   Such an insertion, we observe, for instance, in the labia of some of the
Orchideæ, as in Dendrobium, Sw.

When we mean to signify, generally, that one organ rises
out of another, or is inserted on it, we form a derivative from
the name of the organ which serves as its support. We thus
say, *radicalis*, *caulinus*, *rameus*, *petiolaris*. These denote
that which springs from the root, from the stem, the branches,
or the leaf-stalks, or which is inserted on them ; Flores petiolares of Turnera cuneiformis, (Tab. V Fig. 12.) We also sometimes form words from the names of both the organs,
—from that which is inserted, and that which serves as the
support. We thus say *calyciflorus*, *calycostemon*, (when
the filaments spring from the calyx) ; *gynandrus*, (when the
anthers are set upon the pistillum) ; *rhizanthus*, *thalami-
florus*, (when the flower is set on the *receptacle*.)

We also use the prepositions *epi* and *hypo*, to denote more
precisely the insertion. We thus say, *epipetalus*, *hypophyl-
lus*, and so forth. Lastly, the expressions *dorsalis*, *lateralis*,
basilaris, *terminalis*, are important. We signify by them
that an organ is inserted, on the back, on the sides, on the
base, or on the summit.

We also attend to the situation, as it is in the centre, or
in the axis, (*centralis*, *axilis*), or out of the centre (*excentri-
cus*) ; and also to the tendency towards the centre, or from
the centre, (*centrifugus*, *centripetus*) ; which terms are especially employed respecting the situation of the embryon, and
its radicle in the albuminous substance.

34.

We may consider the insertion or situation of an organ, in regard to the other parts by which it is surrounded. We thus say that a part is inferior (*inferus*), when it is placed below another. If this is the case with the calyx, in respect of the ovarium, it is said to be free (*liber*). But if the ovarium is inferior, in respect of the rim of the calyx, it is then united with the tube of the calyx. In this latter case, the calyx is superior; in the former instance, it is the ovarium. Tab. VIII. Fig. 6, 7, the fruit is superior, the calyx inferior. See also in Tab. VI. Fig. 13.

We are also accustomed to distinguish between *intrafoliaceus* and *extrafoliaceus*. The latter term is used, for example, respecting flower-stalks, which have no relation of position to the leaf-stalks, as happens in many kinds of Solanum. We also say of the stipulæ or appendages of the leaves, that they are *stipulæ intrapetiolares*, when they appear within the leaf-stalk. Flower-stalks are called *petiolos oppositifolios*, when they spring opposite to the leaf-stalks, or the leaves; while, on the contrary, their usual origin is in the axis of the leaves (*axillaris*).

Suprà and *infrà* are used respecting surfaces, *supernè* and *infernè* respecting lines. The former are also used respecting the surfaces of the leaves; the latter, respecting the stalks, to denote their upper or lower part. *Utrinque* is used in both senses: it may thus denote the two surfaces of the leaves, or it may also mean the summit and base of the leaf.

Sursum denotes a direction upwards, towards the summit; for which purpose we also employ *antrorsum:* the opposite of which are *retrorsum, deorsum*. Panicum viride has its *involucella sursum* and P. verticillatum *deorsum hispida*. Valantia Aparine *folia antrorsum*, Galium Aparine *retrorsum aculeata*. *Anticus* and *posticus*, denote position before or behind another part. We also use *anterior* and *posterior*. In Cryptostylis, the labellum is *posticum ;* in Genoplesium, Br. the galea is *antica*.

35.

The position and insertion of the sexual parts, in relation to one another, is of more importance, frequently difficult to investigate, and by no means uniform. In general, we call the stamina *hypogyna*, when, as in the Grasses, (Tab. III. Fig. 7.) they rise from a lower surface than the female parts. Stamina *perigyna* are those which spring from the same plane with the female parts; (Tab. VIII. Fig. 4. Tab. III. Fig. 24. Tab. IV. Fig. 14.) Commonly the former are united with the corolla, pass downwards along with it, and thus show their origin to be in the same plane with the female parts. But in many of the Caryophylleæ, one half of the stamina, in the same plant, are *hypogyna*, the other half *perigyna*. In Silene, Cerastium, Dianthus, and Saponaria, five stamina stand on the *receptacle*, and therefore deeper than the *ovarium*: but the other five are connected with the petals of the corolla, and may be considered as *perigyna*.

Lastly, we call those stamina, or anthers, *epigyna*, which, as in Cleone and in the Gynandriatæ, are united with the pistil, or with the *columna genitalium*; (Tab. IV. Fig. 10—12.)

36.

With respect to the position of organs of the same kind, we must attend to the following expressions.

Parts are said to be *opposita*, when they rise opposite to one another, or lie directly before one another. In the former sense, the word is used respecting leaves which spring from opposite sides of the stem. In its latter acceptation, it is commonly used respecting the petals of the corolla, in so far as they stand directly before the leaves of the calyx; or respecting the stamina, in so far as they stand directly before the parts of the corolla. We also say that the partition of a capsule is opposite (*dissepimentum contrarium* or *oppositum*), when it is placed perpendicularly to the valves of the capsule. On the other hand, the partition is parallel, when, arising out of the suture of the valves, it has the same direction with them, (Tab. VII. Fig. 6.)

C

Star-like (*verticillata, stellata*), when several parts in the same plane, and with different directions, seem to rise from the same point.

Decussated (*decussatus*). The direction of organs is thus designated, when, standing under or above one another, they form with each other a right angle,

Cross-like (*cruciatus* or *cruciformis*). This, on the other hand, consists in the direction towards four opposite sides of parts which lie in a horizontal plane.

Doubled (*geminus* or *geminatus*), when parts, which are distinct, yet stand in the immediate neighbourhood of each other.

Three-together (*ternus*), when the parts stand by threes in the same plain. We perceive from this the meaning also of *quaternus, quinus,* &c.

37.

Alternate (*alternus, alternans*), when the parts either are inserted, like steps, on the two opposite sides, or at least when they stand not directly before, but between other parts. The former occurs in leaves, the latter in the situation of the petals of the corolla with respect to the divisions of the calyx, or of the stamina with respect to the parts of the corolla.

Two-rowed (*distichus*), when, without regard to the opposite or alternate insertion, the planes of the parts lie in one surface.

On the other hand, we use the phrase *in two directions* (*bifariam*), when any property is observed on opposite sides, with the same character. Thus, we say that the stalk of Veronica Chamædrys is *bifariam pilosus*, because a line of hairs springs from both sides of it. Thus, also, we say that the leaves of Lycopodium complanatum are *bifariam connata*, because they grow in two opposite directions. We hence perceive what is meant by *quadrifariam* (in Lycopodium alpinum), *quinquefariam* (in Lycopodium annotinum), *sexfariam* (in Lycopodium dendroideum). It is thus said, respecting the fruit of Nicotiana, *Capsula apice quadrifariam dehiscens.*

In rows (*serialis*), when the parts follow one another, in a certain order or train. We hence understand what is meant by *bi-* and *tri-seriatus*.

Spiral (*spiralis*), when the parts form a spiral line around the common axis; (Tab. III. Fig. 25, Tab. V. Fig. 1.) *Trochlearis* is the same (13.), only applied to solid bodies. *Gyrosus* is used respecting level surfaces, as respecting the fruit of Lichens.

Rose-like (*rosaceus*, or *rosaceo-congestus*), when the parts, by their crowded position, form rosettes, as the leaves of Bryum roseum and Alyssum nebrodense, (Tab. VII. Fig. 5.)

Radiated (*radiatus* and *radians*), when the parts are placed like the spokes of a wheel.

38.

If no order be observed in the position of the parts with respect to one another, they are called *sparse* (*sparsa*). If, in addition to this, they are thickly placed, they are said to be crowded (*conferta*, or *congesta*). But the parts are called *aggregate*, when several of them rise from one point : Intricate (*intricatus*), when the parts are so heaped together that their origin and direction are altogether undistinguishable ; *Hypha Sporotrichi*, Lin.

If they do not originate in exactly one point, but spring up in the neighbourhood of each other, they are then said to be fasciculated (*fasciculata*); and conglomerated (*conglomerata*), when they have no peculiar support, but touch one another.

We also apply the expression compact (*coarctatus*), when the parts, without regard to their origin, stand thick together. We use also the word turf-like (*cæspitosus*), respecting stalks or branches, which seem to stand thick together, and to grow from one point.

When plants grow in distinct patches, they are called *gregariæ*, in opposition to *solitariæ*.

Parts are contiguous (*contiguus*), when their margins seem to meet each other. Continuous (*continuus*), is a part which seems to be one with some other part, or it is an individual organ which goes on without interruption.

Connivent (*connivens*), when the parts, without being connected, are yet bent towards one another, as, for instance, the antheræ of Cyphia serrata; (Tab. VIII. Fig. 3.)

Incumbent (*incumbens*), when a part rests upon the surface of another part, without being united with it, as the radicle on the cotyledons of Erysimum hieracifolium; (Tab. I. Fig. 33.) On the contrary, accumbent (*accumbens*), is used when a part is placed upon the sharp edge of another, as the radicle on the margin of the cotyledons of Sinapis nigra; (Tab. I. Fig. 34.)

39.

The opposite of proximity is expressed by the terms distant (*distans*), remote (*remotus*). Rare (*rarus*) is the opposite of *confertus*: Lax (*laxus*), the opposite of *coarctatus*; and a higher degree of the former, when the parts hang loosely downwards in all directions, is commonly denominated *diffusus* and *flaccidus*.

The opposite of contiguous is discrete (*discretus*); (Tab. VIII. Fig. 3. 4.)

40.

When one part is placed immediately upon another, it is, in general, said to be sessile (*sessilis*). But of this there are several varieties.

One part may form a joint with another; it is then said to be articulated (*articulatus*). It may grow along with it (*connatus*), or it may have a general connection with it (*cohærens*). When plain surfaces pass into one another, they are said to be confluent (*confluentes*), as in the fruit of Lichens.

When one organ, with its lower surface, embraces another, it is called *amplexans*, whence Folia amplexicaulia; and when this lower surface extends itself in the shape of a saddle on both sides, the organ is called riding (*equitans*). When it descends, in the form of a sheath, around the other body, it is called *vaginans*.

When one organ is sunk into another, it is said to be immersed (*immersus*), as the fruits of many Lichens; or bedded and nestling (*nidulans*), as the seeds of the Melastoma in the pulp of the berry.

When, along with the immersion, it projects a little, it is called *emergens*, as the fruit of Lecidea lithyrga, from the crust.   When a part projects a little, without being directly immersed, it is called *prominulus; Nervi foliorum sublus prominuli.*

When a part of one organ runs downwards on the surface of another, it is said to be decurrent (*decurrens*).

When leaves grow in such a manner around the stem, or leaf-stalk, that they seem to constitute one substance with them, they are called *perfoliate* (*perfoliata*), as in Lonicera caprifolium, Bupleurum rotundifolium, and Jungermannia coalita, Hook. (Mus. Exot. ii. t. 123.)

When several parts which are sessile, partly cover each other, they are called *imbricated* (*imbricatus*).

When a part has a stalk, this is variously denoted, by *petiolatus*, in leafy parts ; by *pedunculatus* and *pedicellatus*, in blossoms and fruits; and by *stipitatus*, in other parts.

Peltated (*peltatus*), is a part which has its stalk in the centre, and not on the margin.

Versatile (*versatilis*), is a part which rests in such a manner on the point of another, that it is inserted in one point only, and is easily put in motion ; for instance, the antheræ of the Grasses and other plants.

When one part is so loosely bound to another, that it is held only by one small point or thread, but in every other part is loose, it is called *basi solutus*, as the leaves of Sedum reflexum, saxatile.   The opposite of this is *adnatus*, as in the leaves of Sedum sexangulare.

VI.   *Direction of the Parts.*

41.

The relation of organs, with respect to the horizon, is their *direction.*

A part is called straight, when it proceeds in a straight line. Upright (*erectus*), when it stands more or less perpendicular to the horizon. The higher degree of the straight direction is called *stiff* (*strictus*), in which case a geometrical straight line nearly is described.

Rigid, on the other hand, (*rigidus*), is a part which is inflexible, without having necessarily the straight direction.

The opposite of the straight direction is the bent (*flexuosus*), when the direction is removed on one side or the other from the straight line, (Tab. VI. Fig. 11.) To this belong the subordinate definitions, crooked (*infractus*), when one organ takes suddenly the opposite, or a quite different direction; geniculated (*geniculatus*), when an organ changes its direction, so as to form something like a knee; twisted (*tortus*, *tortilis*), when an organ is twisted round itself, or changes its direction in a variety of ways, (Tab. II. Fig. 13.) *Tortilis* denotes the capability, *tortus* the fixed convolution.

Twining (*volubilis*), when a part winds itself in a spiral line around another part; in which case, we observe whether it is turned to the right or left side (*dextrorsum* or *sinistrorsum*). The former happens in Bryonia and Lonicera periclymenum, the latter in Calystigia sepium. When a part is twisted upon itself, it is said to be snail-shaped (*cochleatus*), (Tab. I. Fig. 35.); and we attend then to the individual turnings (*anfractus*).

When a part clings fast to another, and bends this way and that, but without going round, it is called climbing (*scandens*).

42.

The direction, which is parallel to the horizon, is called *horizontal* (*horizontalis*, *patentissimus*), in opposition to the perpendicular direction (*verticalis*). That which makes an angle with both lines, the perpendicular and the horizontal, is called *oblique* (*obliquus*). But this is subject to the following varieties.

When one organ simply approaches so near to another,

that it has almost the same direction with it, it is said to be *appressus*.

When the upright direction is left (*erectus*), and the part comes more towards the horizontal line, the direction is then said to be spreading (*patulus, patens*), as in the branches of Hirtella glandulosa; (Tab. VII. Fig. 1.)

In this case, an angle of about 45° with the horizon and with the perpendicular is described. But we may continue to call the direction upright, although there should be 20° of variation from the perpendicular.

Divaricated (*divaricatus*), when the direction is intermediate between the spreading and the horizontal, and even the horizontal is frequently so named. But squarrose(*squarrosus*), is usually taken in a different sense, when parts, which lie thick upon one another, raise their upper extremities on all sides.

Divergens expresses the varying direction in general; in particular, it is often taken for *patentissimus*. When long branches diverge, and are divided in a forked manner, they are said to be arm-shaped (*brachiatus*).

When a part is bent with its point towards the horizontal line, it is said to be stooping or nodding (*cernuus, nutans*). The latter is a higher degree of the former, and the direction really changes then into the horizontal.

When an organ, especially a stem, lies upon the ground, it is said to be procumbent (*procumbens*). When it is first somewhat upright, and then is turned down, it is called decumbent (*decumbens*). When it is quite horizontal, it is said to be prostrate (*prostratus*). To this order belong also the terms creeping (*repens*), and rooting (*radicans*); (Tab. VI. Fig. 12.)

An organ is called ascending (*ascendens*), when its lower part lies flat, but its upper part is erect; (Tab. VI. Fig. 12.)

That which lies under the earth, is said to be *hypogæus*, as the cotyledons of the Walnut and Horse-chesnut; *epigæus* is that which comes up above the ground, as the other cotyledons.

43.

When the direction is downwards, and under the horizontal line, it is called *pendulous* (*pendulus*) when we refer to the points of the part; it is called *reflex* (*reflexus, reclinatus,* and *deflexus*), when the direction of a part amounts to about 45° under the horizontal line.

Bent downwards (*declinatus*), is when a part is bent towards the horizon.

Inverted (*inversus*), when the upper part becomes the lower; as when, for example, the embryo in the seed stands with its radicle upwards. To this belongs also the *reversed* direction (*resupinatus*), when the part which commonly is uppermost, is found undermost. Thus the flowers are said to be *resupinate*, when in the Labiatæ the staminæ are forced down, and the lower lip has the form of the upper; and in the Leguminosæ, when the vexillum, which on other occasions forms the upper part, becomes the lower.

When the direction of the parts is altogether to one side, it is said to be partial (*secundus, homomallus, heteromallus*). When a property or form of an organ is observed only on one side, this is expressed by *hinc; Capsula hinc gibba.*

44.

Something has been already said (26.) respecting the direction of surfaces; but we find it necessary to mention the following particulars, as properly belonging to this department.

Complicated (*complicatus*), when a part is folded into itself. *Conduplicatus* expresses the longitudinal folding; to which belongs the term *runcinatus* (26.), only this term is employed in a special sense.

Bent-back (*revolutus*), when the margin or surface of a part is rolled outwards or downwards; (Tab. II. Fig. 14.)

Involute (*involutus*), when the surface or margin of a part is bent inwards. Obvolute, when the parts are rolled round one another. Convolute is nearly what we have already called *snail-shaped;* (Tab. I. Fig. 35., Tab. VII. Fig. 4.) This is also called *circinnatus*, when we speak of threads and fine tubes.

The opposite of all the various bendings is *plain* or *even* (*planus*), (30.)

VII. *Simplicity, or Composition of the Parts.*

45.

A part is called *simple* (*simplex*), either when it is not divided into separate parts, or when it proceeds without interruption ; or, lastly,, when it has certain subordinate parts placed only in one row.

Simple stalks are thus opposed to the branched ; simple lines to those that are articulated ; simple covers, or calyces, to the double or threefold, as also to the scaly.

46.

With respect to composition, we remark the following kinds in the leaves.

A leaf is said to be compound (*compositum*), when it consists, generally, of several distinct parts, which have a common stalk, or point of insertion. A simple leaf may be deeply lobed, without being on that account compound, provided the substance of the leaf is still united in the base. Hence there are transitions from the folium palmatum, or hand-shaped, to the finger-shaped or digitatum. When two leaves stand together on a common leaf-stalk, they are said to be binate (*binatum*), or conjugate (*conjugatum*), as in the genus Zygophyllum. When three of them stand on a common leaf-stalk, the leaf is said to be ternate (*ternatum*), as in clover. When there are five, they form the quinate leaf (*quinatum*). When there are seven, they form the septinate leaf (*septinatum*). The two latter are said to be fingered (*digitatum*).

47.

A leaf is called *pinnated* (*pinnatum*), when it consists of several distinct leaves, which spring along the sides of a common leaf-stalk; (Tab. VI. Fig. 11.) The common leaf-stalk is called the *petiolus communis*, as also the *axis* and *rachis*.

Pinnated leaves are classed according to the position of the individual leaflets (*pinnæ*). When these stand opposite to one another (*oppositè pinnata*), they are reckoned by pairs (*jugum*), and the leaves are said to be two-paired, three-paired, four-paired, and so forth (*bi- tri- quadri-juga*). At other times, however, the leaflets alternate with one another (*alternatim pinnata*). If the summit of the whole leaf terminate with an unpaired leaflet, it is then said to be *impari-pinnatum*: when there is no unpaired leaf on the point, the leaf is said to be abruptly pinnated (*abruptè pinnatum*).

When, between the proper side-leaves, smaller leaflets are placed alternately, the whole leaf is said to be interruptedly pinnate (*interruptè pinnatum*), as in Agrimonia Eupatoria. When the side-leaves run into one another, the leaf is called *decursively pinnate* (*decursivè pinnatum*), as in Scabiosa alpina.

48.

When the common leaf-stalk is divided in two parts, it is said generally to be *doubly compound* (*decompositum*); and when the division of the leaf-stalk is threefold, it is said to be *super decompound* (*supra decompositum*), as in Peucedanum officinale.

Doubly pinnated (*bipinnatum*), is a leaf, of which the common axis is again set forth with pinnated leaves, as in Athyrium Filix fœmina. The leaves of the first order are then called *pinnæ*, or *foliola partialia*; the leaves of the second order *pinnulæ* or *foliola propria*.

Triply pinnate (*triplicato-pinnatum* or *tripinnatum*), is a leaf of which the common axis has a threefold subdivision.

49.

The divisions of the Nerves of the leaves also belong to this department.

Nerves, in general, are those visible continuations of the leaf-stalk, or of the point of insertion, which take place through-out the length of the leaf. The lateral branches of the nerves are called Veins, which thus never run parallel with the axis, but always form an angle with it.

When the nerve is divided at the base, we name the leaf according to the number of the nerves, *three-nerved, five-nerved (tri-nervium, quinque-nervium)* ; but when the side-branches of the principal nerves do not spring directly from the base, but arise first a little above it, so that they have some of the substance of the leaf under them, the leaf is then said to be *triple-nerved, quintuple-nerved (tripli- quintupli-nervium)*.

The veins and nerves often anastomose, or they are united by side-branches (*Venæ anastomosantes, Anastomoses vena-rum*). In the Ferns this is particularly observed.

The Nerves are seen to have leafy processes (*nervi lamel-lati*), in Gymnostomum ovatum, (Tab. IV. Fig. 9.) ; in Po-lytrichum lævigatum Wahlenb., angustatum Brid. tenuirostri Hook., and in some other mosses.

The nerves are said to be excurrent (*excurrentes*), when they go on to the apex. The opposite are interrupted nerves (*nervi ad medium, ad $\frac{2}{3}$ evanidi*). Some also employ the phrase, *Folia ruptinerviæ*.

As the leaf-stalk commonly passes into the central nerve, and two other nerves place themselves on the sides of it, three and five nerved leaves are therefore the most common. Two-nerved leaves appear almost only in the mosses, and most distinctly in Neckera affinis, Hook. (Musc. Exot. ii. t. 122.)

50.

In the division of branches and of stalks, the forked-shape is the most common (*rami, pedunculi dichotomi*). In this case they are always divided into two. In the branches of the Umbellæ, the first division is expressed by *dichotomus*, the second by *bifidus*. Also *pedunculi*, *rami trichotomi* are not unfrequent ; for instance, in the *panicle* of some of the species of Avena.

A simple forked division is expressed by *furcatus*.

51.

With respect to the uninterrupted continuation of an organ, we find its simplicity subject to the following alterations and reverses. We have already remarked (38.), that when a part proceeds uninterruptedly forward, it is called *continuus*. In the Confervæ, there are contracted parts (*stricturæ*), which are exceptions to this simplicity ; (Tab. V. Fig. 10.) In the stem and branches there are knots (*nodi*), swellings produced by a crowding of substance, and which contain within themselves the means of increase. On the base of the leaf-stalk also (*Osteospermum moniliferum*), similar knots appear.

When no indentation takes place at the margin of the part, we have the idea of *smooth-margined* (*integerrimus*), in which case there is thus also no interruption of the progress in a line.

Indentations of the margin are caused by teeth, notches, and cirrhi.

Teeth are, in general, pointed projections on the margin. The rim is called *dentated* (*dentatus*), when there are interstices between these pointed projections ; but when the teeth run into each other, the rim is said to be serrated (*serratus*). Forms of nearly the same kind are, *denticulated* (*denticulatus*), and serrulated (*serrulatus*), (Tab. VIII. Fig. 3.) ; as also, coarsely dentated (*grossè dentatus*), deeply, unequally, equally, doubly, and obsoletely dentated (*profundè, inæqualiter, æqualiter, duplicato*, and *obsoletè dentatus*). We say also, equally, unequally, sharply, hooked, connivent, doubly, obsoletely serrated (*æqualiter, inæqualiter, argutè, uncinato, conniventi, duplicato, obsoletè serratus*).

Notches are blunt rounded teeth. The margin is called *crenated* (*crenatus*), when it has indentations of this sort. Varieties of the crenated are the *finely-crenated* (*crenulatus*), and the *crenated-dentated* (*crenato-dentatus*), when the notch is not completely rounded, but is not properly pointed.

Cilia are hairs, or bristles, which divide the margin. The margin is then called *ciliatus ;* (Tab. VI. Fig. 11.) The hairs frequently rise from sharp teeth, on which account the margin is then said to be *serrato-ciliatus.* The hairs are sometimes so stiff, and at the lower extremities so broad, that they might be taken for spines or thorns. The margin is then said to be *ciliato-aculeatus*, or *spinoso-ciliatus.* Sometimes the hairs have other round bodies, or glands, on their points, the margin is then called *glanduloso-ciliatus ;* (Tab. VII. Fig. 1. 4.)

When fine fringes, in the form of *cilia*, extend themselves from the surface, the idea of *fringed* (*fimbriatus*), is generated.

52.

With respect to the unseparated parts of a surface, we have the following definitions.

Large unseparated parts, which are broad and rounded, are called *lobes* (*lobi*). When they are small and pointed, they are called *fringes* (*laciniæ*).

Hence we use the expression lobed (*lobatus*), when, in general, there are lobes, without denoting their number ; thus, also, *three-lobed* (*trilobus*), four-lobed (*quadrilobus*), and so forth.

When a part has *laciniæ*, it is said, in general, to be cleft (*laciniatus*), when the number of clefts is not mentioned ; but, when these are counted, we observe whether the cleft proceeds to the centre, or almost to the base. In the former case, we say of the part, that it is *tri- quadri- quinque-fidum.* If the clefts go almost to the base, we call the part *tri- quadri- quinque-partitus.* This distinction is chiefly important respecting the calyx.

If the clefts go as far as the central rib, so that the substance of the leaf is divided, we say, that it is *sectus*. We say also, *folia trisecta, ternatim secta*, and we call the cleft parts *segmenta*.

53.

The clefts themselves, or the interstices between the projecting parts, are called *sinus*, when they form curved lines; hence a sinuated leaf (*unciatum*), is that which has bendings of this sort on its margin.

Deep rents on the surface, when they are altogether irregular, give the idea of *torn*, or *rent* (*laceratus*, or *multifidus*); smaller irregular projections and rents render the part eroded (*erosus*).

Angled (*angulatus*), when the margin has projections which are greater than *teeth*, but are not proper lobes. When these angles come out very feebly, and often are undistinguishable, the margin is said to be repand (*repandus*).

54.

Palmated (*palmatus*), is when the surface is lobed, or cleft, and its clefts go commonly in five divisions to the under part of the surface.

Pinnatifid (*pinnatifidus*), when a surface has long parallel lobes, or clefts, on both sides. It coincides frequently with the *decursivè pinnatum* (47.) Bipinnatifid (*bipinnatifidum*), is when either the side-lobes are again *pinnatifid*, or when, in a properly pinnated part, the side-leaves shew this half feathering.

Lyre-shaped (*lyratus*), is a pinnatifid surface, the highest unpaired lobe of which is rounded, and the side-lobes become always the finer the nearer they approach the base.

Runcinate (*runcinatus*), again, is when the uppermost unpaired lobe of a pinnatifid surface is pointed, and the side-lobes hang down.

VIII. *The Manner in which an Organ is Terminated.*

55.

We here attend to the apex of an organ. The termination of an organ, as we formerly remarked (28.), is called its apex (*apex*, seldom *vertex*). This is obtuse (*obtusus*), or rounded (*rotundatus*) (29.), when it approaches, more or less, to the round form. We also say, respecting solid bodies, that they are *thickened at the summit* (*apice incrassatus*). To this belongs partly the club-form, (31.)

It is truncated (*truncatus*), when it seems to form a straight cross-line. It is bitten (*præmorsus*), when a curved line seems to cross it.

It is retuse (*retusus*), when a slight curvature is observable in the middle of the obtuse apex. When a sharp remarkable curvature passes inwards on the obtuse apex, it is said to be emarginated (*emarginatus*).

A slight degree of obtuseness is expressed by *obtusiuscula ;* and the reverse of a sharp, or hairy apex, is called unarmed (*muticus*).

56.

The pointed character is called generally *acutus*, the slighter degree *acutiusculus*. *Acuminatus*, again, denotes a long projecting, highly-tapering apex. When this runs out gradually into an apparently fine spine, it is called *cuspidatus*.

When the apex is somewhat obtuse, and a gentle tapering suddenly takes place at the extremity, it is said to be *apiculatus*. *Mucronatus*, again, denotes a rounded apex, with a herbaceous spine standing on it.

When the long projecting apex is placed somewhat obliquely, it is said to be *rostratus*, *rostellatus*, which frequently appears in the covers of the capsules of mosses.

The apex is also said to be bearded (*aristatus*), when a long projecting bristle terminates it. The *awl-shaped* apex (*apex subulatus*), is easily understood, from what was said

(31.)   In like manner, the pricking-apex (*apex pungens*), needs no further explanation.

IX.   *Duration of Plants, and of their Individual Parts.*

57.

Persistent (*persistens*), is the epithet given to a part, when it continues to exist beyond the time at which, according to the laws of vegetation, it ought to wither or fall.   With respect to leaves, this is also called their property of being perennial (*perennans*), when they are observed to be always green.   *Sempervirens* has the same meaning.

There are also parts, which, towards the time of their probable fall or decay, grow with increased vigour.   We then use the terms *accrescens*, or *auctus*.   Every change from the progress of vegetation, is denoted by the addition of *demum*, or *ætate*: *Apothecia demum angulosa; capsulâ ætate auctâ.*

58.

The opposite of persistent is *decaying*, for which we have, in the Latin technical language, two different terms.

An organ, or part, is said to be *caducus*, when it loosens itself very speedily in a joint at the base, as the calyx of Papaver and Chelidonium; *deciduus*, again, is the term used, when, without releasing itself at one joint, the part falls at the same time with other neighbouring organs.

Decaying (*marcescens, marcidus*), consists in a withering of the part, without a falling off.   The disappearance of a part is expressed by *evanescens*.

59.

With respect to the earlier, cotemporaneous, or later appearance of particular parts, in relation to others, the following expressions are used.   Parts are called *early* (*præcox*), when they shoot out, or come to perfection before others; *coeval* (*coætaneus*), when this happens at the same time;

FOR FORMS AND QUALITIES.   49

late (*serotinus*), when they appear later than others. These definitions are important for distinguishing the Willow tribe.

Again, the female and male parts come to maturity at different times. In a great many plants, the antheræ are sooner ripe than the stigmata or pistillæ. This is called *androgynous dichogamy* ; (Tab. II. Fig. 11. Tab. VI. Fig. 5.) But when the female parts come to perfection sooner than the male, this is called *gynandrous dichogamy.*

60.

Many organs, from internal laws, never attain their perfect state. They vary, in consequence, both in their form and substance, and become unfit for their functions. These are called *abortive* (*abortivus*), and their germs are called *rudimenta.*

61.

With respect to the absolute duration of plants, the following expressions are important.

Very evanescent (*fugacissimus*), when an organ scarcely shews itself before it again disappears ; as happens in various blossoms. As also in Sporidia fugacia Ceratii Pers. Hypha fugax Byssi. By means of some plants, we find the hour of the day, whence they are called *horarii.* If they shew themselves only for one day, they are called *ephemeral* (*ephemeri*). If they appear only in the day-time, they are called *diurni ;* if only during the night, *nocturni ;* during the morning, *matutini ;* mid-day, *meridiani ;* after mid-day, *pomeridiani ;* in the evening, *vespertini.*

62.

Their duration for a month is denoted by *menstruus ;* for two or three months, by *bi- tri-menstres.*

If a plant dies the same year in which it sprung up and blossomed, it is called an *annual* (*planta annua*), for which the sign O is used.

Leaves and shoots of the present year are called *horni ;* of

D

the past year *annotini ;* and those of the year before the
last *bimi.*

63.

When a plant springs up, and grows during the first year,
and during the second puts on fruit and dies, it is called a
biennial (*biennis*), the sign for which is ♂.

If a plant lasts several years, and every year sends out
new matter from its root, it is called *perennial* (*perennis*), the
sign of which is ♃.

CHAP. III.

NAMES OF THE ORGANS.

I. *The Root.*

64.

The root (*radix*), is that part of the plant by which it
descends into the earth. It may be considered as a part of
the stem, which has been changed only by the covering of
earth.

But the root is distinguished from the radicle, or fibrils
of the root (*radiculæ fibrillæ*), which are branches, or fibres,
that descend from the principal root.

65.

A thickened root, in which we can commonly distinguish
the solid kernel from the softer surrounding matter, is called
a *tuber* (*tuber*), (288.)   The forms of these are so extremely
various, that they pass from the common spindle-form to the
perfectly spherical, the turbinated, and other forms ; (Tab.
VI. Fig. 1. 3.)

A bulb (*bulbus*), is a thickened, and commonly a spherical or oval-shaped root, the solid central body of which is contained within scales that lie upon one another, and between which the stem, or shaft, rises ; (289.)

II. *The Stem.*

66.

Under the name Stem (*truncus*), we understand, generally, that part of the plant which rises above the ground, and from which all the other parts are evolved.

In particular, it is called the *stalk* (*caulis*), when it is more or less of a herbaceous nature. A plant which has a stalk is called *caulescens* ; one in which it is wanting, is called *acaulis*.

67.

Tree-like stems (*trunci arborei*), and trees (*arbores*), are those plants which have a simple and woody stem.

In sections of woody stems we distinguish various parts, which commonly lie in concentric layers within one another, namely,

1. The rind (*cortex*), the outer part of which, covered by the epidermis (*epidermis*), is for the most part brown, grey, or of some similar colour ; the inner part is entirely cellular, and of a green hue.

2. The inner bark (*liber*), is an apparently fibrous, whiteish, and very flexible part, which lies under the rind.

3. The soft-wood (*alburnum*), or the layer of young wood, which approaches nearer to the nature of the inner bark, by its brighter colour and greater flexibility.

4. The wood (*lignum*), distinguished by its hardness and cross-joinings, or bundles of rays (*radii corticales*).

5. And, lastly, The pith (*medulla*), apparently of an entirely cellular structure, and in old plants either entirely gone, or only remaining as a thin, almost inorganic, brown kernel ; (291.)

68.

Shrubs (*frutices*), are those plants which send out several woody stems from the same root. For shrubs and trees, we use the sign ♄.

Undershrubs (*suffrutices*), are those, the lower part only of whose stems are woody, but whose upper part, being of a herbaceous nature, dies every year.

69.

The place where the stem and root meet, has received various names. Young calls it the *limes communis*, or *fundus plantæ*. Lamark calls it the *life-knot*. Some denominate this part *rhizoma*, or *root-stock ;* and also *cormus*, and *caudex*. De Candolle calls this part the *neck* (*collum*).

70.

In the different families, different names are used for the stem and its parts.

In the Grasses, and Grassy Plants, it is called the *straw* (*culmus*). In Ferns, Palms, and Fungi, it is called the *stipe* (*stipes*) ; but the latter word is generally used to express different parts.

A leafy stem is called generally a *frond* (*frons*), especially in imperfect plants. The frond of Lichens is either *crusta,* when it is quite uniform, granular, or at least as if some matter had been deposited on it ; or it is called *thallus*, when it is leafy, lobed, or shrubby ; (Tab. II. Fig. 3.)

In the Fungi, we employ the term Hypha, when the stem is very delicate; (Tab. I. Fig. 31. ; Tab. V. Fig. 5.)

71.

Branches (*rami*), are the divisions of the stem. Twigs (*ramuli*), are the last and youngest branches.

Sarments (*sarmenta*), are those branches, or stems, which lie upon the ground, and here and there send out roots.

Shoots (*surculi*), are the stems and branches of Mosses and Jungermanniæ.

Sprouts (*turiones*), are shoots of the present year, which are not completely unfolded.

III. *Buds, Leaves, and Parts connected with them.*

72.

In most plants, especially of the lower kind, there are produced by a crowding together of the constituent parts, what have been named the *germs* (*germina, gongyli*). These are small spheres, or opake grains, which are collected together, and from which new shoots or new individuals arise. When they have grown somewhat larger, they are called *propagines, propagula ;* and their union into something like small plots is named *sorædia.* The layer of cellular texture in which these germs lie, is called *lamina proligera,* especially in the Lichens.

In higher plants the germs press so much upon one another, that they commonly make their appearance enveloped by scales, in the axes of the leaves. They are then called *buds* (*gemmæ*) or *eyes,* (Tab. IV. Fig. 2. 8.)

73.

The forms of buds are extremely various, (304.) Some of them remain hidden under the epidermis, and are then only small knots, composed of compact granular masses, or of the substance of leaves, as in various tropical trees. Most of the buds of trees belonging to the temperate zones, appear as oval, pointed, or angular organs in the axes of the leaves. In the *Prunus depressa, Pursh,* two flower-buds stand one on each side of the leaf-buds. They often take the appearance of actual tubers, and even of small bulbs, as in *Dentaria bulbifera,* and most plants of the same species, where they appear between the blossoms. They are in their simplest form in the Tulip-tree (Tab. IV. Fig. 3. 4.), and consist merely of two flat scales lying upon one another, between which the future leaf appears. At the base of the leaf-stalk, we see the two scales of the buds of the second generation, and we thus often see from three to four generations included.

The scales of which buds are composed, lie commonly in such a manner upon one another, that one covers the half of

each of the two that are below it (*germinatio imbricativa,* in
Salisburia adiantifolia, Tab. IV. Fig. 7.), or they lie *riding*
upon each other (*G. equitativa,* in the Common Ash, Tab. IV.
Fig. 5.) The leaves always lie curled up in the buds, as in
the Elder (Tab. IV. Fig. 2.), and in the Snowball (Tab. IV.
Fig. 8.)

74.

The buds are either leaf-buds (*gemmæ foliiferæ*), when nothing but leaves and leafy shoots spring from them; or they
are fruit-buds (*gemmæ fructiferæ*), which produce both blossoms and fruit.

There is also a remarkable difference in the situation of
the parts which are included in a bud, (304.)

75.

A leaf (*folium*) is a green surface, which, for the most part,
is spread out horizontally.

Leaf-stalk (*petiolus*) is the part by which the leaf is joined to
the stem or branches. There are transitions from the leaf-stalk
to the leaves, when the latter are abortive, and then the leaf-
stalks assume the form of leaves. With De Candolle, we
may call these intermediate forms *phyllodia.* They are seen
most distinctly in the Acaciæ from New Holland, and in
Phyllanthus, (Tab. III. Fig. 1.) We even suspect, that
what are called leaves in Bupleurum, are nothing else but
such intermediate forms.

Transitions-from leaves to roots are also observed in water
plants, when the undermost leaves are much subdivided in
the form of hairs, and thus resemble roots. We observe this
in the Ranunculus of our streams, in Sium latifolium, in Nectris aquatica, and many others.

76.

The axis (*axilla*), is the angle which a leaf or leaf-stalk
forms at its insertion with the stem or branches. *Axillaris*
thus denotes that which springs from the axis of the leaves.

The remains of the leaf-stalks often leave scars or warts on the stem and branches. These are called *cicatrices* or *verrucæ*.

The remains of leaves, and of the scales of buds, are called *ramenta*.

77.

The *sheath* is the cylindrical prolongation of the leaf, by which it wraps itself round the stalk.

The place where the sheath passes into the leaf is called the *opening* (*os vaginæ*), and here a leafy membrane, for the most part white and semi-transparent, is usually found, which in the Grasses is called *ligula*.

In some of the Cyperoideæ and Palms, the interstices of the cleft-sheath are connected by fibrous net-work (*reticulum*). This is the case, among others, in Schœnus ustulatus, capillaceus Thunb. thermalis, and Burmanni, Vahl., which grow together at the Cape.

Stipule (*stipula*) is a leafy part, which grows in the neighbourhood of the leaves, or of the leaf-stalks. There are forms of this kind, however, which have very little of a leafy nature, and are rather membranaceous or feathery, as in Sauvagesia, (Tab. VI. Fig. 12. 13.)

The roll (*ochrea*) is commonly a cylindrical membrane, the upper part of which is open, and which surrounds the leaves or leaf-stalks. It appears as a peculiar organ in the Polygoneæ and Cyperoideæ.

Smaller leaves on the leaf-stalk are called *auriculæ*; hence a Folium auriculatum has these appendages either on the leaf-stalk, or at the base. When there are two appendages bent downwards, one on each side of the base of a digitated leaf, it is said to be pedate (*pedatum*).

Smaller leaflets, under the shoots of the Jungermanniæ, are called *amphigastria*.

79.

Other subordinate parts of plants are called by Linnæus *supports* or *fulcra*, because some of them serve to fasten the

2

plant to others. But all the parts which come under the name
by no means deserve it.

The tendril (*cirrhus*) is a filiform, and for the most part
bent body, by which the plant clings to other objects. It is
distinguished sometimes by its origin out of the leaves or leaf-
stalk (*foliares, petiolares*), sometimes by the number of the
leaves which grow under it. Hence cirrhi diphylli, tetra-
phylli, and so forth.

Absorbent warts (*haustoria*) are spungy bunches, which
supply the place of roots in some parasitic plants, and by which
they attach themselves to other plants.

80.

Among what have been called the armour of plants (*arma*),
we place the spines (*spina*), or woody and sharply pointed
processes, which spring from the wood itself, or generally
from the internal parts. They appear not only on the stem
and branches, but also on the leaves and calyx.

Prickles (*aculei*), again, are similar stiff, and prickly
points, which rise only from the epidermis, and are taken off
along with it.

The awn (*arista*), we have already noticed, (56.) It is
a hair-shaped and stiff prolongation of the body. The bristle
(*seta*), is distinguished only by its smaller length, and in some
instances by its being a continuation of the nerves.

The hook (*hamus, uncus*) is a bristle or prickle, bent at
the point. The double hook (*glochis*) is a bristle or prickle,
with reflex subordinate branches at the point. Hence we see
the meaning of the terms *uncinatus, hamosus,* and *glochi-
datus.*

The opposite of these different kinds of armour is expressed by *inermis* and *muticus*, in reference to the points.

81.

Scales (*squamæ*) are for the most part roundish or pointed
membranaceous parts.

Glands (*glandulæ*) are granular, commonly transparent

bodies, containing peculiar juices. They are of a roundish or bowl shape, (Tab. V. Fig. 12.)

On and near the leaves there arise peculiar flask-shaped organs, which separate fluids, as in Cephalotus and Nepenthes. In this instance, as well as in Sarracenia, where the whole concavity of the leave has such organs, they are observed to have peculiar covers. Willdenow called them *Ascidia*.

Air-bladders (*ampullæ*) appear on Utricularia and Aldrovanda.

IV. *Inflorescence.*

82.

Inflorescence (*inflorescentia*) is all that which belongs to the situation and arrangement of the flower, or the way and manner in which flowers grow. To leave nothing unexplained, we must first settle the idea of a flower (*flos*). This is the name given to the whole apparatus, by which impregnation and propagation are accomplished, although, in common conversation, we give this name only to the coloured coverings of the sexual parts. There are flowers without a corolla, and even without any cover, (Tab. III. Fig. 4, 5.) ; but there can be no flower without sexual organs ; on which account the flowers of the Mosses are doubtful, and in the Ferns and Lichens they cannot be admitted.

83.

The support of the flower is called the *flower-stalk* (*pedunculus*) ; and the small stalks, which are in the neighbourhood of a principal flower-stalk, are called *pedicelli*. The name *rachis* is applied in the case of spikes and panicles, to denote the common stalk.

Marshall of Biberstein uses *thecopodium*, and Hoffman *spermapodophorum*, to denote the receptacle continued downwards.

In many families, the flower-stalk receives other names,

when it is also the fruit-stalk. When it springs immediately from the root, and bears flowers only without leaves, it is called the *shaft* (*scapus*).

In the Mosses the fruit-stalk is called *seta*, in the Lichens *podetium*. In the Gastromyci and Nematomyci, the part which supplies the place of the fruit-stalk is called *stroma*, and also *subiculum* (*cephalophorum*, Nees.) If we consider the fruit-stalk of Calycium as a stroma, there is still another part under the fruit, (*hypostroma*, Mart.)

84.

The name *spike* or *ear* (*spica*), is given to that mode of inflorescence in which stalkless flowers are arranged on a common axis. The spike may be simple or compound. In a simple spike, the lowermost flowers are first evolved, and then follow by degrees those higher up. But when the spike is compound, the evolution takes place in a reversed order.

Spicula in the Grasses, is that mode of inflorescence in which several flowers are contained within a common calyx.

The catkin (*amentum*) is a spike, which, instead of flowers, contains only scales, as in Willow, Hazel, and Poplar.

The spadix is a spike with a thick juicy axis, which contains either very small blossoms, as in Acorus and Saururus, or only sexual parts without any covering, as in Arum and Calla.

Sometimes we call the crowded spikes, whose flowers are separated by coloured bracteæ, *strobili*, as in Origanum. On other occasions this word has a different meaning.

When flowers without stalks, or with short stalks, stand in descending rows around the stem, this is called a *whorl* (*verticillus*). Such is the usual inflorescence of the Labiatæ. Frequently the flowers are only on one side, and form then the half-whorl (*verticilli dimidiati*), as in Medusa officinalis. The flowers of a whorl, however, are not always without stalks.

When stalkless flowers are crowded together on the end of a common stalk, they form a head (*capitulum*) : but when the individual flowers which are thus crowded together have stalks, a fascicule *(fasciculus)* is formed. Of the former, Armeria vulgaris is a common example; and of the latter, Dianthus barbatus.

A ball *(glomerulus)* is an irregular collection of flowers with stalks. We hence speak of *conglomerated flower-stalks (pedunculi glomerati)*, when these are collected together, of different lengths, into one heap.

The associated fruit-capsules of Ferns, upon the back of the frond, form the *sorus*.

A bunch *(racemus)* is an inflorescence, where from one common principal stalk undivided flower-stalks arise. When the lower ranges of these are so much lengthened, and the upper so shortened, that the flowers seem almost to be placed in one horizontal plane, it is called a *corymb (corymbus)*.

The umbel *(umbella)* is that inflorescence, where the subordinate stalks extend themselves in a ray shape, on the summit of a common flower-stalk. When these subordinate stalks are again divided, the umbel is called *compound*.

The panicle *(panicula)* is an inflorescence, where the subordinate stalks of a common principal stalk are again divided. When these are condensed, it is called a *thyrse (thyrsus)*, and when the flowers seem to lie in one plane, it is called a *cyme (cyma)*.

85.

The receptacle *(receptaculum)*, is that expanded part of the fruit-stalk which bears the parts of fructification. This part is also called *discus hypogynus*, when, like a disc, it bears the sexual parts. When it is swelled up, it is called *gymbasis* and *sarcobasis*, (105.) In the compound flowers, the expression *clinanthium* has been lately proposed, to express the same idea. It is of much importance how far we extend this idea of the receptacle, because the separation of the sexual parts is connected with the separation of the receptacle. If we admit the separation of the sexual parts in Euphorbia and some other *Tricoccæ*, we must necessarily regard the small stalk which bears the germen, as a sign of the separation of the receptacle. We also find the two receptacles, which other-

wise are separated, running together in Xanthium hamotha-
lamum; and the attempt at this separation in Flaveria and
Brotera, as also in Calycera, Cav. is very distinctly marked.

86.

Bracteæ are those leafy parts which appear in the neighbourhood of the flowers, and which have either a different
form or a different colour from the other leaves; (Tab. VI.
Fig. 6.) But when they cannot be distinguished from the
other leaves either by the form or colour, they are called,
from their station, *floral leaves (folia floralia)*. In the Anemone, the leaves which stand immediately under the blossom,
have been called *involucres* (*involucra*), although they are
only *folia floralia*.

When the bracteæ are collected together above the flowers,
and contain either abortive blossoms or none, they form the
tuft (*coma*). There is another sense of this word noticed
(25.)

Besides the bracteæ, there is another remarkable part in
Surubea, Aubl. (Tab. V. Fig. 11.), where a club-shaped,
coloured, and forked body sits horizontally on the flower-
stalk, and as it were rides on it. It has been lately called *an-
thoxorynium*. In Ruyschia clusiæfolia, Jacq. Amer. Tab. li.
Fig. 2, there is a similar form, but not cleft.

87.

The spathe (*spatha*), is formed by one or more bracteæ,
which enclose the flowers of the Coronariæ, Irideæ, and other
related plants, and which are either leafy or membranaceous.
The individual bracteæ which compose the spathe, have
been very improperly called *valves* (*valvæ*).

Covers of flowers, which stand at a distance from them, are
called generally *perianthia*.

To this class belongs, in particular, the *involucrum* which
occurs in umbelliferous plants.

If the inflorescence of compound flowers be regarded as
one bunch of flowers, the common cover receives the name of

calyx communis, anthodium, and *periphoranthium,* according to Richard ; and *periclinium* according to Cassini ; (Tab. II. Fig. 2. 3.) The bracteæ which stand on the base of this bunch are then called the *outer calyx* (*calyculus,* or, according to Cassini, *involucrum*).

88.

In the Grasses, the exterior covering of the flowers is called the *glume* (*gluma calycina,* according to Panza, *peristachium*).

In the Ferns, the membrane which covers the fruit, and which in some genera, encompasses it like a bowl, is called the *veil* (*indusium*) ; (Tab. II. Fig. 5.)

In the Mosses, the leafy coverings which surround the apparent sexual parts, are called *perichætium*; (Tab. II. Fig. 5.) The calyptre (*calyptra*), is the interior membranaceous, and often hairy covering of the *ovarium,* which, when the fruit is ripe, bursts in a cross direction, or is longitudinally cleft, and for the most part continues till the opening of the fruit.

Similar calyptræ of the flowers appear in Marcgravia, Ascium, Schreb. and Thylacium, Lour. Also in Calyptranthus, Sw., Eucalyptus, Herit., Endesmia, Br. Pileanthus Labill., and Lecythis, we find deciduous or permanent covers, which pass over the sexual parts.

In sponges there is a soft, open cover, which rises from the root-knot, and is called *wrapper* (*volva*) ; as also the ring (*annulus*) which divides and covers the stalk ; and when these are reduced to threads, there is the *cortina* ; (Tab. I. Fig. 29.)

In the lower fungi, the cover of the germ and seed, which for the most part is spherical, is called the *peridium* ; (Tab. I. Fig. 25. 28., Tab. V. Fig. 7.)

89.

The proper cup (*calyx*) is the external, and commonly green cover of the sexual parts, which can either be easily distinguished from the internal coloured parts, or which passes into them, when it is called *calyx corollinus.* This last we find in the Polygoneæ, Chenopodeæ, and many other plants ; and in Sesuvium, the separation of the two covers is so much in the act of taking place, that they seem to be merely attached to each other.

The separate parts of the calyx are called *sepala.*

V. *The Flower.*

90.

The interior, coloured, and for the most part short-lived cover of the parts of fructification, is called the *corolla* (*corolla*). This, as was formerly mentioned, often passes into the exterior cover, and in particular it is called *corolla calycina,* when it has indeed an integument resembling the calyx ; but is still entirely a corolla. This is the case in the Liliaceæ and Coronariæ. There is often merely a simple appendage to the corolla, in some scattered leaves, as in Aponogeton ; (Tab. II. Fig. 11.)

91.

The corolla consists either of one or of several distinct parts. The division of the corolla may be known by looking at its base, and observing whether its parts are connected with each other or whether they are distinct.

When the parts of the corolla are separate, they are called *petala* ; and from this we perceive the meaning of the terms *di-, tri-, tetra-, penta-,* and *polypetala corolla.*

When the parts of the corolla are connected with each other, they are called lobes (*lobi*), segments (*laciniæ*), or lips (*labia*), which expressions have been partly explained already and will be more fully defined.

92.

A corolla, of which the parts are united, forms a tube (*tubus*) or the hollow cylinder, which unites the parts : the expanded lobes form the border (*limbus*) ; and the junction of this, with the former, is called the throat (*faux*).

93.

In a polypetalous corolla, the smaller part of the petals, which often resembles a stalk, is called the *nail* (*unguis*), and the expanded part is called *lamina*, (28.) When the *ungues* stand thick together, they also form a tube, the entrance to which, in like manner, is called the *throat*. The scales, which in some of these plants protect the entrance, constitute the *corona faucis*, as in Silene.

94.

The corolla is often arranged in several rows ; we have thus an interior and exterior corolla (*corolla interna et externa*), as in the Contortæ, particularly Eustegia, in Sauvagesia, and, as some think, in the Grasses. What in these last has been called, in the Linnæan acceptation, the Corolla, is only an exterior cover, which, in contradistinction to the *gluma calycina*, is named *gluma corollina*. It is divided, like the former, into valves (*valvæ*), of which there are commonly two : they have been lately called *stragula*, and the valves *paleas*. Within this *gluma corollina*, there are found, in most of the Grasses, but not in all of them, two very small, delicate, and transparent leaflets, resembling often a tuft of hairs, and springing immediately from the sexual parts. These seem to form the true corolla. Linnæus called them, falsely, *nectaria*. They have also been called *lodiculæ*; (Tab. III. Fig. 7.), (101.)

95.

We must also pay some attention to the regularity or irregularity of the corolla.

It is impossible to give technical names to the infinitely varied forms which here present themselves. The general varieties of form have also been already noticed (31. 32.), so that we need only to apply them to the corolla. We shall, therefore, notice only at present the distinct forms of the irregular corolla.

Of these the simplest is, undoubtedly, the tongue-shaped (*corolla ligulata*), which, in the Aristolochiæ, changes into the tube-form; in compound flowers, it appears as a half flowret (*semi-flosculus*); and in the Orchideæ, as a small lip (*labellum*). In the latter, the highest upright and open leaves are, notwithstanding their colour, the calyx. The labellum, in the Orchideæ, extends downwards into an obtuse or pointed sack, which is called the *spur* (*calcar*); whilst another small sack (*perula*), is formed by the prolonged base of the calyx.

96.

A very frequent species of irregular corolla, is the ringent and labiated (*ringens* and *labiata*). We not only find this form in what are called the Labiatæ, but in many of the Irideæ, Liliaceæ, and Coronariæ. In this case, the corolla is composed, as it were, of two lips, one higher and the other lower (*labium superius* and *inferius*). The interval between both is here also called the *throat* (*faux*).

If the two lips are so closely set to each other, that we cannot see into the interior of the corolla, such a corolla is said to be personate, or masked (*personata, larvata*). The elevated and arched part of the lower lip, receives the name of the *palate* (*palatum*). The upper lip, especially when it is arched, is called the *helmet* (*galea*).

97.

Compound flowers (*flores compositi*), are those which are crowded together upon a common receptacle, and surrounded by a common cover. They are particularly distinguished from the aggregated flowers (*flores aggregati*), by the following circumstances: that in the former the antheræ are united with the flower, in the latter they are free; that in the former there are five, in the latter four antheræ; that in the former there are two stigmata, in the latter but one; lastly, that in the former the embryo is placed upright in the middle of the albuminous matter, and in the latter it is inverted, whilst the albuminous matter itself is consumed.

In compound flowers, the tube is the principal form; the tongue-form appears to be more imperfect, but very common.

Still less common is the ringent, which, however, has been observed in one whole family in South America.

98.

In the Polypetalous corolla, we also attend, for the purpose of defining its shape, to its resemblance to some generally known form. We thus say that a corolla is rose-shaped, pink-shaped, or lily-shaped (*corolla rosea, caryophyllacea, liliacea*).

To this class belongs also the Papilionaceous corolla (*corolla papilionacea*). It consists of four parts, the upper expanded part, or the *standard* (*vexillum*); the two lateral parts or the *wings* (*alæ*), and the lower boat-shaped part, or the keel (*carina*). These four parts consist sometimes of but one piece, as in clover. In other cases, the keel consists of two parts, or it is entirely wanting, as in Tamarindus, and Amorpha. In many of these papilionaceous flowers, as in Hymenæa, a more perfect resemblance between the parts of the corolla takes place, and they approach, by this means, to the regular form.

99.

We must attend to the situation and folding of the flower before its evolution. These are called *Estivation*. It is with respect to flowers, what the interior structure of the leaf-buds is to the leaves. We observe,

1. An *æstivatio valvaris*, when the parts of the corolla, before evolution, only touch one another with their margins, like the valves of the capsule. We observe this, for instance, in compound flowers.

2. The *æstivatio contorta*. Here the parts of the corolla stand so obliquely, that they cover the margins of each other. This estivation is remarked principally in the family of the Contortæ, which hence derive their name, because, even after their complete evolution, they still retain the oblique position of the parts of the corolla, as may be seen in Vinca, Nerium, and Arduina. We also observe this estivation in Pinks.

E

3. *Æstivatio induplicativa*, when the parts of the corolla are bent inwards, and touch each other with the folds of their margins, as the margins of the valves in the capsules of Violets. We observe this estivation in some of the Clematidæ.

4. *Æstivatio alternativa*, when the parts of the corolla stand in two or more rows, in such a manner that the interior row is covered partially and alternately by the exterior. This is observed in most of the Liliaceæ.

5. *Æstivatio quincuncialis*, when, of five parts, two are exterior, and two interior, and the fifth covers the interior with one of its sides, and is again partially covered by the exterior, as we observe in the *calyces of roses*.

6. *Æstivatio vexillaris*. This takes place in the papilionaceous flowers ; the standard covers the three other parts.

7. *Æstivatio cochlearis*, when one part is larger than the others, and, bending itself into a spoon-shape, it incloses them. This is the case in Aconitum, in some of the Personatæ, and in Antholyza.

8. *Æstivatio imbricativa*, when the parts stand in several rows, and the exterior and shorter parts cover only the base of the interior, as we observe in the common calyces of the compound flowers.

9. *Æstivatio convolutiva*, when the exterior part is bent, and incloses the interior, this again the following, and so forth, as we observe particularly in the *cruciform flowers*.

10. *Æstivatio plicativa*, when all the parts are folded into one another, without any particular order, as we observe in the Poppy, and in the Needhamia of Br.

And, *lastly*, we must observe, what seems to make an essential difference between the corolla and the calyx, that their estivation is completely different ; for example, the *æstivatio* of the calyx of the Garden Pink belongs to No. 5., and that of its corolla to No. 2.

100.

We must yet further notice, with respect to the corolla, its time of full blow (*anthesis*). We mean by this, the point of time when the parts of the corolla, as being the organs of fructification, have completed their evolution. We can determine this point, by observing the emptying of the pollen out of the opening antheræ. As the direction and position of the parts are different, before and after this point of time, we readily perceive the meaning of the expressions *ante,* and *post anthesin.*

VI. *The Nectaries.*

101.

Nectaries (*nectaria*), are all those organs formed within, or near the flower, which secrete a honeyed juice.

This term has been employed too loosely by Linnæus, and his followers, to denote all the parts of a flower, except the corolla and antheræ. Hence the interior double corolla of Narcissus, Sauvagesia, and such like plants, has been frequently taken for nectaries. Also the fine transparent scales, which immediately surround the sexual organs of the Grasses, have been improperly called by this name, (64.)

It is often, indeed, a matter of doubt what parts shall receive this name, especially when we attend to the nectaries of Parnassia and Sauvagesia. These last stand around the corolla, which is at least an unusual position, although not properly contrary to rule, (Tab. VI. Fig. 13.) ; because in Cymbidium alvifolium of Swartz, the nectary is found completely without the calyx, on the base of the ovarium.

In general, we must say that that which secretes honey is a nectary ; on which account neither abortive antheræ, nor false petals, can be designated by this name.

Even situation determines this matter. Commonly we must seek the nectaries in the bottom of the corolla, and they thus stand, for the most part lower than the antheræ, (331.) They frequently are united with the receptacle, and often with the ovarium ; (Tab. III. Fig. 10. 13. 14. ; Tab. IV Fig. 18.) Not unfrequently they are united with the filaments ; but they can scarcely appear higher than the antheræ.

E 2

102.

Beside the proper organs for secreting the honey, there are other parts which preserve it, the *nectarothecæ.* These are cavities, sacks, or spurs (*calcar*). These parts belong often essentially to the corolla.

There are other organs which serve for protecting the honey. These are called *nectarilymata,* and are formed either by tufts of hairs, as in the Geranium, or by scales and subordinate leaves, as in Phylica, (Tab. II. Fig. 15.) ; or, lastly, by the situation and direction of the petals themselves.

Lastly, We must not overlook the *nectarostigmata.* These are, for the most part, coloured parts, lines, or spots, which lead to the proper nectaries, as we see them marked out in Pelargonia especially.

VII. *Sexual Parts.*

103.

Sexual parts are those organs which serve for the propagation of the plant.

As in all the higher forms of organized nature, we observe two sets of organs, the one of which, as being the active and impregnating, are called the *male organs,* and the other, as being more passive and adapted for being impregnated, are called the *female parts ;* we distinguish also, according to this idea, the male and female parts of plants.

The time at which plants arrive at the full exercise of their functions is called *puberty ;* before this time, they are called *impuberes,* and afterwards *effœtæ.*

Dichogamy consists in that arrangement, by which the sexual organs come not at once, but after one another, to maturity. The dichogamy is *androgynous,* when the antheræ come first ; and *gynandrous,* when the stigmata come soonest to maturity. The former is the case in Tropæolum, the latter in Euphorbia, (331.)

A flower is called *neutral* (*neuter*), when no sexual organs are produced in it ; it is called *hermaphrodite* (*herma-*

phroditus), when both the sexual parts are contained in the same cover, and for these last we employ the sign ☿. A plant is called *androgynous* (*androgynus*), when the male and female parts are separated from one another, but grow upon the same common stalk, in the same ear, in the same bunch, and so on.

A plant, again, is called *monœcius*, which contains male and female flowers separated from one another, but upon the same plant; it is called *diœcius*, when the separate sexual organs appear upon different plants; and, lastly, it is called *polygamus*, when sometimes male, sometimes female, and sometimes hermaphrodite blossoms appear.

104.

As the female parts appear first, we must begin our account with them.

The Germen (*germen, ovarium*), is the rudiment of the future fruit. It is distinguished into the Simple and Compound. In the Cherry, for example, the germen is simple; in Sage, it is made up of four compartments. In the germen we discover the beginnings of the future seed, like small eggs (*ovula*), which are frequently more numerous than the perfect seed.

105.

The germen rests upon the bottom of the calyx, or it is supported by a fruit-stalk. From this, or from the calyx, there often arises a fleshy elevated support, which is called, in general, *gynobasis*, (Tab. III. Fig. 17.) ; and when this support, during the ripening of the fruit, swells powerfully, it is called *sarcobasis;* (Tab. I. Fig. 36.) To this belongs the juicy swelling of the receptacle in the Strawberry, and the related Genera; (Tab. III. Fig. 22.) In the umbelliferous plants, Hoffman calls this part *stylopodium*.

106.

The pistil (*pistillum, stylus*), is the part which proceeds upwards from the germen, or it is the prolongation of this organ, and bears the scar, or stigma. The pistillum is often entirely wanting, as in the Poppy: it often rises, not from the top of the germen, but from its base, and on its sides, as in the Labiatæ and Hirtella, (Tab. VII. Fig. 3.): it is seldom hollow, commonly it resembles a solid pillar, which might be confounded with the filaments, if we did not attend to the usual central position of the pistillum, and to its strength, which is usually somewhat greater than that of the filaments.

The stigma (*stigma*), is that part of the pistillum which has a soft spungy structure, and is destined to the reception of the impregnating principle. It is by no means always found on the top of the pistillum, for in the Caryophylleæ it is placed longitudinally on the side of the pistillum. In the Iris, it forms a small fold under each of the three divisions of the pistillum, which in this species resembles the petals of the corolla. The softer and more spungy the surface is, and the richer in fine warty matter, the more certainly may we regard it as a true stigma. Nor is it necessary, with Richard, to give to the stigma of the Orchideæ a peculiar name, *gynizus*, (properly *gynixus*), because the idea upon which this name is founded is applicable to all families. The stigma in the Lobeliæ has a peculiar veil (*indusium*), which covers it, before it has attained its perfect state, (Tab. II. Fig. 23).

107.

The male part consists, in common plants, of two organs, namely, the filament (*filamentum*), and the anther (*anthera*); and these, when they are taken in connection, constitute the stamina. The filaments have the same origin with the corolla; and, in innumerable plants, are closely united with it. In the Canna, Scitamineæ, Calothalamus Labill., and some species of Thalictrum, they evince, by their colouring and breadth, their approach to the nature of the corolla.

The fertilising dust (*pollen*), is contained in the antheræ, and has, for the most part, a resemblance to small globules, but it often varies from this form.

When an anther consists of several horizontal compartments, the cellular texture, which connects the compartments, is called *connective* (*connectivum*) ; (Tab. II. Fig. 2. 3.)

There are remarkable variations from these forms, however. In the Contortæ, particularly the Asclepiadeæ, we observe two bodies of a club-shape, and waxy nature, which are bound together by a peculiar knot, and stick under the folds of the *common column of impregnation*. In the Orchideæ, too, we find some masses of a granular substance (*massa granulosa, Listeria, Epipactis*), or of a substance composed of globules of a definite number, (from two to four, Limodorum, Tab. IV. Fig. 11.). These masses are often united by distinct threads, but they are always connected into pairs, by means of a small spherical body (*retinaculum*). These masses of pollen are deposited in peculiar cavities of the common pillar of impregnation (*columna genitalium*, or *clinandrium gynostemii*, Richard), and are commonly covered by a projecting part of the pillar (*rostellum gynostemii*), (Tab. IV. Fig. 12.) This pillar, which supports both the anthers and the stigma, has two longitudinal appendages, which seem like so many abortive filaments (*staminodium*, Richard). The mass of pollen, also, is frequently divided, in these plants, into two longitudinal valves (*massæ sectiles*, Orchis), and are connected with the retinaculum by particular tails (*caudicula*).

There are other anthers of a compound form, some of which seem to be more completely unfolded than the others. Thus, Melastoma contains five large anthers, coloured like the corolla, which bend down the filaments, that seem to have joints in their centre, into a curved line. Five other yellow coloured anthers have no jointed filaments. Similar forms are observed in Cassia, Hofmannseggia Cav. Anthonotha Pal. Beauv., in Solanum cornutum, heterodoxon, Fontanesianum, and rostratum Dunal.

Abortive anthers are called *effœtæ*, as in the Heterostemon Desfont.

VIII. *The Fruit and Seed.*

108.

The fruit (*fructus*), in a general sense, is every thing that contains seed.

Fruits are hence usually divided into simple (*simplices*), compound (*compositi*, or *carpella*), that is, when a single flower-stem, having several pistills, produces several fruits, as the Ranunculus, Clematidæ, and Thalictra; and aggregate (*aggregati*), when the fruits of several flowers are aggregated into one common fruit, as is the case with some of the Urticeæ, Anoneæ, and with the Mulberry. The term *carpidium* has been proposed for this kind of growth.

In the fruit, it has been usual to distinguish more particularly the *pericarpium* from the proper seed. The former is the cover by which the latter is surrounded.

109.

According as the pericarps are too thin, simple, and small, to be distinguished from the seeds, or separate themselves more obviously from them, we apply the phrases of *naked seeds* (*semina nuda*), or seeds surrounded by a cover (*pericarpus tecta*). But more correct observation teaches us, that no seed is wholly naked, or destitute of a covering; on which account those called *angiospermia* must embrace but a very small proportion.

What have been called *perfectly naked seeds*, are only such as are surrounded with a simple covering of a peculiar kind. These are now called *caryopses* (*caryopsis*), as in the Grasses.

Achenium, again, is an apparently naked seed, which yet, beside its proper cover, has a calyx overspreading it, as is the case with the Compositæ, and partly with the Umbellatæ; (Tab. I. Fig. 14. 15.; Tab. VIII. Fig. 8.) Both these apparently naked seeds are called, by De Candolle, *carpella*.

When the seed is loosely surrounded by its cover, a bladder (*utriculus*) is formed, as in the Amaranths, and the

Plantago species. In the Geranium, the loose bladder springing up laterally terminates in a bill-shaped appendage, which
fixes itself to the pistil; (Tab. I. Fig. 17.) Cocculi in cau-
dam longam terminati, Gært. Linnæus called this part, improperly, *arillus.*

When a fruit of this kind is furnished with a membranaceous wing, it is called *samara*, as in the Elm, and partly in
the Maple.

On these simple fruits appear appendages, which serve
for their dispersion. *Small chains* (*catenulæ*), are attached to the seeds of the Jungermanniæ, Marchantiæ, and
Targionia hypophylla, (Tab. III. Fig. 8.) Thread-shaped
appendages, under the name of *tails* (*cauda*), are found in
the fruit of Clematis and Puccinia. Hairs, which spring
from the base, form the tuft (*coma*), as in Epilobium; (Tab.
I. Fig. 13.)

Further, the pappus (*pappus*) is an important part, being
the remnant of the covering calyx, which still continues in the
achenia of the Aggregatæ and Compositæ. It is bristly
(*setaceus*), when it consists of stiff hairs; (Tab. I. Fig. 6.)
hairy (*pilosus, capillaris*), when it consists of soft long hairs;
awned (*aristatus*), when the bristles are thick below, and
long; plumose (*plumosus*), when the hairs are beset with
smaller hairs; pencil-shaped (*pencillatus*), when small hairs
stand on the top of the hairs, (Tab. I. Fig. 12.); chaffy
(*paleaceus*), when dry membranes crown the seed; (Tab.
VIII. Fig. 8.)

In the caryopses and achenia of the Umbelliferous plants,
the peculiar sap-vessels, under the external membrane, are
called *vittas.* The ribs of the fruit are called *juga*, or *costæ;*
the hollows, or small valleys between them, are called *val-
liculæ.*

110.

Pericarps are divided according to their compartments,
their valves, and their partitions.

Compartments (*loculi*), are the chambers, or divisions of
the vessel containing the seed; hence we say, bilocular, trilo-

cular (*bilocularis, trilocularis*). A kind of compartment, which opens with a certain elasticity, is called *coccum*, whence the Euphorbiæ, and similar plants, are called *tricoccæ*, because their fruits have three such compartments.

Valves (*valvæ*), are the side-pieces, or exterior walls of the compartments, into which the pericarp is divided. Hence a fruit is called *bivalve, trivalve* (*bivalvis, trivalvis*).

Partitions (*dissepimentum*), are the interior walls of the compartments. They are often formed, as in the Andromedæ, by the valves turned inwards. They are then called *dissepimenta valvaria,* and *valvas septiferas.*

We must also notice the suture (*sutura*). This is the place in which two valves are united. The suture properly, therefore, points out the springing of the fruit from valves; but the genus Eucledium Br. has distinct sutures, and yet the fruit does not spring; and Bunias, in the unripe fruit, shews sutures which disappear during the ripening The opening of the valves often takes place from above to the middle of the fruit (*semivalves fructus*); they often also open from below, as in the Orchideæ and Triglochin.

111.

An uncommonly important part is that which has been called the *placenta* (*placenta*, or *receptaculum*), which either stands like a free column in the middle, or is formed by the thickening of the partitions, or even by the bending in of the valves; (Tab. I. Fig. 21). At the same time, every column that is observable in the fruit is not the placenta. In the capsules of the Mosses, for instance, we observe an entirely free central column; but the seeds are fastened to the walls of the capsule. The way and manner in which the placenta is formed is called *placentation.*

In the Umbellatæ, Hoffman has called the column *spermapodium.*

112.

A nut (*nux*), is a fruit with a shell, which does not burst of itself. This shell is often surrounded by an external tough covering, which is called *naucum.* Almonds and Hazel-nuts present examples of this.

But if the nut be surrounded by a juicy or fleshy covering, it is called a *drupe* (*drupa*), as in Plums and Cherries, and Myoporum ; (Tab. III. Fig. 21.)

113.

A berry (*bacca*) is a juicy fruit, which contains one or more seeds imbedded in the sap or juice. Grapes and Gooseberries are common examples; (Tab. I. Fig. 35.)

114.

A legume (*legumen*) is a long fruit with two valves, the seeds of which are fixed on one and the same suture, but alternately upon the two valves; (Tab. I. Fig. 16.) The legume has commonly but one compartment; in Astragalus it has two, and in Kennedia, Vent. it has several.

A loment (*lomentum*) is a legume, which is divided crossways into cells, as in Hippocrepis and Ornithopus; (Tab. I. Fig. 18.)

A silique (*siliqua*) is a long, two-valved fruit, the seeds of which are fixed to both sutures, as in Rape, Cabbage, and Stocks; (Tab. I. Fig. 38.)

A small silique (*silicula*) is properly a silique, which is not much longer than broad, as in Thlaspi and Camelina, (Tab. I. Fig. 7.) ; but improperly the nuts of Bunias and Crambe also are so called, although they have neither sutures nor valves.

The follicle (*folliculus*) is a long, one-valved fruit, which opens only in one suture. It is found in the Contortæ, in Pæonia, Cimicifuga, and Butomus.

Ehrhart has given the name *Pyxidium* to an utriculus with one seed, which bursts crosswise, as in Plantago and Amaranthus. We also find the phrase *Capsula circumscissa* applied to this.

115.

A capsule (*capsula*) is every dry fruit which does not fall under the preceding or following article.

When it is surrounded by a fleshy covering, it is called an *apple* (*pomum*) ; (Tab. I. Fig. 23.) We must distinguish the pumpkin (*pepo*) from this kind. The latter name is given to a fleshy fruit, the seeds of which are fastened in the interior circumference. Also the fruit of the Agrumæ (*Aurantium*, De Cand.) is a peculiar, fleshy, and inflated fruit, which can easily be divided into several membranaceous compartments.

116.

Fruits are often collected together in numbers. The Umbellatæ and Rubiaceæ bear double *achenia*; the Labiatæ and Asperifoliæ bear a fourfold fruit; and five stand together in the Geranium.

Aggregated fruits are found in a great many plants, where, by the swelling of the receptacle, their union is promoted. In the Annoneæ, for example, the single seeded kernels (*pyrenæ* or *acini*) are collected together in the swollen receptacle. In the Fig, a soft fleshy covering unites together many *caryopses*. In the Mulberry several juicy *utriculi* are united into one.

The strobile (*strobilus*) of the Pines and Proteæ consists also of utriculi, which stick under bracteæ that are very involved and hard, and which together form a ball ; (Tab. I. Fig. 9.) When these bracteæ swell and flow together, they form the *galbulus* of the Cypress, Thuia, and even of the Juniper; (Tab. I. Fig. 8.)

117.

The fruits of imperfect plants must also be considered. In Ferns the membranaceous and spherical reservoir of the seed is called the *capsule*. In the true Ferns, we observe a jointed ring (*annulus*), by the elasticity of which the capsule is thrown off ; (Tab. II. Fig. 9.) In the Mosses also, the fruit has been called a Capsule, although some use the expression *theca* for it. We distinguish also the *operculum*, which, when the fruit is fully ripe, loosens itself all round. This loosening is often assisted by a fringed ring (*annulus fimbriatus*), which is placed horizontally between the operculum and capsule, and by its elasticity throws off the operculum.

The mouth of the Moss capsules (*os, stoma*) is the upper, circular part, which is either naked (*nudum*), or is furnished with teeth, cirrhi, and membranes, which arise from the prolongation of the capsular partitions, and are called *peristomium*; (Tab. II. Fig. 7 8.) Sometimes also a membrane, which in Polytrichum and some others is called *epiphragma,* passes across the mouth of the capsule.

118.

In Lichens, the whole *thallus* is capable of producing granular germs, destined for propagation. Yet there are *apothecia*, which contain apparent, frequently twin-seeds, in peculiar layers; (Tab. II. Fig. 2.) Formerly the various forms of these apothecia were furnished with peculiar names, which, however, are no longer in use. *Lirella* is a linear longitudinally opening apothecium, as in Opegrapha.

Tricæ are closed, twisted seed-beds, of a black colour.

Thalamia are close round seed-beds, in the substance of the leaf, surrounded by a peculiar membrane, within which the seeds are enclosed in peculiar bags.

Tubercula are close, roundish or spherical seed-beds, which project from the leaf.

Cephalodia are highly coloured, roundish, open, and commonly stalked apothecia, covered with a seed-bed, which passes off like a powder.

Orbillæ are flat, slightly coloured, open fruits, without a raised margin, covered with a thin seed-bed.

Scutellæ are open, circular, hollowed fruits, the margin of which is formed by the substance of the leaf.

Patellæ are open, flat or elevated fruits, without a raised margin.

119.

In the Fungi, the fruit is generally called *peridia*; (Tab. I. Fig. 25.) More particularly, however, the bladders which contain the seeds or germs (*sporæ*), are called *sporidia.* They are also called *thecæ sporophoræ*; (Tab. I. Fig. 30., Tab. II. Fig. 1.) Their reservoir, in certain groups of Fungi, is called *perithecium*; and when the peridia are included in a distinct case, this latter is called *sporangium*; (Tab. V Fig. 7.) When the lines or hairs on which the sporæ of Fungi sit, are collected together in tufts, this sort of tuft is called *capillitium*; (Tab. I. Fig. 28.)

In the proper Sponges, the germ-bladders, or seed-bladders, form a peculiar covering, or a layer, which is called *hymenium*; (Tab. I. Fig. 30., Tab. II. Fig. 1.)

120.

On the seed itself, we remark, in the first place, what has been called the *umbilicus, hilum, cicatricula,* or a hollow part commonly found in the base, but often also on the sides, by which the seeds are fastened, and from which the germ proceeds; (Tab. I. Fig. 4. 6.)

The umbilicus, in many plants, particularly in the Leguminosæ, is covered by a warty substance, which is called *strophiolus* or *strophiola.* In Urania and Strelitzia, this strophiolus is a heap of beautifully coloured, intermingled, and stiff hairs; (Tab. I. Fig. 10.) From the umbilicus proceeds the *funiculus umbilicalis* or *podospermium*, a thread which effects the insertion of the seed.

A leafy or solid expansion of the funiculus umbilicalis frequently surrounds the seed, and is known by the name of *arillus.* In the Nutmeg, the mace is nothing else but this arillus. In the Oxalideæ, this membrane has a certain elasticity, by means of which the seeds are pushed forth. In Euonymus, and in fresh Coffee-beans, this arillus can be very distinctly seen.

Small hooks (*retinacula*) are observed on the seeds of the Acantheæ, by whose elasticity the opening of the capsule and the ejection of the seed is favoured; (Tab. I. Fig. 37.)

In the seeds of many of the Leguminous plants, a small cavity appears under the umbilicus, called *micropyle*, but its use is unknown.

Chalaza is the place in the interior membrane of the seed, where the funiculus umbilicalis passes into the seed. Sometimes this is at a distance from the umbilicus, and it is even, as in the seed of the Citron (Tab. 1. Fig. 1. 2.), placed opposite to it.

121.

When we open the seed, we find the embryon either surrounded by albuminous substance (*albumen, perispermium, endospermium*), or this substance lies in the centre of the curved embryon, as in some of the Polygoneæ and Caryophylleæ; or it lies on the side of it, as in the Grasses and in the grains of Corn; (Tab. I. Fig. 3. 11. 20.) But this substance is often entirely wanting, that is, when the future plant being already perfectly evolved, fills the seed; (Tab. I. Fig. 1. 2.)

Seeds which have albuminous substance are called *albuminosæ*, and those which want it *exalbuminosæ*.

In some imperfect seeds, there is an intermediate body between the albuminous substance and the embryon, which in the Grasses is called *scutellum* (Tab. I. Fig. 11.), and in the Scitamineæ *vitellus*, (Tab. I. Fig. 3.) In the latter it surrounds the whole embryon.

The embryon, or future plant, is either *unevolved*, when it resembles a small point, or a short thread; or it is *evolved*, and then we distinguish on it the two seed lobes (*cotyledones*), the plant itself (*plumula*), and the root (*rostellum* or *radicula*); (Tab. I. Fig. 1, 2. 5.) The direction of the last, commonly towards the umbilicus, may, however, be either upwards or downwards, according as the seed is placed. In the former case the embryon is called *inverted* (*inversus*), because it has the opposite direction with respect to the fruit, but not with respect to the seed. In the latter case, the embryon is said to be erect (*erectus*), (195.)

PART II.

TAXONOMY,

OR THE

THEORY OF CLASSIFICATION.

CHAP. I.

GENERAL OBSERVATIONS

122.

THIRTY thousand species of plants are at present known upon the earth. This number might be encreased to fifty thousand, if all the plants which are still undescribed in the great collections were known. And if we suppose the central regions of Asia, Africa, and New Holland, to have been once as well explored, as many of the countries of Europe have already been, we may consider it as extremely probable that there are above a hundred thousand species of plants upon the Earth. Every one of these species has its native Country, its Name, Form, Properties, and Uses. The knowledge of these must have an important influence, both upon the developement of the human mind, and upon the progress of trade and useful arts. But who shall clear up for us this immense study ? To what guide shall we trust ourselves in this frightful labyrinth ? How shall we able, not only to become acquainted with the particular Natural History of each plant, but to find out what others who have preceded us have observed respecting the plant that is before us, and

F

to know whether or not the observed form be an entirely new one, which no individual before us had observed ? This most important service is performed for us by what has been called the *Method;* that is to say, the Scientific Arrangement and Division of Plants, either according to one common principle, or according to Families and Groups, the common marks of which have been learnt.

Botanists have always been so much convinced of the importance and utility of such an arrangement, that they have regarded the knowledge of the laws of this arrangement as the highest object of their exertions.

123.

As long as a small number only of plants were known, the necessity of classification was not felt. But the more that native plants have been studied, since the beginning of the sixteenth century, the more has the necessity of such a method pressed itself upon our attention ; and however imperfect the first attempts of Lobelius and Bauhin were, every unprejudiced person must confess, that the principle upon which they proceeded, that, namely, of arranging plants, as Nature has done, is the only right principle.

In general, the various methods may be divided into the *Empirical* and *Scientific.* The former are at the same time the most ancient. They are founded, not upon nature and upon essential forms, but upon accidental things. The Alphabetical Arrangement of Plants,—or their division according to their Uses, as when, for instance, the edible vegetables, the orchard-trees, the forest-trees, and the ornamental plants, are collected into distinct assemblages,—these are some of the *Empirical* Methods.

Although *Scientific* Classifications have a reference to the nature of the objects, there is, however, a multitude of views which may furnish the foundation of such a division. No person has pushed further the attempt to find out, and even in some degree to complete, the manifold methods, according to the various properties and parts of plants, than the immortal *Michael Adanson,* who has proposed no fewer than five-

and-sixty different classifications. Among these there are some which are founded simply upon the Stature, others upon the Thickness or Substance, and others again upon the Colour, the Smell, the Taste, and similar properties.

124.

In assuming the parts and properties of plants, as principles of Classification, we must, in every instance, make a distinction between essential parts, and those that are accidental or less essential. The former are such organs as have an intimate connection with the purpose of vegetation : less essential forms, again, are those which have a more distant connection with the purpose of vegetation. If we seek this purpose in the propagation of plants,—and it seems undoubtedly to consist in this,—then the seed and fruit, and also the flower, which in most plants precedes the fruit, are the parts which furnish the most important ground of Scientific Classification.

But, with respect to these, we may proceed in two ways. We may regard the determinate relations of these essential parts as the only Principle; and without considering other properties and distant organs, we may employ the former only as the principle of classification. In this case, we sketch an Artificial System ; that is to say, an arrangement of plants according to one common principle.

But we may consider the relations of essential parts, in their joint connection with other organs and their properties; and we may proceed in this way so far, as universally to avail ourselves of resemblances and agreements, without binding ourselves exactly to one and the same leading principle. We then follow a *Natural Method*, which cannot be called a System, because it is destitute of unity of principle.

CHAP. II.

ARTIFICIAL CLASSIFICATION.

125.

If, now, we would either study the nature of plants themselves, or would learn their uses, in both cases, we feel the necessity of having names assigned to them; because without names, we can neither make ourselves intelligible to others, nor find out what others have remarked concerning them. The Nomenclature of Plants is the first object of the artificial arrangement. The second is to give them their place in some order or other, and beside plants that are already known; because without this, the bare knowledge of names would be completely useless.

126.

When we find a plant, the simplest way of discovering its name, and finding its place in the system, is to look into the great Registers, or Scientific Catalogues, unless, in a completely empirical way, we betake ourselves to the turning over of plates, which consumes much time, and yet often does not lead to our object. But in order to be able to use those scientific catalogues, or artificial systems, we must know the principles according to which they are formed; we must possess the art, or have acquired the address, of regarding, with respect to every plant, only the relations; and of keeping in our eye the organs, upon whose diversity the artificial system proceeds.

127.

The Artificial Method, or the System, must necessarily assume, as the grounds of classification, such parts only as are invariable; so that all those things which are changed by propagation, are justly excluded from the principles of the system. The Duration of plants, their Stature, their Taste and Smell, even sometimes their Colours, and also their situation and time of flowering, are all things and relations which we must consider variable; whilst, on the other hand, the Forms and Numerical Proportions of the parts of fructification, are seldom subject to change. These, therefore, must constitute the principle of classification.

128.

We must especially employ, as the grounds of classification, such organs, as, besides their constancy and invariableness, are also found in the greatest number of plants; and, when it is possible, such parts too as are easily observed, and which appear at the same time.

But the two latter requisites are of less importance than the first; and cannot always, according to the nature of things, be obtained. There are a great many plants, whose essential parts are so small, and lie so hidden, that they can be discovered only by the aided eyes, and after previous, often troublesome, preparation. The finer differences in the structure of the Mosses, Fungi, and other imperfect plants, cannot be observed without powerful magnifying glasses. The situation of the embryon and of its parts in the seed, can only be exhibited by particular preparation. But as these relations are among the most stedfast and important, we must avoid no labour to become acquainted with them; and no system can be reproached for paying regard to those parts and relations.

The same remark applies to the want of cotemporaneous growth. It is impossible to determine a plant with certainty, before its vegetation has come to maturity; because then only are all the essential parts unfolded, and a plant which has been observed for years, without being seen to bloom, or to carry fruit, cannot be defined with certainty, and in a scientific manner. Especially the ripening of the fruit is to be waited for; because, as we shall see, important

changes often take place in the fruit and seed, when they
are passing from an unripe into a mature state.

129.

The permanent relations of the parts which constitute the
foundation of the system, must be strongly and distinctly expressed in the nomenclature. Those designations of properties
and relations, which are also called Characters, must, as much as
possible, be positive, and must not consist merely in negation,
or in the assignation of absent properties, (for we may wish
to denote, by definite oppositions, what the characters exclude). But even in this case, the absent properties may
easily be expressed positively. When, for instance, the chaffy leaves of the receptacle are taken into the one character,
the *naked receptacle* properly takes its place in the other, as
the opposite of the former. If in one Genus it be important
to notice that the antheræ *grow together*, then the other Genus is distinguished by having its organs standing *free*.

130.

An artificial system, in order to be useful, must have as
many subdivisions as are demanded by the essential differences of the principal organs. Too few divisions oblige us
to place together too great a number of different plants, and
thus the investigation of particular plants is made very difficult. But the arrangement must be made according to certain general properties or relations of the parts, in order that,
by seeking out the general division, we may pass with greater
ease to the particulars which it contains.

131.

If we examine by these principles the artificial systems
which have been hitherto devised, we shall find the most celebrated of them, that which Linnæus proposed, to possess a
decided superiority, not only because it is consistently derived
from one simple principle, but also because the author of it,
by means of a new nomenclature, has given to his terms the
greatest distinctness of meaning.

But it will be necessary to present the whole system in one Tabular view, before we proceed to pass judgment upon it.

View of the Linnæan System.

I. Plants whose parts of fructification are manifest, *Phæne-rogamia.*

  A. Antheræ and pistilla upon the same receptacle, *Mo-noclinia.*

    * Antheræ and filamenta free.

      a. Filaments of equal length, *Isostemones.*

1. One anther,	Class 1.	MONANDRIA.
2. Two anthers,	2.	DIANDRIA.
3 Three,	3.	TRIANDRIA.
4. Four,	4.	TETRANDRIA.
5. Five,	5.	PENTANDRIA.
6. Six,	6.	HEXANDRIA.
7. Seven,	7.	HEPTANDRIA;

        (Tab. VI. Fig. 7.)

8. Eight,	8.	OCTANDRIA.
9. Nine,	9.	ENNEANDRIA.
10. Ten,	10.	DECANDRIA.
11. Twelve to twenty,	11.	DODECANDRIA.

    12. Twenty and more anthers, but the filaments upon the sides of the calyx, Class 12. ICO-SANDRIA; (Tab. III. Fig. 22.)

    13. Twenty and more anthers, but the filaments on the receptacle or corolla, Class 13. PO-LYANDRIA.

      b. Filaments of unequal length, *Anisostemones.*

    14. Two longer than the other two, Class 14. DIDYNAMIA; (Tab. III. Fig. 22., Tab. IV. Fig. 16.)

    15. Four longer than the two remaining ones, Class 15 TETRADYNAMIA.

   ** Filaments united

    16. Into one bundle, Class 16. MONADELPHIA (Tab. III. Fig. 11., Tab. IV. Fig. 14.)

17. Into two bundles, or one free, the others united,
    Class 17. DIADELPHIA.
18. In more than two bundles, Class 18. POLY-
    ADELPHIA.
*** Anthers united
    19. Among themselves, Class 19. SYNGENESIA;
        (Tab. III. Fig. 2.)
    20. With the pistillum, Class 20. GYNANDRIA;
        (Tab. IV. Fig. 10.)
B. Antheræ and pistilla on different receptacles, *Diclinia*.
    21. On the same plant, Class 21. MONŒCIA.
        (Tab. III. Fig. 1.)
    22. On different plants, Class 22. DIŒCIA.
    23. Sometimes separate, sometimes united, Class
        23. POLYGAMIA.
    II. Plants whose parts of fructification are hidden, or
which want them. Class 24. CRYPTOGAMIA.

132.

From this view it is evident, that the relations of the
parts of fructification afford the foundation of arrangement.
These relations consist in the Number, the Insertion, the Difference of Length, the Union and the Separation of the male
parts, as well considered in reference to themselves, as in
respect to the female parts. A preference is given to the
male parts, which is less founded in nature, than apparently
rendered necessary by the circumstance that, besides being
commonly in greater number, they present also more varieties.
Nature seems, upon the whole, to have given less constancy
to the Numerical proportion, than to the Forms, the Situation,
the Union, and the different Lengths of the sexual parts.
The finding of plants in those catalogues which have been
arranged after this system, is very much aided by the simplicity of the principle; yet, in such researches, we must attend
to all the other relations, beside the number of the parts.

133.

In judging of this system, we cannot speak of its want of
conformity to nature, because, from the very fact of its being

an artificial system, it renounces the pretension of presenting naturally related groups. Therefore, it exposes this system to no well founded reproach, that the affinities of Nature are torn asunder, and the most dissimilar genera of plants brought together. The Grasses are thus scattered through several classes : the Labiatæ are found in two; the Chenopodeæ, the Rubiaceæ, the Palms, and many other families, in several classes. Every artificial system, from the very circumstance of its assuming one simple principle, founded on the relations of a few essential parts, must depart from Nature.

134.

But the first well founded objection which may be made to this celebrated system, consists in this,—that in many of the classes more regard is paid to natural affinities, than the artificial structure of the system, and the unity of its principle permit. If all Monadelphous plants must be referred to the sixteenth Class, so must also a great number of the Diadelphous ; and the Melieæ and the Malpighieæ must be transferred from the tenth, and even from the eighth and fifth Classes, into the sixteenth. If attention had not been paid to natural affinity, all the single flowered plants, having their anthers united, ought to stand in the nineteenth Class.

135.

A second objection, and one of the most important, is, that a greater value is placed upon numerical proportions, than is ever observed to be justified by Nature. There are genera of plants, such as *Valeriana, Stellaria, Rhexia,* and innumerable others, which observe so little steadiness in the numerical proportions of the male parts, that Linnæus must necessarily have been perplexed, when he wished to assign to these genera a definite class. He used, in such cases, to fall upon three plans.

In some genera, he remarked, in the *first* place, what numerical proportion was established in most of the species. When, for example, among nine *Convallariæ,* which were known to him, he observed six or seven species to have six anthers, and the other two or three species to have only four, he placed the genus in the Sixth Class, but referred, in what he called the Key to his Classes, or in the Preliminary General Index, when speaking of the fourth class, to the *Genus Convallaria* in the sixth. By this means the investigation was completely facilitated. But it has happened, that later discoveries have exhibited a multitude of species, in which the subordinate numerical proportion has become the prevailing one, in so much, that we must now transfer many plants from the class in which Linnæus placed them into another. This has happened with *Verbena*, which Linnæus placed in the second class, but which later botanists have justly placed in the fourteenth. Thus also *Boerhavia*, which Linnæus placed in the first class, is now transferred, on account of the greater number of its species which have two anthers, into the second. With respect to *Rhexia*, it is doubtful whether it belongs to the eighth or tenth class, because we find nearly as many species with eight as with ten anthers.

In the *second* place, Linnæus was accustomed to observe, among several species of the same genus, which of them was most common, and was produced in greatest abundance; that, according to these circumstances, he might determine its place in the system. Thus he decided with respect to *Lythrum*, the most common species of which, namely *Lythrum salicaria*, has from twelve to fifteen anthers, whereas, in other cases, the number is far smaller. But even here Linnæus was not consistent; at least the most common species of *Euonymus* is almost always observed to have four anthers, although the genus stands in the fifth class. Defective observation, also, often led him into mistakes, which have been propagated to our time. Thus *Ruppia* stands in the fourth Linnæan Class, although no such numerical proportion can be discovered in it. We thus also find *Calla* in the seventh class, and yet the number of the filaments is wholly indeterminate and fluctuating. We shall speak of *Euphorbia* upon another occasion, when we shall shew that it holds a very unsuitable place in the eleventh class.

Lastly, The founder of this system, when he remarked a fluctuating numerical proportion, was accustomed to set such a value upon the first flower, (*flos primarius*), that according to it he determined the place in his system. Thus the Garden Rose has always ten filaments in the first flower, and in the others eight; hence Linnæus placed it in the tenth class. When the flower-top of *Adoxa* unfolds itself, the first blossom is seen to have eight anthers, the following ones ten, and Linnæus accordingly attached it to the eight class.

136.

A third well founded objection which may be made to this system, is founded on the value which is laid in it on the difference of sex; a difference which, considered in itself, is so fluctuating, that we see it strikingly displayed, not only in species of the same genus, but even frequently in different plants which belong to the same species. The whole three-and-twentieth class, and a great part of the genera which stand in the one-and-twentieth and two-and-twentieth, shew a separation of the sexual parts, which is completely destitute of constancy. Thus some of the species of Hordeum are polygamous, others hermaphrodite. The same thing happens with respect to the different species of Acer. In numberless genera, which Linnæus referred to other classes, we find monœcious, diœcious, and polygamous species, as the common examples of *Rumex* and *Rhamnus* shew.

137.

Beside these difficulties and deficiencies which the Classes of the Linnæan system present, it cannot be denied, that some well founded objections may be made with respect to the Orders which it admits. In the *first* place, it fails here completely in unity, since sometimes one and sometimes another principle of the order is assumed. For the most part, attention is paid to the number of the Pistils, but instead of pistils, sometimes the germen, and not unfrequently also the stigma is taken, so that we often are not in a situation to find the Order. Not to mention, that in the fourteenth class

2

the fruit is improperly called *naked*, because it presents Caryopses with a simple covering, and frequently even Nuts. In the fifteenth class, the idea of a *silicula* causes some difficulties in the first Order, because in many genera we observe true nuts without the power of bursting.

Similar exceptions may be made to the orders of the nineteenth class. No doubt this arrangement promises much at first, because in the first order of the nineteenth class, *Syngenesia, Polygamia æqualis*, the flowers have all equally good seed, and are all hermaphrodites. In the second order, *S. P. superflua*, the flowers on the margin are only female, those in the centre hermaphrodite, but they are all equally fruitful. In the third order, *S. P. frustranea*, the flowers on the margin are neutral, or the pistillum is abortive, but the flowers in the centre are hermaphrodite, and alone bear perfect seed. In the fourth order, *S. P. necessaria*, the relation is exactly reversed : the central flowers, which are commonly male, are abortive, and only those on the circumference, which are for the most part female, bear perfect seeds. *Lastly*, in the fifth order, *S. P. segregata*, every flowret has also its separate calyx, although they all stand upon a common receptacle.

To this division it may be objected, that very often the *frustranea*, in the same genus, coincides with the *æqualis* ; for in *Bidens* and *Centaurea*, for example, a neutral ray often makes its appearance, and often fails. The same thing happens in some genera of the *superflua*, where *Anthemis, Anacyclus*, and *Pyrethrum*, sometimes lose the ray. But these objections affect every other division of this class in an equal degree; for the Cynareæ pass into the Radiatæ, and *vice versa*. In like manner we might notice, respecting the *segregata*, that what has been called their *peculiar calyx*, is often nothing else but chaffy leaves, as we also see them in many genera of the remaining orders.

138.

These, and similar defects, induced the founder of this system, during his later years, to think of some improvements on it.

This necessity seems to be most urgent in regard to the sexual parts, as the principle of the classes from the twenty-first to the twenty-third. The later scholars of Linnæus assure us, that he wished to leave out the twenty-third class as altogether unnecessary. It was reserved, however, for Sir James Edward Smith, the worthy inheritor of the Linnæan treasures, to propose the happiest alteration. This consists in recognising the difference of sex as then only essential, when it manifests itself by an actual difference in the structure of the female, male, and hermaphrodite flowers; when thus the male-blossoms, as in the Oak, stand in catkins, but the female flowers are insulated. This difference, not of sex alone, but also of forms, is called by Smith *diclinia ;* and he accordingly rejects from the twenty-first, twenty-second, and twenty-third classes, all the genera, in which no such correspondence between the difference of sex and the difference of form takes place. Hence among others, the genera *Acer, Veratrum, Hydrocharis, Stratiotes, Sagittaria,* and so forth, are placed much rather according to the number of their filaments, than according to their difference of sex.

139.

Less fortunate attempts to improve the Linnæan system have been made by Thunberg, Suckow, Rebentish, and lately by Claude Richard. According to Thunberg's plan, the Gynandria is removed, and is brought, with peculiar impropriety, under the second class. He also rejects Monœcia, Diœcia, and Polygamia, whilst he classes all these plants according to the numbers of their anthers. Others have thrown away the Monadelphia and Diadelphia, and have acknowledged the numerical proportions as the only principle of classification. But by these means the difficulties have only been made greater, and some classes' have been overloaded with genera, so that their investigation has been rendered much more troublesome.

Richard's later proposal is founded partly on a pretended better division of the Polyandria and Icosandria, from which he separates a new class, Hysterandria, in which the filaments stand on the germen, and are therefore *epigynous ;* partly on the assignation of new names to the Syngenesia, Polygamia, and Cryptogamia, namely, Synantheria, Anomalæcia, and Agamia, which are very superfluous. And, *lastly,* he forms from the Linnæan Monogamous plants in the Syngenesia, a peculiar class which he calls *Symphysandria.* All these changes seem to fail of answering their purpose.

140.

Beside the Linnæan, there are some other artificial systems, the most important and best known of which must be here noticed. The nearest in principle and value to the Linnæan, is that which *Gleditch* proposed in 1764. It is founded entirely on the situation of the filaments, and the more therefore may be said for it, the more constant this situation and insertion are. The filaments, according to this system, stand either on the receptacle, (*thalamostemones*), or in the corolla (*petalostemones*), or on the calyx (*calycostemones*), or, lastly, on the pistil (*stylostemones*), or no filaments are found, or they are hidden. Nevertheless, the small number of divisions which this system allows, are a disadvantage to it; and a multitude of subordinate divisions must again be made, according to the Linnæan system, by which means the value of the former is much diminished.

However, *Borckhausen* (1792), and *Mönch* (1794), on the principle of the system of Gleditch, have published peculiarly acute treatises, which evince as much a fine talent for observation, as for just criticism.

141.

Older attempts to form artificial systems have been founded on other essential parts beside the male organs. *Cæsalpinus* (1583), *Morison* (1683), *Paul Hermann* (1690), and *Boerhaave* (1720), divided plants simply according to the differences of their fruit, without regarding natural affinities. They were called *Fructistæ.* To these belongs *Joseph Gærtner,* in later times, although he has never completed the plan of his system.

Others, as *Rivinus* (1690), *Ludwig* (1750), and *Tournefort* (1700), regard the corolla only in classification; whilst *Rivinus* attended to the number and regularity of the parts, *Tournefort* to the general form of the corolla, and *John Ray* (1682), connected the fruit with the corolla. All these *Corollistæ*, as they were named, paid, however, a constant regard to the natural affinities.

It is scarcely worth while to mention the unsuccessful attempts of *Antony Magnol* (1720), to class plants according to their calyx, and of *Sauvage* (1751), to arrange them according to their leaves. We are far from considering every artificial system as just as good as another, because we find some fault with them all. Our opinion on the contrary is, that the Linnæan system, with the improvements which *Smith* has proposed, is best adapted to the instruction of beginners.

CHAP. III.

ON THE MUTUAL CONNECTIONS OF PLANTS.

I. *Idea of Species.*

142.

By Species (*species*), we understand a number of plants, which agree with one another in invariable marks.

In this matter every thing depends upon the idea of invariableness. When an organ, or a property of it, is changed neither by difference of soil, of climate, or of treatment, nor by continued breeding, this organ or property is said to be invariable. When, for instance, we have remarked during centuries, that the Centifolia has always unarmed leaf-stalks, we say correctly, that this property of the Centifolia is invariable.

When we express these invariable properties in words, we give the Specific Character (*character specificus*).

143.

This idea proceeds on the supposition, that the species which we know, have existed as long as the earth has had its present form. No doubt there were, in the preceding state of our globe, other species of plants, which have now perished, and the remains of which we still find in impressions in shale, slate-clay, and other flœtz rocks. Whether the present species, which often resemble these, have arisen from them;—whether the great revolutions on the surface of the earth, which we read in the Book of Nature, contributed to these transitions,—we know not. What we know is, that from as early a time as the human race has left memorials of its existence upon the earth, the separate species of plants have maintained the same properties invariably

To be sure, we frequently speak of the transitions and crossings of species; and it cannot be denied that something of this kind does occur, though without affecting the idea of species which we have proposed. We must, therefore, understand this difference.

144.

We perceive the *Transitions of a Species*, when it loses or changes the properties, which we had considered as invariable in the character. Thus, it would be a transition, if we had stated as an invariable character of winter wheat (*Triticum hybernum*), that it was biennial, and had an ear without awns; and if we should remark, that by frequent reproduction, and by very different treatment, it began to assume awns, and, when sown in spring, came to maturity during the same summer.

But this shews only that our idea of the difference between the two kinds of grains had been incorrect; for it is the universal rule, that the character does not constitute the species, but the species the character. Species, then, only appear to undergo transitions, when we have considered an organ or a property as invariable which is not so.

The case is similar with respect to the *Crossing of Species*. By this we understand such changes as arise from the mutual impregnation of two related species. It cannot be denied a kind of bastards are thus produced: they are called *hybrid plants* (*plantæ hybridæ*). They occur most commonly among plants which are cultivated. Thus, in the genus *Pyrus*, *Prunus*, as also among the species of Grain, and in the kitchen vegetables, there are a considerable number of real hybrids, which do not lose their properties even by reproduction. But most of these plants are productions of art, and Nature seems to prevent the mutual impregnation of related species in more ways than one, although these are not completely understood by us, (332.)

145.

All properties of plants which are subject to change, form either a Subspecies (*subspecies*), or a Variety (*varietas*). By the former we understand such forms as continue indeed during some reproductions, but at last, by a greater difference of soil, of climate, and of treatment, are either lost or changed. When the different Cabbage species receive the same treatment in the same climate, they continue to be frequently reproduced, without changing their appearance. But we cannot on this account maintain, that Cauliflower would retain the same favourite form in very different climates, and under a complete change of treatment. It at last changes so much, that it can scarcely be distinguished from the Common Cabbage. This, therefore, is a subspecies. Varieties again do not retain their forms during reproduction. The variable colours,—the very variable taste, and other properties of the kitchen vegetables, the ornamental plants, and the fruit-trees, shew what varieties are; and the Scientific Botanist must therefore be particularly attentive to distinguish permanent species from the variable subspecies, degenerate plants, and varieties.

146.

To this discrimination belongs, above all things, a careful, continued, and unprejudiced observation of the whole vegeta-

G

tion of the same plant during its different ages, and amidst the most different circumstances which have an influence on it. When, for instance, in the common *Lotus corniculatus,* on whatever soil it may grow, we uniformly observe that it has a solid stem, even and erect divisions of the calyx, and expanded filaments, we must of necessity distinguish, as a particular species from it, another form which grows in bogs and in watery meadows, which has a much higher, and always hollow stalk, the divisions of its calyx spread out into a star-shape and hairy, and which has uniformly thin filaments; and we must name this latter species either *Lotus uliginosus* with Schkuhr, or *Lotus major* with Scopoli and Smith. As, on the other hand, the *Pimpinella Saxifraga* grows sometimes quite smooth, and sometimes, in woods and shady meadows, considerably hairy ; as it displays sometimes simple and small stem-leaves, sometimes half and even doubly pinnated leaves ; and as these forms vary according to the situation of the plant and during reproduction, we cannot regard these forms by any means as distinct species, but we must view them as corruptions.

We see, that, in order to decide respecting the idea of a species, an observation of many years, and of much accuracy, is often required; and that the cultivation of plants, from the most different climates, in botanical gardens, is in the highest degree necessary for their discrimination.

II.   *Idea of a Genus.*

147

By a *Genus* we understand the sum of the species which agree in certain constant properties of the essential parts. When we compare several species of Roses with one another, we soon find that they have all certain common marks, and that it is hence an easy matter for every person, who perceives their sum of corresponding properties, to say that he has a Rose before him.

148.

But we must confess, that there is an important difference in genera, according as they are founded on an agreement in the properties of most of the parts, or only in the marks of a few essential organs. When the former is the case, we recognise a *Natural* Genus. We then observe correspondences in structure, in external appearance, in situation; often in the form of the roots, the leaves, the buds, in the subordinate parts, or in the armour and supports; sometimes even in the composition of the sap, in the colours, in the smell and taste. Such natural genera are, for example, those of the Rose, Wheat, Stocks, Willows, and innumerable others.

Artificial Genera, again, are those which, though they want a correspondence in external appearance, shew the same formation of the essential parts. Whenever the preference has been given to the organs of fructification as the essential organs, we must then be permitted to connect the agreement of these with the idea of a genus. Indeed, such genera are by no means adapted to strike the eyes of every person; nay, in a Natural arrangement of plants, we might even very possibly overlook them. But when Art has once reared a system, she must necessarily assume constant differences in the essential parts, as foundations for the discrimination of genera. If the innumerable plants which are known to us as Umbellatæ, and which agree more or less in their external marks, were not marked out as peculiar genera by such fine and scanty characters of the essential parts, it would be contrary to all scientific ideas, to unite them into one common genus, which would then include within it innumerable species, with important differences in the fruit, and in the other essential parts.

149 a.

There are here two opposite errors to be avoided. If we fall into the one, we then seek to simplify all things, and collect the most different forms into a few great genera. The less progress the knowledge of plants has made, the more are we disposed to lose ourselves in this error. Indeed, when we attend to the number of species which Linnæus knew, we

cannot assert that he would have formed fewer genera than
his followers have made ; but as he put less value upon the
fruit in particular than was proper, many of his genera are
too comprehensive, not to be separated with advantage. Whoever should receive the *Fumaria* or *Polypodium* of Linnæus
even now, in their entire latitude, would evidently overlook
the most important and essential differences, and strive for
simplicity at the expence of Science, and even of Nature.

The second error leads to too fine a discrimination of insignificant marks, which might well furnish the foundation for
constructing a species, but can never be approved of as generic
characters, unless we would ultimately make as many genera
as there are species.

149 *b*.

We are naturally led to ask, how we may most securely
avoid both these errors.   It cannot be denied, that the power
of separating important from unimportant circumstances, and
of being as little misled by fancy as by refinement, in connecting or separating things which ought not to be connected or
separated, is a talent which is either born with some individuals, or acquired to a great extent by practice. This talent
belongs to some Botanists in so high a degree, that we may
safely trust ourselves to their glance, and to their discrimination.   Others, with the best intentions, have never been
able to acquire the talent, especially as they seem to believe
that refinement is the only requisite for the construction of
species and genera.

150.

In the first place, in order to avoid these errors, Genera
must be founded on such characters, as, compared with one
another, have an evidently uniform value.   Thus, when a
certain number of genera have already been distinguished from
one another by the difference of one character, we may with
propriety avail ourselves of new differences of the same, to
construct new genera ; since the organ or property must, in
all similar cases, have a uniform value.   If we have thus
once begun to distinguish the Umbellatæ by the form of their

fruit, and the Syngenesistæ by their pappus, we cannot, in the former case, associate a winged with a solid fruit, nor, in the latter, a pinnated pappus with one that is bristly.

It is much to be wished, that in all classes, or in all tribes of plants, we had such characters as have a constant value, and which could hence be employed in the formation of new genera. But that this is by no means the case, we see, in particular, from the numerical proportion in very many genera and families; that is to say, this proportion is often of so little value, or it fluctuates so much, that we cannot employ it for the discrimination of genera. It is also obvious, that the character, taken from one organ, cannot be applied to several families. Important as the inflorescence is in the Grasses, we cannot use it for the discrimination of genera in other families.

151.

The second rule, by following which we may avoid the above-mentioned errors, is, that as much as possible natural genera should be constructed, and, when that cannot be accomplished, artificial genera; but that the variation of form and of proportion in one part, ought not to be considered as sufficient for the construction of a genus, unless this difference be also expressed by other marks. " *The character does not constitute the genus*," is a very wise saying of the founder of Scientific Botany. A genus is not on this account firmly established, because one or another difference of structure occurs in individual parts, but because the plants actually exhibit striking differences in their whole vegetation. We might easily, by giving to a genus a very circumscribed character, be able to separate from it all the species, and to unite them to new genera, which have not that circumscribed character; but it is the business of Art, to evolve a more comprehensive and generally availing character, so as to shew the value of the principle, That the genus forms the character.

152.

Lastly, we must remark, that the further the knowledge of plants is extended, the more do we find that many great ge-

nera, in the natural arrangement, are not correctly placed,—
that they ought much rather to be considered as Families, and
their subdivisions raised to the rank of genera. This has happened in our days with the genera *Lichen, Fucus,* and *Pro-
tea,* which have justly been expanded into several genera, according to fixed characters.

153.

Many genera consist of such numberless species, that it is
much to be wished they were separated into a greater number
of genera. But so long as Nature shews a correspondence
in essential parts, we dare not separate what she has united.
The genera *Aster, Erica, Mesembryanthemum, Salvia,* and
some others, thus justly remain undivided, and we must only
endeavour to arrange the species in such a manner as is
required by their natural correspondence.

This arrangement into divisions or sections, is as necessary
in numerous genera, as the arrangement of the genera themselves according to a certain principle of affinity. Not only
the outward view of the arrangement, but the investigation
of it, is generally relieved by this means. Many of these sections, if we were to carry our distinctions to a great length,
especially in the genera *Convallaria* and *Polygonum,* might
become as many genera, especially as they are separated from
each other not merely by numerical proportion of parts, but
by other marks. Sometimes we give peculiar names to these
sections, as the genus *Polygonum* is divided into Atraphexoi-
dæ, Bistortæ, Persicariæ, *Polygona,* and Helxinæ. ·Otherwise
these sections are denoted by the signs §, * or †.

III.   *Idea of Tribes and Families.*

154.

As in the animal kingdom there are related genera which
form tribes, as the Cetaceous Animals, the Marsh Fowls, and
the Graminivorous Animals, we also find in the vegetable
world a multitude of related genera, which together form

either *Tribes* (*tribus*) or *Families*.   Every person admits
that the different species of Grain, the Palms, Ferns, and
common Fruit-trees, are such Tribes or Families.

155.

But we distinguish these two ideas from one another in
this way ; that by a *Tribe* we understand a smaller number
of related genera ; a *Family*, again, denotes the sum of all
the genera, which agree in one or more essential parts : a family may thus consist of several tribes.   If, for example, we
assume all the species of Grasses as one family, then this consists of the Hordeaceæ, the flowers of which are placed on a
spike,—of the Avenaceæ, which flower in panicles, and have
a twisted awn,—and so forth.   The names of these tribes, as
the examples we have given shew, are commonly borrowed
from the principal genus, to which a termination is given, expressive of the resemblance.   We hence say, the Cyperoideæ,
Orchideæ, Junceæ, Aroideæ, Jasmineæ, Gentianeæ, and so
on.   We proceed, in the same manner, in giving names to
Families.   But these are, with as much propriety, denoted,
usually, by general names, which have a relation to the principal property.   We thus derive the Ferns, Sarmentaceæ,
Coronariæ, Labiatæ, Asperifoliæ, Cruciatæ, *Compositæ*, *Aggregatæ*.

156.

The laws which are followed in the construction of Genera, have the same importance in the establishment of Tribes
and Families ; for, fundamentally, the latter are nothing but
genera, and our only concern, therefore, is to separate them
from one another by constant characters, and to be able to
mark out their connections with one another.

CHAP. IV.

ON THE NATURAL ARRANGEMENT IN GENERAL.

157.

The solution of the last mentioned problem, that, namely, of marking out the connections of families with one another, and of so arranging them with respect to each other as Nature has arranged them, is the object of Method, or the Ideal after which Science is incessantly striving, and to which she has recently approached nearer than she ever did before, without having yet perhaps completely reached it.

To present more distinctly the meaning of this problem, let us think of two nearly related families, for instance, the Musci frondosi and the Musci hepatici. Along with general correspondences, we find also differences, which must be so marked, that we may perceive which of the two families stands higher or lower,—which of the two shews at once the greatest number of organs, and the greatest perfection in these. Let us next ask further, respecting these families, to what others they stand contiguous. In the foregoing examples, we shall see the Musci frondosi passing into the Lycopodeæ, and bordering on the Ferns ; the Musci hepatici, in several forms, approach the Homallophyllæ, and through them the Lichens ; the Lichens pass in several shapes into the Fungi, which at last resolve themselves into the most imperfect of organised bodies, the *Coniomyci* and *Nematomyci*.

158.

When we have in this manner denoted the connections of families and their differences, we are frequently led to intermediate forms, which still more strongly establish the alliance. Between the *Musci frondosi* and *hepatici* is the *Andreaca* ; between the *Musci frondosi* and the Ferns are some species

of *Trichomanes*; between the *Musci hepatici* and the Lichens
are *Riccia* and *Endocarpon*; between the Lichens and Fungi
are *Calicium*, *Stilbum*, and *Opegrapha*;—all evidently intermediate forms. If we could shew similar intermediate forms
throughout the whole of the vegetable kingdom, it might then
be likened to a chain, the links of which were in every instance
connected, and where no proper separation of the parts could
be perceived. On the one hand, this account seems to be
daily more and more confirmed by recent observations. The
Compositæ and Lobelieæ have lately been united by the genus
Brunonia, in such a manner, that this genus may be considered as their intermediate form.

But, on the other hand, we must recollect, that although
Nature makes no leaps, yet she does not appear to proceed
uninterruptedly from inferior to higher degrees of perfection,
but that her forms are repeated in several families; and if we
take the whole together, we shall commonly find, that of two
nearly related families, the one is the more perfect in many
respects, but that in other respects again it is by much the
least perfect. If, for instance, we compare the Musci hepatici with the Ferns, the latter, by their frequently shrubby growth, resembling that of Palms, and still more by the
complete state of their spiral vessels and slits, shew a considerable degree of perfection. The Musci frondosi, again, although
they want these distinctions, shew not only doubly formed
sexual parts, but the leaves which cover these parts are often coloured in the manner of a corolla. These are marks of a higher
perfection, which is wanting in the Ferns. In the same manner, we shall find throughout the whole vegetable kingdom,
that the growth with two separate cotyledons, and the consumption of the albuminous substance, by the formation of
the embryon, are proofs of a higher perfection. But there
are not only tribes and families, which, along with their more
perfect formation in other respects, yet retain the albuminous
substance, as the Caryophylleæ, Portulaceæ, and Aizoidæ,
but among what have been called the lower plants we often
remark a degree of completeness and perfection in the forms,

as in the Scitamineæ, Orchideæ, and Coronariæ, which must often lead us into error in arranging these vegetable tribes.

159.

To this must be added, that there are a multitude of separate genera, and even of entire families, which cannot at present be arranged with respect to others. The genera *Begonia, Cynomorium, Datisca,* and *Nepenthes,* for instance, are completely unknown in their affinities; and it thus remains doubtful, whether the Caryophylleæ are not more properly connected with the Chenopodeæ, than with the Liliaceæ and Myrteæ.

160.

We here already perceive one of the difficulties which the Natural Order presents in instruction. For as several genera stand quite insulated in this order,—as others have some properties of one family, and others have the marks of a different family,—as the situation of families with respect to one another is not pointed out by nature, but is the work of human genius,—we must confess that such a method, from its uncertainty and difficulty, is by no means fitted for beginners, however elevating to the human mind the study of those alliances may be,—however much it sets all the powers of the mind in activity,—and however much it is to be wished that this species of knowledge should constantly be making progress.

161.

The difficulty of the natural method is still greater, when we look for some bond, or, what is the same thing, for some common form, which may unite the natural families with each other, and lead us to their arrangement. If we take this leading principle from one or a few ever so essential parts, we in fact do nothing else but connect an artificial system with the natural arrangement, and this arrangement itself can no longer be called a natural one. We may employ, as our principle, the form and situation of the embryon, the shape of the fruit, the insertion of the filaments, or any other relations of the organs we please, but such a principle cannot be employed throughout but in a constrained and artificial manner. This objection strikes even at the very celebrated, very ingenious, and in many respects the immortal work, which Jussieu has published. Hence there remains no other plan but that of arranging the tribes and families according to the sum of related characters in the greater number of their parts. But as, in this operation, a great deal depends upon a peculiar glance of the observer, the objects of which frequently cannot be distinctly stated in words, fluctuation and uncertainty are here unavoidable; and the more completely all the marks are collected, the more impossible is it to impress upon the mind of the learner the sum of these characters.

In the last place, the comprehensive survey of the Natural Method necessarily requires the knowledge of such genera, tribes, and families, as are only accessible to him who has either accomplished himself by travel in foreign countries, or is in possession of a very rich collection of plants, and at the same time has access to one of the best botanic gardens. Without these assistances one could scarcely become acquainted with the Anoneæ, the Guttiferæ, the Sapoteæ, and many other families.

162.

If with all this we compare the artificial system, nothing is more easy than, during the very first lesson, to produce examples from all the classes; nothing more easy than to obtain a view of the subdivisions of this system. It would be difficult to find any other system better adapted to instruction than the Linnæan. But besides the objections which were before stated, the study of this system, if the attention of the student be limited wholly to it, has the great disadvantage, that it produces a partial and confined way of viewing things, which must necessarily estrange the mind from the higher object of science. Accustomed to view the relations of the sexual parts as the most important of all others, and to consider plants only under this point of view, we come to believe, in the end, that this is the object of science. We become satisfied to thrust the plants that come in our way into the ranks which the Linnæan Classes furnish, and we neglect the relations of the remaining organs. The mind becomes unaccustomed to consider Nature in her greater relations, and a real distortion seizes the understanding, from the habit of assembling together the most dissimilar things, and of separating from one another the most closely related bodies.

To avoid those evils, we must begin, as soon as we have obtained a competent knowledge of the common plants according to the Linnæan system, to study the Natural Method. The Cryptogamous Plants, as they have been called, force us as it were to have recourse to these ideas of natural affinity. Because here, where no artificial system is of any avail, we must necessarily pay regard to general relations, to resemblance of structure and of outward appearance, and to the sum of the other marks, in order to be able to arrange these plants. As every person perceives the necessity of a natural arrangement of these lower plants, wherefore should the study of relationship be confined entirely to these, and not be extended also to the higher plants?

163.

That Linnæus himself viewed his artificial system in a right manner, is evident from a great multitude of passages, where he rates the value of the natural method very high, and views this arrangement as the last object of Botany;— where he expressly asserts, that only imperfectly instructed botanists set a small value upon this method, but that all accomplished natural historians regard it as the highest aim of their labours;—where, when he was only thirty years of age, he promises to dedicate his whole life to the formation and perfecting of the natural method;—where he solicits all able botanists to make common cause with him in securing the great purpose of a scientific knowledge of the affinities of plants; and where he expressly says, respecting the artificial arrangement, that it is merely an expedient of necessity, and must, on all occasions, yield to the natural arrangement. We find those passages as well in the *Classibus Plantarum*, t. 484. and 487. as in the *Philosophia Botanica*, § 77. Even in the latter days of his life, he read ingenious lectures respecting what he named *Ordines naturales*, which lectures have been published by Giseke, at Hamburgh, 1792. These natural orders are formed without any particular bond, and Linnæus used to compare them with the different sections upon a land chart. According to this idea, we find in Giseke's edition such a land chart, where indeed very many regions stand quite insulated, but where also the conterminous boundaries of many are correctly marked. That the Palms are conterminous, through the genera *Cycas*, *Zamia*, and *Nyssa*, with the Ferns, and on the other side with the Hydrocharideæ; that the latter are related through the Junceæ with the Calamariæ, and Cyperoidæ, and on the other side with the Ensatæ or Irideæ; that the latter are conterminous with the Orchideæ, and these with the Scitamineæ; all this, and a good deal more, is very correctly stated in that chart.

Even in his artificial system, the putting together of genera, according to their natural affinities,—a peculiarity of his system which we have already blamed,—is a proof of his predilection for the natural arrangement.

164.

Even during the life of Linnæus, Michael Adanson proposed a natural method in his *Familles des Plantes*, Paris 1763, which contains an inexhaustible treasure of observations respecting the essential characters of families, and of innumerable genera. It wants indeed a principle of arrangement, many of the genera are imperfectly constructed, and, with capricious obstinacy, are improperly named; but the series of families is for the most part derived from nature. Thus to the Boragineæ or Asperifoliæ, succeed the Labiatæ, then the Verbeneæ, the Personatæ, the Solaneæ, the Jasmineæ, the Anagallideæ, the Salicariæ, the Portulaceæ, the Sedeæ, the Alsineæ, and so forth. Besides, in this arrangement, attention is paid to every thing,—even to the finest parts of the fruit and seed ; and those recent times, the improvements of which have been derived from the labours of Adanson, have, in many instances, only been able to confirm his observations.

165.

By Joseph Gärtner's single work on Fruits and Seeds, an entirely new light has been cast upon the natural method, by which a multitude of obscurities have been cleared up, and affinities have been discovered, where they were before sought for in vain.   In this way, and by treading in the footsteps of Jussieu, De Candolle, Richard, Batsch, Correa de Serra, and Robert Brown, we may hope to afford essential advantage to science, whilst we are searching out every where relations and affinities.

It cannot but happen, that the farther progress we make, the greater number of families will be discovered ; because in particular tribes we shall ever be noticing such distinct assemblages as separate themselves properly from the families to which they had formerly been ascribed.   If Jussieu reckoned only a hundred families, we must now be acquainted with nearly a hundred and fifty.   The Dillenieæ, Pittosporeæ, Tremandreæ, Combreteæ, Cunoniæ, Rhizophoreæ, Haloxageæ, Atherospermeæ, Hackhouseæ, are examples of later families, which Brown has established by well-founded investigations.

166.

In a few words we shall notice the analytical method which Lamark proposed, forty years ago, as a middle path for avoiding the inconveniencies of the artificial system, as well as the difficulties of the natural method.   He remarked, that the procedure of the human mind, in the investigation of plants, was of such a kind as leads us to divide the whole vegetable kingdom into two principal departments, the characters of the one of which are always completely exclusive of the marks of the other.   Each of these two may again be separated into other two divisions, and this mode of dividing may be continued, until we have at last only two species to compare together and to distinguish. Whilst in this plan we are confined by no method, we actually exhaust the region of possibilities, at the same time that we are studying realities. Yet the circumstantiality and discursiveness of these investigations are objections to the plan. No person has pursued it with more ability and success, in later times, than Gaudin, in his *Agrostologia Helvetica*, 1811. When, for example, he wishes to teach us to distinguish the species of *Festuca*, he first attends to the leaves, whether they are all bristly, or whether the stalk-leaves be smooth ; then to the integument of the leaves, whether it be very short and truncated, or very prominent ; next he attends to the spiculæ, whether they be oval or oblong, with awns or without them ; lastly, to the awns themselves, whether they be as long as the spiculæ, or longer.

For facilitating the diagnosis of individual species, the analytical method is very useful ; but it requires too great an expence of time and labour to be used on every occasion in the examination of plants. If, for instance, we wish to determine a Myrtle, we must first ask whether it belongs to the division of plants with distinct sexual parts, or with those that are hidden ; whether it be monoclinous or diclinous ; whether the sexual parts stand free or are united, to what organ they are fixed, in what number they are present ; whether the seeds have albumen or not ; what situation the plumula has with respect to the umbilicus and the cotyledons ; still further, how the corolla is fashioned, how its estivation takes place, whether the plant be a tree or an herb, what is the formation of its leaves, and their situation in the buds, and so forth. The answer to these questions depends on inquiries, which always lead to certain, comprehensive, and useful knowledge. They are also applicable to every system, but for elementary instruction they are in every case too discursive.

CHAP. V.

THEORY OF NATURAL CLASSIFICATION.

167.

To the Theory of Natural Classification belong essentially the three following particulars. In the *first* place, we must be acquainted with the relative importance which belongs to organs, compared with one another ; in the *second* place, we must know the circumstances which might lead the observer to mistake the true nature of organs ; and, in the *third* place, we must be able to estimate the importance which may be attached to each of the points of view, under which an organ may be considered.

I. *Comparison of Organs.*

168.

As in organic bodies every part has its relative import-ance, so this importance can have a reference only to the function for which it is destined, and not to things to which it stands in no relation. In classification, therefore, the de-gree of importance of every organ, can be estimated only in relation to those organs which have a reference to the same function.

169.

But as the functions of the vegetable kingdom are of two kinds, namely, nourishment, which relates to the maintenance of the individual, and propagation, which relates to the main-tenance of the species, so each of these functions, considered by itself, must have an equal importance ; and a classification, which is founded upon one of these two great functions of the vegetable kingdom, must necessarily be as natural as that which has a reference to the other function.

This principle, indeed, is opposed by the commonly received idea, that the parts of fructification are properly the most essential and important. It is opposed by the general practice of the founders of systems, who regard only the parts of propagation, and derive from them the principles of arrangement; but this happens principally because we usually find fewer differences in the organs of nourishment, and because, in order to discover these differences, dissection is previously required. Hence, in order to proceed securely to work, and not to throw unnecessary difficulties into the study, we must view the Organs of Fructification as the chief basis of Classification.

170.

As in an organised body all the parts have a mutual influence, and are connected with one another, important differences in the organs of one function must of necessity draw along with them differences in the structure of the parts which belong to another function; and it is a very important principle, that the entire structure of plants is different when important alterations are found in the seed. Plants, whose seeds contain an unevolved embryon and rich albumen, have a completely different internal structure of stem, leaves, flowers, and commonly even a different numerical proportion, from plants, whose seeds contain a completely developed embryon, and have little or no albuminous substance.

In many instances this has even an influence on the nature of the secreted juices, because these derive their character only from the nature of the separate organs. Thus the coloured juices in the Hypericeæ and Guttiferæ, the milky juice of the Euphorbiæ, and the aromatic ingredients of the Labiatæ, are well known examples of this influence of natural affinity even on the constitution of juices. The theory of medicines may, in this view, derive advantages from the natural arrangement of plants.

H

But it is a matter of consistency, when we admit the nature of the secreted juices among the characters, to understand the existence of peculiar organs and vessels, which secrete and contain these juices. If we have not yet completely ascertained the internal structure of these organs, we are justified in assuming it, when, instead of the organs themselves, we exhibit their productions, namely, the nature of their secreted juices

171.

But it is not enough to know, that the organs deserve more attention than their products; in every function we must know the means to be employed for estimating the importance of the organs. Sometimes reason, and sometimes observation, are these means.

Reasoning can only be employed when we know the use of an organ; then we obtain, by a simple exercise of reason, an idea of its importance. If we attend to the organs of propagation, these are evidently of more importance than their integuments. If we compare together the sexual parts, both of these parts are alike indispensable; but the male organs perform their part for but a very short time, and in this respect may be compared to the stigma, which disappears after impregnation. But as the female organ, beside this short-lived part, contains another, for which all the other parts are constructed, it is plain that the female organ is of more importance than the male, and that the part of it which lasts is of more importance than that which passes away. As, farther, in the permanent female organ, the integument, or the fruit, may be separated from the seed, and exists but for it, the seed has thus a higher value than the fruit. And when, lastly, we divide the seed into the embryon and the albuminous substance, or, in the want of the latter, when we distinguish between the cotyledons and the young plant, then the latter has a greater value than the other parts of the seed. If these conclusions are just, we have then the following degrees in the importance of these organs. In the 1*st* place stands the embryon, the ultimate object of the whole vegetation. 2. The parts of the seed. 3. The fruit. 4. The filaments and an-thers, the latter of which must be more important than the former, because the former exist but for them. 5. The nec-taries, which, when they are present, essentially promote fructification. 6. The interior cover of the sexual parts, or the corolla. 7. The calyx, or exterior cover.

172.

There is a second mean of determining the value of or-gans, which, however, although very instructive in certain re-spects, is liable to more objections. We may consider any part of fructification as having so much the greater import-ance, according to the number of species in which it is found. By following this rule, we obtain nearly the same results as from the former; because the object of vegetation is in all cases affected, when the same means are not in operation. The seed, or germ-grain, is universally present, even in the lowest plants. The individual parts of the seed cannot be so com-monly distinguished. All plants have not fruit; but more shew fruit than sexual organs. Of these organs, the female parts, even those which are shortest lived, are found in more plants than the male parts; since even in the Homalo-phyllæ, and Musci hepatici, there are pistils and stigmata, where no anthers have yet been shewn. Anthers occur more commonly than filaments. We dare not decide whether nec-taries or corollæ appear most frequently; since as many flowers wanting the corolla have nectaries, as there are nec-taries without the former organ. As little can we affirm that the corolla appears more frequently than the calyx, because, in numberless cases these integuments pass into each other.

173.

A third mean of judging respecting the importance of or-gans, consists in observing how far a certain organ is more or less constantly united with the structure of definite and gene-rally received vegetable tribes. If, for example, we had to determine whether the stipulæ or spines are most import-ant, we must give the preference to the former, because there is a multitude of families in which the stipulæ are a constant appendage of vegetation; for example, the Rubiaceæ, Malvaceæ, Leguminosæ, Amentaceæ; whilst spines may be present or wanting in a great many families, without making any perceptible difference; for instance, in the Rosaceæ, Leguminosæ, and so forth. In the same manner, the constant absence of an organ, in certain families, is of more moment than its accidental appearance, when it is wanting in other related forms. Thus the Grasses, so far as we know, have never nectaries, as they also never have compound leaves.

II. *On the Means which Nature affords for enabling us to know Organs, and thereby to avoid mistakes.*

174.

In all inquiries into the history of plants, it is an object of the highest importance that we should have a correct idea respecting the purport and nature of any organ, because then only can we flatter ourselves with the idea of having obtained a proper insight into the economy of plants. This department of the study has made remarkable progress in our days, since men began to free themselves from the fetters of the schools, and from the prejudices of authority. The great Founder of Scientific Botany frequently mistook the nature of organs, especially when he saw nectaries, where none are and never can be. And it is not long since it was believed, that in the Ferns there were anthers in very various organs, which, however, have a very different purpose.

The first thing that we have to do, in order to become acquainted with the nature of an organ, is to endeavour to find out whether it really performs the functions, to which, by its form, it seems to be destined.

Even when the form varies, we must be determined by the function to assign to the organ a definite nature. It is well known how different is the form of nectaries, and how great a variety takes place in the structure of filaments. If, however, the function be really the same, we must explain their nature upon a common principle. But he who permits himself to be led, by the mere similarity of form, to suppose that, therefore, the organs are alike, has fallen into a mistake which may give occasion to important errors.

Experience teaches us, that when, from some peculiarity of structure, a function cannot be performed by the organ commonly destined for that purpose, another organ supplies its place. As the proboscis of the Elephant performs the part of a hand; as the tail of the Kangaroo, although fashioned like other tails, serves the animals as a bone; so in the Acacia of New Holland, the leaf-stalks supply the place of leaves. Thus, also, those leaves of water plants, which grow under the water, are divided in the manner of roots, and seem to perform a similar function.

175.

There is yet another law to be understood, to enable us to judge properly respecting the Nature of Organs. In innumerable instances, there appear forms similar to those which are connected with a definite function, but which do not fulfil that function; and Nature seems, in these instances, as in the animal kingdom, to produce forms which are completely useless, merely for the sake of a harmonious and symmetrical structure. The appearance of filaments with empty anthers in flowers that are altogether female, and of female parts in flowers wholly male; the structure of filaments in other forms, where they resemble nectaries; the false nectarothecæ in such Orchideæ as have no nectaries; these all are formations which can only be explained by the law of nature we are now illustrating.

The third mean of knowing the Nature of an Organ, consists in the dissection of its structure; for which purpose powerful magnifying glasses are frequently necessary. If we wish to ascertain the existence of the integument possessing the nature of the calyx, we must observe the continuation of the epidermis with its slits in that integument, (312.) A glandular and fleshy structure determines respecting the nature of nectaries, as a multitude of short absorbent warts leads to the belief of the existence of the stigma.

By studying these relations, by carefully comparing the
structure with the function, we arrive at a sure knowledge of
the nature of organs.

176.

Mistakes in this respect are committed, when we overlook
three things, namely, the Abortion, Alteration, and Union of
Organs.   Respecting all these three phenomena of nature
we must now give a more particular account.

A. *Of Abortive Organs.*

177.

The abortive state of an organ is often the consequence of
an imperfect evolution.   The cause lies not unfrequently in
unfavourable weather, in an unfertile soil, and in other accidents.   We thus see that fruit is not completely formed,
when the soil is arid.   But the circumstance of most importance is, that abortion arises very often from a law of Nature, in consequence of which one part is evolved at the expence of another; and this latter part, therefore, remains imperfect.   We every day see a remarkable example of this in
many species of Violets, where some blossoms, in which the
corolla is chiefly unfolded, fall off, without leaving any fruit,
whilst other flowers, which had not a corolla, set perfect fruit.
The beautiful colouring of the bracteæ of *Buginvillæa*, leaves
no distinct evolution to the corolla ; and in our Melampyrum
nemorosum and cristatum, the flowers do not expand on the
upper parts, where the beautifully coloured bracteæ seem to
supply their place.   It is a law of Nature, which remains
constant even in entire families, that half of the filaments
are abortive, or that the loculi of the fruit, originally destined undoubtedly for the reception of seed, remain partly
empty.   In the Acantheæ and Sapoteæ, we find the phenomena of abortive filaments as a law of these families.   In the
Rhamneæ, Palmæ, Sapindeæ, and many other families, the
abortion of the individual loculi of the fruit is constant ;
and who can have compared our *Gaura biennis*, in the

state of the unripe germen, and of the ripe fruit, without re-
marking, that the former contains several ovula, whilst the
latter always incloses but one seed. In the Siliculosæ, as in
Crambe, *Cakile*, and *Myagrum*, we always find empty abortive loculi. Is it further necessary to multiply examples,
by calling to recollection the common Snow-ball, and *Hydrangea hortensis*, where the abortive state of the sexual parts
affords opportunity to the evolution of the beautifully coloured integuments of the flower ?

178.

A question naturally arises, in what way we may avoid the
mistakes to which such abortion may lead. For this purpose,
we propose the following means. 1. We must examine the organs in all their proportions, even in those that are most different. It cannot but happen, that during such examination a return to the symmetrical or natural structure will be observed.
These symmetrical proportions are often the natural ones,
and art has merely produced the abortive state, by evolving
one organ at the expence of another. The Snowball of our
gardens bears perfect flowers when it grows wild. It is frequently art itself, or a luxuriant growth, which produces
this return to the original structure. As, for example, the
appendages to the filaments of Sage are abortive filaments,
the plant, like other Labiatæ, should properly have four filaments. In fact, we observe, that, in many species of this genus, the oblique processes of the filaments carry anther-shaped
bodies; and in *Salvia glutinosa*, we find these, during moist
summers, passing into true anthers. When we compare with
these *Stachytarpheta* and *Westringia*, we see very distinctly
the two abortive filaments, which also, in the last named genus, bear empty anthers.

But in every case it is important to observe the organs at
their first appearance, whilst they are as yet unrestrained in
their structure. Thus, we must examine (and this is one of
the most important rules for the botanist) the loculi of the ovarium in its unripe state, in order to determine respecting the
true nature of the ripe fruit. When we examine a spine in

its first appearance, we perceive that it was intended to have been a branch, which has only proved abortive, and which, notwithstanding, even as a spine, bears leaves, and, in *Euphorbia heptagona*, sometimes flowers and fruit.

179.

2. To avoid mistakes, we also often follow analogy or induction. When a form is common to several families, we cannot mistake its sign, even when it seems to fail. Thus, as the Orchideæ are related to the Scitamineæ and Irideæ, we find in the former the sign of the three male organs, in the two lateral appendages of the column of fructification, or the Staminodia of Richard. In like manner, the two fibres, which *Gratiola* carries beside the fruitful filaments, will appear to be abortive filaments, when we compare them with the other Scrophulariæ, which sometimes have four fruitful filaments, and sometimes, in addition to these, even a fifth.

The same analogy which leads us from genus to genus, explains also one species by another. As in the *Leea* we consider the cleft scales, with which the filaments are intermingled, to be abortive filaments, because in the nearly related genus *Melia* ten filaments have perfect anthers; we also conclude that *Polygonum amphibium*, which has only five, and *Polygonum persicaria*, which has only six filaments, have lost the rest by abortion, because several species of the same genus possess eight filaments.

180.

From abortion principally arise the many irregularities in the structure of plants ; because we may suppose that the original formation of natural bodies is regular. When we thus find unimportant irregularities in the organs of a plant, we may suspect that there are plants in which these irregularities assume a more marked aspect,—that there are others where these organs are completely abortive,—and others in which complete regularity takes place. The usual form of the papilionaceous flower is very irregular in *Dimorpha* ; the vexillum and carina in *Amorpha* are completely abortive ; and in

Tamarindus, Hymenæa, Parkinsonia, and even *Cassia,* the papilionaceous blossom assumes the appearance of pretty regular four-leaved flowers.

181.

Another effect of abortion is, that an organ, which has completely altered its form becomes incapable of performing its function. It is then either altogether superfluous, and remains only as an ornament (175.), or it performs the function of another organ, whose form it has taken,—as the expanded filaments of *Canna* and *Thalictrum petaloideum* supply the place of the petals of the corolla.

The usual effect of abortion, in which another part is enlarged at the expence of the abortive part, has been partly noticed already. We thus see the fruit swell and become better flavoured, when the seed is abortive. Thus, in the Acacia of New Holland, the compound form of the leaves is only remarked during the earliest growth of the plant. Afterwards the leaves become abortive, and in their stead the leaf-stalk is evolved to such a degree, that, along with the form, it assumes also the function of leaves.

In fact, attention is every day more carefully directed to the laws of abortion; because by them we are able to explain a great multitude of phenomena in the vegetable world, and of otherwise incomprehensible varieties of form ; and because excellent applications of these laws can be made to the physiology of animal bodies.

B. *On Change and Degeneration of Parts.*

182.

It is an important law of the whole vegetable kingdom, that from every individual part of a plant, every other may be evolved. In animal bodies of a perfect structure, this is not possible in the natural state, because the internal structure of the individual organs is much more compound and various. It is only in the sickly state of the higher animals that muscle is changed into cellular texture and fat; or the common membrane of the vessels into the substance of bone. But in plants this change takes place the more readily, the nearer the organs stand to one another in the process of formation; so that the roots may become stem and branches,—these again may be changed into roots,—leaves may become leaf-stalks, and the reverse,—the calyx may become corolla, and the filaments may change to petals; and even in some cases the fruit may again push out leaves, and from the receptacle new branches may spring up.

183.

In the formation of one part from another, Nature proceeds in such a way, that she first renders the parts compact and simple, when she intends to give them an extended and compound form. The root-leaves are naturally more simple than the following stem-leaves, and these again are the simpler the nearer they stand to the flower. When hindrances to this compacting of the part occur, the consequent extension must also undergo a change, and hence we see many alterations arise.

184.

These alterations, when we attend to the substance of the part, may be arranged under the five following classes.

1. The parts become prickly, as happens in the branches of Fruit-trees,—the stipulæ of the Acacia,—in innumerable leaves,—in the calyx even,—and, in one instance, the *Cerviera De Cand.* a genus of the family of the Rubiaceæ, in the petals of the corolla.

2. The parts become lengthened out into a flexible fibre, which, when it is curved, is usually called *cirrhus*. It is well known, that the leaf-stalks of Vetches, and of the species of *Lathyrus*, are thus extended into cirrhi; as also that the flower-stalk of the Vine and of the Passion-flower uniformly undergo this alteration. The transition of the stipulæ in *Smilax*, and of the nerves of the leaves in *Flagellaria* and *Nepenthes*, into cirrhi, is less

2

known. Even the petals of the *Strophanthus De Cand.* and the filaments of *Hirtella* resemble twisted threads.

3. The usual fibrous or rounded form of particular parts, may also become of a leafy nature, and thus conceal their original form. The branches of *Phyllanthus*, and even of the *Cactus* species, are of a leafy structure ; and, in the former, the flowers seem to be set on the margin of the leaves, whilst it is only the expanded and leafy flower-stalks which bear them. We have already remarked, that the nerves of the leaves of some species of Polytrichum are fashioned into lamellæ. In the same manner, it is not unusual for the filaments to expand themselves, and, in the Irideæ, even the pistils from this expansion assume the appearance of petals.

4. Parts that are naturally green and juicy, often become dry and membranaceous (*scariosus*). This is most frequently observed in the calyx-scales of compound flowers, and in the calyx of the single florets, which we consider as a pappus.

5. Lastly, It very often happens, that membranaceous or leafy parts become fleshy, as we every day see in the receptacle of the Strawberry and Raspberry. Thus, also, the one-seeded berries of the Anoneæ unite by the swelling of the receptacle into a single juicy fruit ; and, in like manner, the strobilus of the Juniper, by the swelling of the scales and their union, takes the form of a juicy berry. In the *Hovenia dulcis,* even the flower-stalk, after the time of blossoming, becomes juicy, well flavoured, and of a beautiful red colour.

C. *On the Union of Organs.*

185.

Every person knows that there are instances in which two fruits, two branches, and even two trees, are united with one another. But there is a law of nature which regularly produces this junction ; namely, when similar organs have a disposition not only to hang together, but when they cannot pro-

ceed in their growth without passing into each other. Thus
we see leaves grow together, which, in the Loniceræ, constitute
under the flower an entire leaf, so that we say they have com
pletely united. Thus we every day see in the Umbellatæ, two
fruits united, which only separate when they are fully ripe. In
the same manner, we find the parts of the involucra of many
Umbellatæ growing together, especially in the species *Bupleurum* and *Seseli*, so as to lead us to the conclusion, that the calyces of compound flowers are nothing else but united involucra,—as is proved by their transition into leaves in the calyx
of *Buphthalmum*, in the *Acmella* of *Richard*, in the *Georgia,
Wild.* and in the *Sigesbeckia*. The distinction between *Tragopogon* and *Urospermum, Scop.* also leads us to the same
conclusion, because in the latter the parts of the calyx are
united into one body, whilst in *Tragopogon* they are disjoined.
In these examples, therefore, it is evident that a union of the
leaflets of involucrum takes place, rather than that the calyx
deserves to be considered as a single and independent body.

186.

What has now been said of the calyx is applicable also to
the corolla. There is a regular union of the usually-sepa-
rated parts of the corolla, of which we have obvious examples
in some papilionaceous flowers, particularly in *Trifolium,*
and even in the vexillum of the *Lotus*. In like manner,
many monopetalous flowers may be regarded as made up
of parts that originally stood free; and it is evident in what
manner the forms of *Cyphia* and *Phyteuma* pass into imi-
tations of the corolla in the other Campanuleæ.

187.

A similar disposition to united growth belongs to filaments,
whose analogy to petals strikes every person upon an attentive consideration. When we consider, that among papilionaceous flowers and leguminous plants, the numerous tribe of So-
phoreæ have free standing filaments, whilst the tribe of Spar-
teæ have them united into one bundle, both tribes will appear to be related to the Diadelphous plants, because we observe frequent transitions from the one tribe to the other.

Crotalaria and *Abrus* shew, in the empty cleft on the back of the cylinder of the filament, this disposition to the separation of the one filament, which, in proper diadelphous plants, we see standing single, whilst in *Dipteryx*, *Schreb.*, the junction of the filaments is incomplete, and in *Dalea* and *Petalostemon* they are less united with each other than with the parts of the corolla. Are not these instances proof sufficient that the union of the filaments presupposes an original disunion of them? Do we not every day, in many of Caryophylleæ, perceive the filaments standing free, whilst in others (*Dianthus*, *Saponaria*, *Silene*, and *Agrostemma*,) they are attached to a ring surrounding the pistil, or are united with the petals? The tribe of Chenopodeæ, to which belong *Illecebrum*, *Herniaria*, *Gomphrena*, and others, shew the same thing. The filaments in *Iresine*, *Paronychia*, *Tournef.*, *Anychia*, *Herniaria*, and *Bosea*, stand free. In *Achyranthes*, *Illecebrum*, *Gomphrena*, and others, they are united.

188.

What has been said of petals and filaments is also applicable to pistils; that is to say, we are often forced, when we see a simple pistil, to consider it as made up of several : otherwise it would be difficult to understand in what way many plants have several pistils, whilst some nearly related to them have but one. It may here become a question, whether *Mespilus monogyna* be the only original species, or whether there be not many other species, which have from two to three, four and five pistils. When in most of the Grasses we regularly perceive two pistils, it is extremely probable, that the few, as *Nardus*, *Cenchrus*, *Lygeum*, and *Spartina*, which are monogynous, have their one pistil made up of two that are united. This is still farther confirmed by observing, that the number of the pistils corresponds with the number of loculi in the germen. If there be but one pistil, the number of loculi in the germen corresponds with the number of stigmata ; and it is probable that originally there were as many pistils as stigmata. Lastly, in a cross section of a strong pistil, as for instance in any of the Cactus species, we observe

2

that there is not merely one canal in its axis to the germen, but that several such passages stand in a circle, and probably represent as many united pistilla.   When we observe how the four caryopses of the Labiatæ and Asperifoliæ surround the single pistil, we cannot but believe that the latter part is made up of four individual pistils, especially when we attend to the almost complete separation of the stigma,— the distinct coherence of two individual pistils in the *Perilla*,— the deep disunion of them in the *Thymbra*, in *Echium* and *Echiochilon Desfont.*,— and, above all, when we attend to the four divisions of the stigma in *Cleonia* and *Coldenia*.

189.

Lastly, we must apply this idea of an union of parts even to the fruit.   What we mean is, that caryopses, in indefinite numbers, are more original than capsules with many loculi : that the latter must be considered rather as a collection of united simple capsules, and that when, in a simple fruit, we observe a certain irregularity, we may safely suspect that originally there were several loculi, but that they have become abortive and united.

We may prove this account by striking instances.   Every person knows that the fundamental genus of the Ranunculeæ, from which they take their name, contains caryopses in indefinite numbers, which are also found in *Myosurus*, *Anemone*, *Thalictrum*, and so forth, but which pass in *Xanthorrhiza* into the single-seeded utriculus, and in *Aconitum* and *Pæonia* into capsules which open laterally.   On account of the relationship of *Nigella* to those plants in other respects, we must consider its fruit also as single capsules, which open laterally, and which are only a little united at their lower parts.   Now, when we observe a capsule with five loculi in *Vallea Mut.*, which is evidently a Ranunculus, nothing is more natural than the conclusion, that in this case, as well as in *Nigella*, there have been five simple capsules which have here undergone a union.   If we attend to the fundamental genus, from which the Diosmeæ have their name, we find it carrying five pointed caryopses, surrounded by an *arillus*, which bursts by an internal elasticity. The *Melicope Forst.* has a similar fruit, only that, in this case, the capsules have but one seed. In the *Corræa Sm.* we find a capsule of many loculi, derived undoubtedly from four single capsules which have been united; (Tab. III. Fig. 9. 10.) We all know that the fundamental genus of the Malvaceæ contains numerous capsules placed in a circle. The same arrangement takes place in *Sida* and *Lavatera. Hibiscus*, again, has a capsule of many loculi, which is only distinguished from the former fruits by the circumstance, that in the one case the seeds are separated by a simple, and in the other by a double partition. In illustration of the last stated principle, I produce the striking instance of the *Linnæa.* Its relation to the *Lonicera, Hallera, Schradera*, and especially to the *Triosteum*, and even to the *Sambucus*, is obvious; and it belongs, therefore, to the Caprifoliæ. Now, these have, for the most part, fruit with three loculi and three seeds. *Linnæa* also carries a berry with three loculi, but when it is fully ripe, we always find but one seed, and the other two loculi are united with the one which bears this seed. The Polygaleæ have commonly a germen of two loculi, but the ripe fruit contains only one seed. The case is the same in the *Phillyrea* and *Rytidea Decand.* Three or four seeds are here united, so that the situation of the embryon is changed and pressed to one side.

Lastly, The opinion of Brown does not seem to be without foundation, that the caryopses of the syngenesious plants, which we always see simple, have originally been double; because, in some species, we see very distinctly two lateral funiculi umbilicales, which would not be there if the seed were simple. The divided pistil points, as we have formerly hinted, to the same conclusion; (Neue Entdeck. I. s. 171. 172.)

190.

Respecting the Union of Organs we may establish the following laws. The importance of this union of the parts of fructification increases with the difficulties that are presented to its accomplishment, because the more numerous the obstacles which are to be surmounted, the more powerful must be the cause which overcomes them.   But these difficulties may be founded either in the consistence of the organs, or in their degree of analogy.   Fleshy parts are easily united.   Such a junction is accordingly of no great importance ; and when, therefore, a capsule has sometimes the consistence of a berry, as in *Hypericum, Androsæmum,* and *Bacciferum,* we are not justified in considering this phenomenon to be of so much consequence that a generic distinction can be founded on it.

The analogy of parts facilitate their junction.   That similar parts should be connected is of so little consequence, that we perceive the filaments, in particular, to be connected at their base innumerable times, without on that account arranging the plants in the Monadelphia Class.   In the same manner, filaments and petals, calyx and corolla, may very well be united, without any great difficulties being surmounted.   But between the germen and the corolla, between the corolla and the filaments, there is no particular analogy, and hence we find these parts seldom united.   But when they are, this phenomenon is of considerable importance.   In the same manner we consider the junction of the calyx with the fruit, and the union of the two sexual parts, as at least more important than the union of the nectaries with the filaments and sexual parts.

191.

The union of different organs of fructification is the more important, the more intimately it is connected with very great changes in the general symmetry.   This observation limits some of the foregoing remarks.   With respect to the connection between the corolla and the calyx in particular, this may take place in two ways.   If the petals alternate with the parts of the calyx, then they are commonly united only at their base, and the general symmetry is not thereby destroyed.   But if the petals stand directly before the parts of the calyx, then the junction may be complete, of which we see a remarkable example in *Sesuvium.*   In this case the general symmetry is completely changed, and this phenomenon is of very great importance.

If the filaments do not cling to the corollar integument of the calyx, then the calyx cannot be united with the germen. But if the filaments are fixed in the corollar integument of the calyx, then the calyx may either stand free, as in the Melastomeæ, or may be united with the germen, as in the Pomaceæ.

When similar parts are united at their base, they always shew a disposition to be more expanded at their lower part; and we may safely presume that they are not united, when we observe that they become narrower towards the base. Hence the Myrsineæ, although they have a very deeply divided corolla, are not at all inclined to produce polypetalous flowers.

192.

Lastly, The difference of numerical proportions may often be explained. both from the abortion and from the union of parts. When we compare two related plants, one of which has five, the other ten filaments, either the former has taken its character from an abortion of filaments, or the latter from the union of two flowers. Although the Grasses have commonly three anthers, we sometimes see Grasses of the Hexandria Class, as if they had sprung from the union of two belonging to the Class Triandria; an idea which is strikingly confirmed by the structure of the *Ehrharta*. In like manner we often remark, that plants which have usually six filaments, exhibit twelve of them, as happens in *Lythrum*. On the other hand, Hexandria Plants may lose a part of their filaments, and appear as if they belonged to Tetrandria, of which the genus *Convallaria* is a striking instance. Those of the Class Tetradynamia, are perhaps originally of the Class Decandria, and the four petals are to be regarded as altered filaments, as in *Thlaspi Bursa* we sometimes observe this return to the original structure.

I

III. *On the different Points of View under which an Organ,
     or a System of Organs, may be considered.*

193.

In order to seize these points of view, we must examine
as well the internal symmetry of an organ, as its relations to
other organs, or to the whole plant. Our attention, therefore, must be directed to the presence or absence of the organ,—to its situation and relative position,—to its numerical
proportion,—to its size,—its external form,—its duration,—its
uses,—and to its sensible properties.

194.

In this examination, the most important matter is, that we
should be convinced of the presence or absence of an organ.
We must endeavour to avoid the errors which may arise
from abortion or union of parts. Especially we must put
more value on the positive characters, which express the presence of an organ, than on the negative, which indicate its
absence.

195.

The situation and relative position of parts is their most important point of view. This situation may either be considered simply with reference to the place of insertion ; or with reference to the dissimilar organs which are attached to the
same place ; or, lastly, with reference to the similar organs
which arise in different places.

With respect to the first of these ; the essential situation of
every organ is always so defined, that we understand its true
support ; that is to say, we understand what is the part by
which it is nourished, or from which it springs. To give an
instance,—the situation of the embryo must not be considered
in relation to the fruit, but to the umbilicus, or to the point
where the funiculus umbilicalis is inserted. In this sense
we observe that almost all embryos direct their radicle towards the funiculus umbilicalis. When we say, therefore,

that the radicle is directed upwards or downwards, it is the same thing as if we said, that the seed stands upright, or hangs, in relation to the fruit, downwards; (121.)   Properly, therefore, this character has a reference, not to the situation of the embryo, but to the situation of the seed, and its value is not of the first but of the second rank.

The situation of the parts of the flower must always be referred to the receptacle, because they arise from it.   But it is often very difficult to determine the true position of these parts on the receptacle; and we sometimes only conjecture it from the mutual junctions which take place, because we naturally take for granted, that organs stand nearer to one another at their origin, the greater disposition they shew to unite together.

196.

Still more important than their absolute situation is the position of different organs with respect to one another.   We must not only know that the filaments stand on the receptacle, but also, whether they alternate with the petals, or are set opposite to them, and stand before them.   Especially, it is of much consequence to consider the position of the parts of the corolla and calyx with respect to each other, because in this matter Nature preserves a very particular regularity. The most usual case, is that in which the filaments stand before the parts of the calyx, and alternate with the petals and with the loculi of the fruit.   There is no known instance, however, in which the loculi of the fruit, the filaments, and the parts of the corolla and calyx, all stand directly before or behind one another.   The instances are rare in which the parts of the corolla and calyx stand before one another, and the filaments alternate with them ; or in which the filaments, and the parts of the corolla, stand before one another, and alternate with the parts of the calyx.   The position of the stem-leaves also depends on the same principles. The alternate position of the leaves is peculiar originally to the imperfect and albuminous plants,—the opposite and

whorl-shaped position belongs to the higher, or to the albuminous plants.   But there are many exceptions to this.

197.

It is very difficult to establish any fixed principles respecting the value of numerical proportion.   On the one hand,
Nature shews a regularity which cannot be mistaken in the
numerical proportions of many parts, as, for instance, in the
impregnating organs of the Orchideæ and Liliaceæ.   On the
other hand, she seems sometimes to sport in such a manner
with numbers, that in many genera we scarcely find the same
number of filaments and pistils in all the species.   Astonishing changes often take place, in this respect, from abortion
and union of parts.   We shall endeavour to state some
rules on this subject.

198a.

Numerical proportion appears to be more steadfast, and
consequently more important, the more scanty the number
is.   It is on this account that the numerical proportion
of the anthers remains so steadfast in the Scitamineæ, Orchideæ, and Grasses, and also in the Labiatæ, because they
have only one, three, or four.   Dodecandria Plants observe
much seldomer the same numerical proportion than those of
the Class Hexandria.   But there are exceptions also to this
rule, of which *Valeriana* and *Boerhavia* are well known instances.

In the organs of impregnation, unity seems chiefly to belong, as a character, to the pistil.   In most of the other parts
of fructification, unity appears only as a consequence of abortion or union.   This has been already remarked in the
Orchideæ; it is equally evident respecting the Scitamineæ,
because on both sides of the principal filaments, these commonly have two filiform processes, which, on account of the
resemblance of the plant in other respects to the Irideæ,
lead us to suspect that they are properly two abortive filaments.

198 b.

To know the true and absolute number of the organs of a plant, we must trace them, with the help of the theory of abortion and union, to the original type of the tribe or family to which the plant belongs. When, for instance, I remark that the Primuleæ have almost all five parts in their corolla and calyx, and five filaments, I cannot be led into any mistake, although in the *Trientalis* and *Tovaria* I commonly observe seven filaments. The original type is five, to which our *Trientalis* sometimes returns. Also, the four dissimilar filaments of *Lindernia, Limosella,* and *Centunculus,* cannot occasion any mistake, since these plants appear as Primuleæ in all other respects. The fifth filament is here abortive, as it is also in the Personatæ, Acantheæ, and Bignoniæ. Among the Campanuleæ, whose original number is five, we yet find *Canarina* with six, and *Michauxia* with eight parts.

199.

We perceive from this, that it is a matter of more consequence to know the relative number, than the absolute number of parts. It is of more importance, in particular tribes, to know, that the number of the filaments is twice or three times the number of the parts of the corolla and the calyx, than to be able to state the precise number. This numerical proportion, which the different parts maintain with respect to one another, has often also an influence on the divisions of the fruit, and even on the number of the seed. Thus in *Alyssum* we observe a simple division of the silicula, four seeds, the same number of parts in the calyx and corolla, and one and a half times as many filaments. But in many genera and families the number is altogether indefinite, especially in the Ranunculeæ and Magnoliæ.

200.

In all regular flowers the relative number of the parts of every system must be the first object of our examination; but in irregular flowers we must begin by examining the absolute

numbers of every system, and, by deriving from them, the relative number.   Because, as irregular flowers arise from the
union or abortion of regular petals, the whole number is
evidently lessened, and can no longer stand in a fixed relation
to the parts of the calyx, or to the filaments.   Therefore
the absolute number is here the first object of examination.

201.

The number of petals has a fixed proportion to the number of the parts of the calyx, when each of these systems has
but one row of parts.   The relative numbers are less remarkable and less applicable in these two organs, when they stand
in several rows.   But in some cases this relation may be expressed by multiplication.   In *Nymphæa alba* we reckon four
divisions of the calyx, four times four pistils, four times five
petals in two rows, and four times twelve anthers in four
rows.

But as the petals themselves, as in *Delphinium, Nymphæa,
Calycanthus*, and several other genera, are evidently altered
filaments, the numbers of each row of the filaments maintain
a fixed proportion to the parts of the corolla and calyx.

202.

It is difficult to state the numerical proportion of the loculi of the fruit.   If the ovaria are placed in the shape of a whorl
around an imaginary axis, they often maintain a definite proportion to the parts of the calyx and corolla, as in the Geraniums, Diosmeæ, and Junceæ.   But when the ovaria are
arranged in heads or ears, their number is for the most part
quite indefinite, and maintains at least no proportion to the
parts of the corolla and calyx.   We see this very distinctly in
the Ranunculeæ.

Now, as we have already remarked that capsules of many
loculi are to be considered as individual capsules that have
been joined together, what has been already said respecting the whorl-shaped position of individual ovaria, is applicable to capsules of several loculi, when the pistil stands
in their centre.   But if the pistil be on one side, this is pro-

bably a consequence of abortion or union, as we have already remarked of the Labiatæ, and as in the *Gleditschia triacantha* and *Spartium scoparium* we often actually observe two pistils more or less united.

203.

The absolute size of organs is a very insignificant circumstance in the theory of classification. A more important consideration is the proportional size of similar or dissimilar parts. We may assume the following rule as a fundamental law of vegetation. The parts of one and the same system are by nature equal in size, and only become unequal from changes, which are more or less intimately connected with the general nature of plants. We have already noticed, that the Labiatæ and Cruciform plants shew an abortion and alteration of parts ; hence proceeds the want of uniformity in the length of the filaments.

But, in general, the regularity or irregularity, and also the similarity or dissimilarity, of the size of parts, depend very much upon their position on the stem. If a flower stands alone at the tip of a branch, where no other flower hinders its evolution, it will necessarily be regular, even when it belongs to a family with irregular flowers. *Parnassia* and *Sauvagesia* have regular flowers, although they belong to the Resedeæ with irregular flowers. *Asarum* stands among the Aristolochiæ, because it has always a stalk with but one flower. But let us make the supposition, that around and near this blossom others arise, it will then become a whorl, a head, an ear, or an umbel. The uniformity is now overturned. The central blossom continues regular, but those on the margin must be irregular, as we every day see in the Umbellatæ, the Aggregatæ, and in the compound flowers. Hence the most irregular flowers never stand single, and never on the tip of the branch, if we except some of the Orchideæ. When in the Labiatæ flowers appear on the tip of a branch, even they sometimes are regular, as is the case in *Teucrium campanulatum* and some species of *Galeopsis*.

204.

From these considerations it follows, that in the theory of classification all these irregular forms must be traced back to their regular primitive form, even although this should be of but rare occurrence. If we compare the Solaneæ with the Personatæ, the latter appear, from the irregularity of their flowers, and the unequal number of their filaments, to be completely distinguished from the former. But if we examine the fruit, the placentation, the situation of the embryon, and the other relations of the flower, we find the greatest agreement, and the transitions from *Nicotiana* to *Hyoscyamus*, to *Verbascum*, and to *Celsia*, strike every person ; for in *Verbascum* the lobes of the corolla are often irregular, and the filaments of unequal length. If we change but one filament in *Verbascum*, it becomes *Celsia*.

In general, we must observe, that an irregularity of parts seldom appears in one organ, without being also apparent in others. But the fruit is frequently an exception to this.

205.

The unequal length of the filaments is nevertheless sometimes connected with perfect regularity in the other parts, of which we see daily instances in *Phlox* and *Oxalis*. Another law prevails here, namely, that the parts of a system are unfolded successively, and not at once. Hence in *Phlox* in particular, there are some only of the filaments which have the necessary length for enabling the anthers to impregnate the stigma. In *Oxalis*, there is more regularity, because exactly one half of the filaments is longer than the other half.

206.

With respect to the connection of parts, the rule is, that all organs which are united with their supports are persistent, and all those which loosen themselves at their base are deciduous. In like manner, related parts, which have a suture, open themselves ; whilst those which have no suture remain shut. With respect to the internal structure of parts, we find that the interruption of continuous structure arises or the most part merely from chasms in the cellular texture. Hence it appears, that such interruption is not of any great moment. But in so far as the cellular texture is a part of the whole body, and a great regularity is observed in this interrupted or uninterrupted connection, such structure belongs to the most important characters. It is as important a character, that nuts and stony fruits do not burst in valves, as that the calyx of the Papavereæ is deciduous.

207.

The uses of organs and their sensible marks, as their colours, smell, and taste, can only be so far important in Taxonomy, as we can thence draw conclusions respecting the internal structure of parts.

IV. *On the Determination of Value of Characters.*

208.

The determination of the value of characters, is in general very simple. In this matter, we may take it as a rule, that the value of a character stands in a compound proportion to the importance of the organ, and to the point of view in which we consider it. If the question relates to a single organ, the characters stand in simple proportion to its variations. If the question relates to a single variation, the characters are in the same proportion as the importance of the organs. If we employ both these elements, their union may afford similar or dissimilar results. Characters are similar in three cases. 1. When the same variation takes place with respect to two organs of the same rank in one or two functions. 2. When two variations of the same rank take place with respect to one or two organs also of the same rank. 3. When the degree of importance of an organ is correctly balanced by the value of the variation. If, for instance, I compare the embryon, considered under its least important point of view, namely, its sensible marks, with the nectary, under its most important point of view, which is its existence, then I obtain two analogous results according to the theory, which may be employed also in practice; because a character which, considered by itself, has a small value, may assume a very important value, if we consider it with respect to the whole organization of a tribe or family. This character then becomes the sign of a permanent internal variation of structure, which must be of great consequence. The form of the leaves is in itself of small value in classification; but in the Grasses, for instance, it has a very important significancy, because it is inseparably connected with the internal organization. If we knew the internal structure of all plants, these relations in many families, for instance in the Rubiaceæ, the Leguminosæ, and the Ferns, would be completely cleared up.

209.

If, in a tribe, we observe certain plants which are nearly related to each other by their general aspect as well as by their other characters, but which are distinguished from one another by one mark, this mark cannot be of any great consequence. In the Saxifrageæ, the superior or inferior position of the fruit, when the other characters correspond, is of just as little consequence as the upright position of the embryon in the Berberideæ, where it is often also inverted. The Aggregatæ, again, are distinguished from Compound flowers by this permanent character, among others, that in the former the embryon is inverted, in the latter it is upright.

CHAP. VI.

NATURAL ARRANGEMENT OF FAMILIES.

210.

I. Plants of a cellular Structure. Scarcely proper Seeds. Propagated by Sporæ.

    Fam. 1. Fungi. (Introd. II. Fig. 32.)

        a. Conyomici, Nees von Esenbeck.

        b. Nematomyci.

        c. Goniomyci.
        d. Gastromyci.
        c. Spongiæ.
        f. Myeolomyci.
   2. Lichenes; (Anleit. II. Vid. 51—64.)
   3. Algæ;   (Do.   33—51.)
   4. Homallophyllæ; (Do.   64—66.)
   5. Musci hepatici; True seeds. Double sex-
         ual parts.
   6. Musci frondosi;   (Do.   73—89.)

II. Plants with Spiral Vessels and Slits. True Seeds.
The Sexual Parts not Double.

  Fam. 7. Filices; (Anleit. II. Vid. 89—104.)
        8. Pteroidæ;   (Do.   104—107.)
        9. Lycopodeæ; Uncommon sexual parts.
                  (Do.   107—110.)
     10. Rhizospermæ; (Do.   110—114.)
     11. Naiadæ;   (Do.   114—122.)

III. Plants with the Sexual Parts obvious, and of the
usual Form. The Spiral Vessels dispersed through the
Stem. The Embryon unevolved in the Albuminous Substance. The number three prevailing.

  Fam. 12. Aroideæ; (Anleit. II. Vid. 122—128.)
      13. Cyperoidæ;   (Do.   129—137.)
      14. Grasses;   (Do.   137—184.)
      15. Restiaceæ and Junceæ; (Do. 184—195.)
      16. Palmæ;   (Do.   195—209.)
      17. Sarmentaceæ, (Dioscoreæ, Smilacinæ, Br.)
                 (Do.   219—231.)
      18. Coronariæ; (Liliaceæ, Amaryllideæ.)
                 (Do.   231—256.)
      19. Irideæ;   (Do.   256—261.)
      20. Hydrocharidæ; (Do.   263.)
      21. Alismeæ; De Cand. (Do. 266.)
      22. Scitamineæ;   (Do.   270—277.)
      23. Orchideæ;   (Do.   280—298.)
      24. Museæ;   (Do.   278—279.)

IV. Plants with Sexual Parts obvious, and of the usual Form. Spiral Vessels in Concentric Rings. The Embryon more or less evolved. Numerical proportion variable.

A. Simple Floral Cover.

Fam, 25. STYLIDEÆ; (Anleit. II. Vid. 298—300.)
 26. ARISTOLOCHIÆ; (Do.   300—302.)
 27. POLYGONEÆ; (Do.   303—307.)
 28. CHENOPODEÆ; (Do.   307—320.)
 29. SANTALEÆ; (Do.   320—323.)
 30. THYMELŒÆ; (Do.   323—329.)
 31. PIPEREÆ; (Do.   123.)
 32. STROBILIFERÆ; (Do.   209—219.)
 33. AMENTACEÆ; (Do.   344—353.)
 34. URTICEÆ; (Do.   353—362.)
 35. TRICOCCÆ; (Do.   363—375.)
 36. PROTEACEÆ; (Do.   329—339.)
 37. LAURINÆ; (Do.   339—342.)
 38. MYRISTICEÆ; (Do.   342—344.)
 39. PLANTAGINEÆ; (Do.   376—377.)
 40. NYCTAGINÆ; (Do.   377—382.)

B. Double Floral Cover. Number five prevailing.

 a. The Petals united.
 41 PRIMULEÆ; (Do.   383—390.)
 42. PERSONATÆ; (Do.   390—406.)
 43. ACANTHEÆ; (Do.   407—411.)
 44. BIGNONIÆ; (Do.   412—418.)
 45. VITICEÆ; (Do.   418—426.)
 46. LABIATÆ; (Do.   427—443.)
 47. ASPERIFOLIÆ; (Do.   444—452.)
 48. SOLANEÆ; (Do.   452—460.)
 49. CONVOLVULEÆ; (Do.   460—468.)
 50. JASMINEÆ; (Do.   468—471.)
 51. GENTIANEÆ; (Do.   471—479.)
 52. CONTORTÆ; (Do.   479—496.)
 53. SAPOTEÆ; (Do.   497—502.)
 54. STYRACEÆ; (Do.   505—507.)
 55. ERICEÆ; (Do.   513—517.)

56. CAMPANULEÆ; (Anleit. ii. vid. 522—525.)
57. COMPOSITÆ;   (Do.   527—583.)
58. AGGREGATÆ;   (Do.   583—588.)
59. VALERIANEÆ;   (Do.   589—590.)
60. CUCURBITACEÆ;   (Do.   591—595.)
61. PASSIFLOREÆ;   (Do.   595.)
62. CAPRIFOLIÆ;   (Do.   617—623.)

b. The petals more or less free.

63. RHODODENDREÆ;   (Do.   509—513.)
64. EPACRIDÆ;   (Do.   617—522.)
65. LOBELIÆ;   (Do.   525—527.)
66. RUBIACEÆ;   (Do.   596—617.)
67. UMBELLIFEROUS PLANTS; (Do. 623—645.)
68. SAXIFRAGEÆ;   (Do.   646—650.)
69. TEREBINTHACEÆ;   (Do.   650—658.)
70. RHAMNEÆ;   (Do.   658—666.)
71. DIOSMEÆ;   (Do.   666—669.)
72. BERBERIDEÆ;   (Do.   670—672.)
73. RUTACEÆ;   (Do.   672—675.)
74. MENISPERMEÆ;   (Do.   675—678.)
75. ANONEÆ;   (Do.   678—680.)
76. MAGNOLIÆ;   (Do.   680—682.)
77. MELIÆ;   (Do.   682—690.)
78. MALPIGHIÆ;   (Do.   690—693.)
79. AHORNÆ;   (Do.   693—695.)
80. SAPINDEÆ;   (Do.   695—700.)
81. ONAGRÆ;   (Do.   700—707.)
82. SALICARIÆ;   (Do.   707—711.)
83. CRUCIFORM PLANTS; (Do.   711—724.)
84. PAPAVEREÆ;   (Do.   725—730.)
85. RANUNCULEÆ;   (Do.   730—736.)
86. POLYGALEÆ;   (Do.   736—740.)
87. LEGUMINOUS PLANTS; (Do.   740—773.)
88. CAPPARIDÆ;   (Do.   774—777.)
89. GUTTIFERÆ;   (Do.   779—789.)
90. AGRUMÆ;   (Do.   789—793.)
91. GERANIEÆ;   (Do.   793—797.)
92. MALVACEÆ;   (Do.   797—806.)

93. BUTHNEREÆ; (Anleit. II. vid. 806—811.)
94. OCHNEÆ;   (Do.   811—813.)
95. DILLENIEÆ;   (Do.   813—815.)
96. TILIACEÆ;   (Do.   815—821.)
97. HERMANNIEÆ;   (Do.   821—824.)
98. CHLANACEÆ;   (Do.   824—825.)
99. CISTEÆ;   (Do.   825—826.)
100. RESEDEÆ, De Cand.; (Do.   777—778.)
101. IONIDIÆ;   (Do.   827—829.)
102. CARYOPHYLLEÆ;   (Do.   829—839.)
103. PORTULACEÆ;   (Do.   839—842.)
104. AIZOIDÆ;   (Do.   842—845.)
105. CEREÆ;   (Do.   845—847.)
106. LOASEÆ;   (Do.   847—848.)
107. MYRTEÆ;   (Do.   849—854.)
108. SEDEÆ;   (Do.   854—856.)
109. MELASTOMÆ;   (Do.   856—859.)
110. ROSACEÆ;   (Do.   859—872.)

PART III.

PHYTOGRAPHY,

OR

DESCRIPTIVE BOTANY.

CHAP. I.

ON THE NAMES OF PLANTS.

211.

THE original names of plants are those by which they have
been designated in every country by common use. The most
ancient writers on botany have availed themselves of these
names only ; and as the common use of language is subject
to no rules, the same names were formerly given to the most
different plants, when they had only a remote or accidental
resemblance, and quite different names were given to nearly related plants, to those even which belong to the same
genus. As more plants were discovered, greater perplexity
must have been felt respecting their names. A remedy for
this perplexity was endeavoured to be obtained, by furnishing the names of known plants with such additions, as
might fit them for denoting new plants which bore a resemblance to them. Hence the names *Caryophyllus, Lysimachia, Consolida regalis, Nasturtium, Auricula muris,*
and so forth, were not only assigned to an innumerable multitude of different plants, but were accompanied with verbose
definitions of a more particular character, which on the one
hand rendered it impossible to fix them in the memory, and

on the other enlarged beyond all bounds the size of botanical writings, and, as caprice rather than law ruled in this matter, multiplied the synonymes of every plant to an endless length.   In the middle of the seventeenth century, indeed, Joachim Young, an ingenious philosopher in Hamburgh, first attempted to introduce order into this chaos, by giving laws, founded on correct views, to the nomenclature; but his writings became known for the first time almost a century after his death, (*Joach. Yungii Opuscula botanico-physica, Coburgi,* 1747, 4to.); and the herculean labours of Caspar Bauhin, respecting the older synonymes (*Pinax Theatri botanici, Basil,* 1671, 4to.), corrected and improved by Morison, (*Præludia botanica ; Hallucinationes C. Bauhinii in Pinace, Lond.* 1669, 12mo.), continued till the eighteenth century to be the only guiding clew in the labyrinth of botanical nomenclature.

212.

Linnæus earned for himself immortal honour, by inventing what he called Trivial Names, in addition to the Generic Names which several earlier writers (Ray, Plumier, Tournefort), had established according to correct principles.   In this way every Species of plant was now designated by only two invariable names, which could be easily retained, and by means of which the acquisition of the science must have been much facilitated.

To this nomenclature of Linnæus, it has indeed been objected, not without reason, that it serves only for enabling us to retain the name of a plant, without denoting its essential properties.   Hence Haller, and others, proposed various plans for expressing the characters of plants in their names. But these attempts failed, and, at any rate, could lay no claim to general approbation, because their object was less to facilitate the study, than to express a preconceived meaning. The Linnæan Nomenclature must ever endure, because, with about a thousand trivial names, and from two to three thousand and a half of generic names, we are able correctly to designate more than fifty thousand different species of plants.

But we must now point out the laws, according to which both the generic and trivial names have been invented.

I. *Of the Generic Name.*

213.

The Generic name should be a substantive, and the Trivial name an adjective. Hence adjectives, as generic names, are objectionable. We allow many at present, because custom has consecrated them, (*Scabiosa, Gloriosa, Impatiens, Fontinalis*), but to form new names on this principle is not permitted.

214.

Generic names, which have been employed by the most ancient classical writers, are always to be preserved, provided they do not stand in direct opposition to the other rules of nomenclature, (*Betula, Samolus, Humulus,* and so forth.)

215.

The best generic names are those which express the character of the genus in a single well-formed word. They are compounded of Greek or Latin words, (*Epilobium, Ceratocarpus, Lithospermum, Tragopogon*). This rule is not exactly attended to, when we attempt to express, in the generic name, such peculiarities as are not immediately connected with the essential generic character. These have often an allusion to the general aspect of plants, to their situation, colours, and other properties. (*Lychnis, Stratiotes, Lonchitis, Adoxa, Mimulus, Hydrocharis, Potamogeton*).

216.

Generic names should contain positive information. Hence all those are exceptionable which are founded upon a resemblance to other genera, and which express this resemblance by diminutives, or by syllables, either prefixed or added.

K

(*Ionidium, Ampelopsis, Ricinoides, Acetosella, Lupinaster, Orchidocarpus, Pseudorchis*).

On these grounds generic names are to be rejected which have been formed by transposition from others. *Galphimia* instead of *Malpighia*, *Tepesia* for *Petesia*, and *Mahernia* for *Hermannia*, have already been admitted; but *Meoschium*, formed from *Ischœmum*, instead of *Ischœmum*, is not to be endured.

Names which have the same sound with others already in use, are to be avoided. (*Picria Lour.* and *Picris*, *Castelia Cav.* and *Castela*, *Turp.*; *Dysodia*, *Willd.* and *Dysodium*, *Rich.*)

217.

Generic names must be of Greek or Latin origin, because these are the learned languages of which botanists avail themselves. The original national names are therefore exceptionable, which may be called barbarous, in so far as they have no Latin termination. If this rule is not observed, we may commit as ridiculous mistakes as those into which Adanson fell, who adopted German, Dutch, and other names of plants, as generic. (*Gansblum, Kolman, Chanterel, Amberboi, Kreidek, Rulac, Hond-bessen.*) However, several French botanists follow him in this respect, by assuming American, African, and other barbarous names, as generic. (*Harongana, Lam. Icacoria, Aubl. Paypayrola, Aubl.*)

However, in this respect, we must not be too strict, but must endure such original names as have either been consecrated by custom, or which have a Latin or Greek sound. (*Coffea, Thea, Musa, Cadia, Scorzonera.*) Linnæus called these names *quasi modo genita.*

The formation of generic names must be guided by the laws of the Latin and Greek languages, and therefore all those are exceptionable whose composition is ungrammatical. (*Genosiris, Calyxhymenia, Aixtoxicon.*)

Hence hybrid names, compounded from Greek and Latin, are exceptionable. (*Caturus, Laurophyllus, Alternanthera.*)

218.

Generic names should consist only of one word, otherwise the additional trivial name would make up three names. Yet we have adopted many names formed on a different principle, because they have been in immemorial use; (*Rosmarinus, Cornucopiæ, Sempervivum.*)

219.

Generic names should designate definite genera of plants. They ought not, therefore, to be family names. (*Gramen, Filix, Lichen.*) These names, too, ought not to be taken from other sciences and arts, especially not from other parts of natural history, (*Naias, Elephas, Natrix, Buprestis*): but even here long use has its privilege; (*Heliotropium, Hyacinthus, Pastinaca, Taxus*).

220.

As we are often at a loss for generic names, it is allowable to borrow an allegorical name from mythology. This practice is allowable, but not to be imitated; (*Adonis, Narcissus, Danais, Urania, Hecatea*). To denote his perplexity, Linnæus *called* a plant *Quisqualis*.

221.

From the earliest times it has been a custom to honour the merits of great promoters of botany, by naming plants after them. When neither flattery, nor other private views, lead to this practice, it may be justified; (*Mithridatca, Eupatorium, Cliffortia, Josephinia, Munchausia*). But the practice is very reprehensible, when later botanists have sought, by means of it, to do an agreeable service to their superiors, or to make themselves acceptable to them; (*Ferdinanda, Napoleona, Bonapartea, Theodora, Carludovica, Alexandra*).

222.

To preserve, in this way, the memory of meritorious botanists, is a laudable custom, which must, however, be intelligently adopted, and not misused. We must be able, in the first place, so to change the names as that they shall be easily pronounced; (*Gundelia* for *Gundelsheimeria*, *Crassinia* for *Krascheninikovia*, *Goodenia* for *Goodenoughia*). The French syllables *de* and *du*, *le* and *la*, are also to be omitted; (*HERITERIA*, *Fontanesia*). But if these syllables are a part of the name, they must be adopted into the generic name; (*Duhamelia*, *Lapeyrousia*, *Desvauxia*). It is only to be regretted, that we are often at a loss how to pronounce these names, and the study cannot surely be said to be facilitated, when it is intended that names such as *Knightia*, *Knaultonia*, *Palafoxia*, *Munnozia*, should be pronounced as the English and Spaniards pronounce them.

There is a further evil in this kind of nomenclature, that no person can form to himself any conception of the plant, when he hears it called by a name which has been formed from that of some botanist. No person can have an idea attached to many names, borrowed from men who are either altogether unknown, or have been distinguished but in a small circle. There is a multitude of such names, the derivation of which it is almost impossible to state. It is improper to immortalize, in this way, men who have been known in other departments of study, or who have gained but a doubtful reputation. Adanson named *Carrichtera*, after the wild enthusiast *Carrichter* of Reckingen. *Oribasia*, *Æginetia*, *Podaliria*, *Machaonia*, *Hippocratia*, *Avicennia*, *Averrhoa*, *Fernelia*, *Chaptalia*, are also objectionable on the same grounds.

This honour may be shewn not only to botanical writers, but also to celebrated travellers, who, by extending our knowledge of the earth, have also enriched the science of plants; as also to artists, who have distinguished themselves by excellent representations of plants; on which account the names *Cookia*, *Buginvillæa*, *Magellana*, *Bauera*, *Ehretia*, *Turpinia*, *Redutea*, may be justified.

II. *Of Trivial Names.*

223.

A fundamental rule in the construction of a trivial name, is, that it be a descriptive word, short, and derived from the Latin, or at any rate from the Greek language.

Latin trivial names are the best; but many properties which we wish to designate, cannot well be expressed in Latin; in which case we betake ourselves to Greek. *Micranthus* may be used indiscriminately with *parviflorus*, and *macrophyllus* with *longifolius*; but *macrostemon* and *isostemon* are evidently better than *longistamineus* and *æqualistamineus*. The practice of those is to be blamed who employ Greek too frequently in trivial names, when we have equally significant Latin terms in use; because the latter language is more generally understood than the former. *Cycloselis* is a superfluous expression, since we have *orbicularis* and *cincinnatus*.

224.

A second leading rule is, that the trivial names must be as little varied as possible, so that synonymes may not become infinite. The first trivial name must remain, even although a better might easily be had. But there is an exception to this, when the writers who first employed the trivial name denoted several different species of plants by it, or considered what was merely a subspecies to be a species. This has happened with *Ballota nigra*, under which name Linnæus included a completely different species during his better days, from those which it had formerly denoted.

It is impossible at present not to urge the rule, that no trivial name, but the Linnæan, ought to be pronounced or written, without adding the authority to it; because it happens that different plants have the same trivial name, and because, without adding the authority, no person can know where to find more information respecting the plant. *Neckera splachnoides*, Schwagr. is a quite different plant from *N. spl. Sm.* : *Panicum fasciculatum Sw.* different from that of

Lam. ; *Solanum scabrum Lam. Jacqu. Zuccagn. Mill. Ruiz et Pav, Vahl.*, and *Dunal,* are six completely different plants.

225.

Trivial names should be expressive; they must either express the specific character, or denote generally some striking property, such as the general aspect of the plant,—its resemblance to other plants,—its native country,—its situation,—its time of flowering,—the duration of its life,—its smell,—taste,—and also its uses.

Resemblance to other plants is expressed by the termination *oides*, (which, however, must not be added to any Latin name, as *muscoides, riparioides*) ; also by *formis, pseudo,* or by the name of the plants to which the species bears a resemblance ; (*Begonia Urticæ, Veronica Anagallis, Satureia Thymbra*).

226.

We see that the trivial name is commonly indeed an adjective, but may sometimes be a substantive, and is then written with a large initial letter. If it is an adjective, it must conform itself to the gender of the generic name, and is always written small, unless the adjective be formed from the name of the discoverer or propagator of the plant ; (*Carex arenaria, Euphorbia Gerardiana*). When the trivial name is a substantive, we must not be offended if it sometimes has the termination of an adjective, and does not correspond with the gender of the generic name; (*Ilex Aquifolium, Erysimum Alliaria*).

We even allow trivial names to be sometimes barbarous, because less strictness prevails here than with respect to generic names; (*Centaurea Crupina, Robinia Chamlagu*).

227.

Trivial names should be short. We therefore avoid, as much as possible, those which consist of two separate words, or which are very much compounded and difficult to be pronounced. Some that are in common use are allowed, which, however, we willingly abbreviate, as *Thlaspi Bursa Pastoris*, *Hedysarum Caput galli, Lysimachia Linum stellatum.*

III. *Delineation of Characters.*

228.

Next to nomenclature, the delineation of characters is the most important part of phytography. But we must understand the method of delineating, and of correctly expressing both generic and specific characters. This demands the observation of some general rules, which we are about to state. The most essential requisite in a good character is; that it be expressed in the generally understood technical language, and neither contain undefined terms, nor comparisons with other plants, or with other objects of nature and art. Comparison and metaphorical expressions are objectionable on this ground, that the character must give a positive acquaintance with the object. But from this rule those terms of comparison are excepted, which are of common use in the artificial language.

229.

It is also a principal maxim in the delineation of characters, that we should abstain from the introduction of all accidental and non-essential things, because these lead only to confusion, and are by no means invariable.

Into the character of genus and species we must admit only those marks which distinguish plants from those that are related to them. This may be done without instituting comparisons, which belong much rather to the description and diagnosis. The sum of the marks which are found in related plants, must always be before us whilst we are delineating the character ; on which account it is impossible to give the character of a new genus or species, unless we know the plants which have a similar nature.

A. *On Generic Characters.*

230.

The generic character (*character genericus*) is the expression of the peculiar and invariable marks by which a genus of plants is distinguished from all others. This character may be of three kinds.

1. It may contain a complete enumeration of all and each of the marks which are found in the whole plant. It is then called a *Natural Character*. Such a character must always be very circumstantial, and it is so much the more difficult to give it correctly, that in the organs of nourishment non-essential things may easily be confounded with such as are essential.

2. The *Artificial Character* (*character artificialis*). This contains only the marks of the organs of fructification, or, where these are not distinct, the marks of those parts which are subservient to propagation.

3. The *Diagnostic Character* (*character factitius, diagnosis*). This consists in the selection of such marks only as serve to distinguish one genus from another. This last character facilitates, indeed, the acquisition of the knowledge of plants, and renders it more sure; but when a genus is represented by itself, this character cannot possibly be sufficient. When, for instance, we merely state respecting *Bromus,* that the awn springs from the ridge of the valve, we can distinguish *Bromus* from *Festuca* by this means, but the whole character of the former demands something more, and we must, for this purpose, pay attention both to the inflorescence, and to the number of flowers on the ear.

231.

Every generic character must state shortly and distinctly the common marks which belong invariably to all species of the same genus. No generic character, therefore, can be formed, till we have compared together all the species. Inattention to this rule is the reason why very many of the older characters are entirely useless, because they apply to but one or a few species.   But if there be a correspondence in most of the species, a difference of one or two species among a great number may take place, without inducing us to reject, on that account, the generic character.   Commonly this difference is taken into the character, especially a difference of numerical proportion.   As, for instance, in the genus *Amaranthus*, almost as many triandria as pentandria species appear, it is expressly stated in the character that there are either three or five anthers.

232.

The generic character of the higher plants is borrowed solely from the organs of fructification.   And we begin with the inflorescence ; we then proceed to the covers of the sexual parts ; next to those parts themselves, and to the nectaries ; after these, to the fruit ; and, lastly, to the embryon.   All invariable peculiarities of these organs, which are common to all the species, compose the Generic Character.   The general aspect of the genus,—the formation of the root, stem, leaves, and other parts,—the composition of the peculiar juices,—the smell and taste,—are all things which are no doubt important, and which may lead us to the formation of a genus, but they cannot be taken into the generic character itself, unless we would make it more circumstantial and natural.   If, for instance, we observe a herbaceous plant, which in its organs of fructification shews almost the entire character of another genus, all the species of which are trees, we are not allowed, at once, on this account to view that herbaceous plant as a peculiar genus, and to assume this difference into the character ; but this circumstance induces us to institute a closer examination of the essential parts, that in them also we may discover marks of difference which correspond with the general aspect.

According to the gradation in the value of organs, which was noticed in Sect. 171. and 172., we must not overlook even the smallest parts of the fruit, but must minutely examine the situation of the embryon,—the structure of the cotyledons, of

2

the vitellus, and other parts of the seed. Every person, indeed, cannot institute such an examination, because it presupposes not only the presence of the ripe fruit, but peculiar practice and dexterity. But these things must be noticed in the generic character, unless the genus is not distinguished from its related genera by its structure in these parts.

233.

In the lower plants, where few or no organs of fructification are found, we the more readily betake ourselves to other parts, and borrow the character from them; because in many families the whole plant contains within itself the germs of propagation. Neither in the Lichens, therefore, nor in the Algæ, and least of all in the Fungi, is it objectionable that we admit the structure of the whole plant into the character. We are even permitted, in these cases, to admit such differences into the generic character, as can only be discovered by the most powerful magnifiers.

234.

The shorter a generic character is, it is so much the better; because it is intended to impress itself upon the memory. On this account the diagnostic characters are chiefly to be recommended to beginners, although in them there is a reference to related genera. It must be understood, however, that natural characters are excepted from this rule.

The generic character is put in the nominative, and in it we follow the order in which the parts are successively unfolded.

B. *On Specific Characters.*

235.

The *Specific Character* ought to be the expression of all the invariable marks by which one species is distinguished from all others, (142.) It is called also the *Phrase.*

It follows from this, that the character of a species which is the only one of the genus, cannot be delineated. Such plants are rather explained by description.

236.

The specific character derives its elements from every part of the plant, the properties of which are invariable. The nature of the root,—the qualities of the stem and branches,—the forms of the leaves,—the armour and supports,—the form of the calyx and corolla,—the relations of the nectaries, filaments, pistils, ovaria, and fruit,—these are the true elements of which the specific character is properly compounded. Even the integuments of the parts must not be neglected, in as far as they are constant.

237.

On the other hand, neither colour nor the size of parts, neither smell nor taste, neither the situation nor the frequency of plants, belong properly to these elements. Nevertheless, use may be made even of these things under certain circumstances.

Colour, in the first place, is taken into consideration, when it is not only constant, but when, among a few characters, it is also the most remarkable ; on which account, in the lower plants particularly, the colour is very correctly defined. In more perfect plants, we usually state merely that the parts are differently coloured, commonly by the words *coloratus*, *maculatus*, and such like.

The measure and size of parts do not indeed properly belong to the specific character, in so far as it is positively expressed, because these qualities are subject to change. But the relative size, or the proportion of the parts to one another, is usually viewed as one of the most important ingredients of the specific character. That the corolla projects above the calyx, or is shorter than it,—that the filaments are longer than the pistil, or the reverse,—that the leaf-stalks exceed the leaves in length,—all these things must be taken into the specific character. It is also usual, when an organ, in rela-

tion to others, is very long, very large, or very short and
small, to express this simply by a superlative. We thus say,
pedunculi longissimi, calyx maximus, and so on.

Situation can only be taken into the specific character
of the lowest organised bodies, as the *Algæ* and *Fungi*. In
these we sometimes even take the profusion, or the insulated growth of plants, as a specific distinction.

238.

The specific character is put in the ablative, and this is
done for the sake of brevity. But we must endeavour to
avoid putting two ablatives, one after the other, because this
may occasion mistakes. Instead of *corolla calyce majore,* we
say, with more propriety, *corolla calycem excedente.* Instead
of *petiolis pedunculis brevioribus,* it is better to say *pedunculis petiolos superantibus.*

239.

The specific character must be compounded only of the
usual artificial terms. In general, it must contain neither unexplained words, nor allegories, metaphors, nor any other
comparisons but those which are customary.

240.

The specific character must be positive, and therefore all
negative expressions in it are objectionable. These last, proceeding upon comparisons with the nearest related species,
may very easily be expressed positively. Instead of *non ra-
mosus,* we say *simplex ;* instead of *non tortilis, strictus,* and
so forth.

241.

The specific character must be as easily comprehended,
and as short as possible. For giving an easy view of it,
it is important that the organs, whose properties are to be
stated, should always be placed first ; that these properties should not be separated by signs, and that the organs only should be separated by commas. Brevity requires
that all particles shall be avoided as much as possible. The

particles *insigniter*, *maxime* and such like, are replaced by superlatives. *Subinde*, *raro*, *nonnunquam*, and such like, are better expressed by the syllable *sub*, prefixed to the adjective. It is of importance to brevity, that a property which belongs to several organs should not be repeated with each of them, but put at the end after the organs have been connected by the enclitical conjunction *que*, *(pedunculis petiolisque aculeatis.)*

But the richer in species any genus is, the more necessary is a circumstantial character.

242.

The order in which the elements of the specific character follow one another is, that the properties which belong to most of the species should be placed first, or that certain leading parts should be selected, in which the differences lie. These, in the case of Roses, are the *ovaria*; and, in the case of Pinks, the lobes of the *calyx*.

IV. *Descriptions of Plants.*

243.

Good and complete descriptions of plants (*adumbrationes*) may be compared to excellent pictures, and in one respect they are even preferable to them, namely, in that, at less expence, they exhibit all the relations of parts as correctly to the imagination, as pictures present objects to the external sense. Hence he who reads with attention the description of a plant which is utterly unknown, can represent to himself its image so perfectly, that when he happens to see the plant, he instantly recognises it. For this high value, descriptions are indebted to an observance of certain fixed rules, which we are now to state more particularly.

244.

A good description must, in the first place, be complete, that is to say, it must so comprehend the whole of the essential parts and their relations, that nothing shall be omitted, which is necessary for a complete representation of the peculiarities of the plant. Superficial descriptions,—of which kind were those given by the first writers on botany,—have for the most part a reference only to what was most striking in the general habit of the plant,—to its obvious colours, its size, and other properties, sometimes merely accidental. It is often very difficult to guess from such descriptions, what is the plant which the writer had in view. This difficulty is so much the greater in the old writers, because their variable nomenclature, and the distance of the countries, whose plants are described, also prevent us from forming a correct judgment.

245.

But descriptions may also be too full, when they express common properties and such as belong to many species, or when they dwell too much on the peculiarities of unessential parts. In this case, the reader of such descriptions is perplexed by their too great exactness : he knows not in the end which among the innumerable marks is the most distinguished, and which are those that deserve most attention. To maintain the happy middle course between too great circumstantiality, and too rapid brevity, requires the union of an acute talent for observation, genius, and sound judgment,—talents which are seldom acquired, but are commonly innate, and which constitute the proper botanical genius. Placing in the back ground, or neglecting properties of less importance, we give a prominent place to those which are subservient to the knowledge and discrimination of plants. Such descriptions are always the most instructive; but when a person has no other view, but that of delineating one plant, he is very apt to be led into a useless prolixity.

246.

The order in which parts are described, is that pointed out by their growth. But we often begin with the general aspect (*habitus*), in order to present the image of the plant

2

before the eyes of the reader at the very first. Otherwise we begin with the root, with the bulbs and tubers; we then proceed to the stem and branches; next to the leaves, leafy appendages, armour, and other subordinate parts; we then describe the integument; we next pass to the inflorescence, calyx, corolla, nectaries, male and female sexual parts; and lastly, we dissect the fruit and seed, with all their properties. Some good writers have reversed this order; at least, have spoken first of the flower. But the reasons for maintaining this order preponderate.

247.

In these descriptions, attention is paid to every thing that is observable in the plant; that is to say, not only the specific character, but a complete picture of the plant must be given, on which account an exact delineation of forms, dimensions, integuments, and colour, smell, taste, and other particulars, may be introduced. But whatever serves to distinguish the plant, must be made especially conspicuous; on which account the diagnosis of two related species, carried through all the parts, is often much more important than the most careful and circumstantial description. But both methods can be easily united, provided in our description we make the discriminating marks particularly prominent.

248.

In description, we must on all occasions employ the usual artificial language; and when forms are so changed that the usual expressions are not proper to denote them, it is better to give a more extended description, than to designate these forms by new and uncommon words.

In descriptions that are quite complete, it is of use to arrange the individual organs at intervals under one another, and to write or mark in some other way the names of these organs. To save space, the parts likewise are often successively arranged, their names also being always expressed in writing or print. Individual properties are in this case se-

parated by commas, which does not happen in drawing up the
specific characters. Colon and Semi-colon are made use of,
when the parts of the principal organs are to be described ;
for instance, in speaking of leaflets, after the principal leaf ;
of petals, after the corolla in general ; of the partitions and
valves of the fruit, after the fruit itself.

249.

Having finished the description, we proceed to give an account of the situation and duration of the plant, as also of
the use which is made of it in arts and trades.   In assigning the station, correctness is chiefly to be recommended,
not only for facilitating the finding of plants, but also because the nature of a plant, and its discrimination from related species, depend partly upon this.   The culture of
plants in Botanic Gardens may derive the most important advantages from such exact descriptions of their stations.   No
advantage is derived from knowing that the plant grows in Africa, America, or New Holland ; but it is of the utmost importance to know under what degree of latitude, at what height
above the level of the sea, in what soil, and amidst what circumstances it grows.   From these descriptions, which scarcely any person has given so carefully as Humboldt, both the
natural historian of the vegetable world and the botanical gar-
dener may receive directions.

In stating the duration of plants, we often commit mistakes,
especially when plants have been reared in gardens ; because
many tropical plants, which in their native country are perennial, become annuals in our climate, from causes which we
cannot explain.   Hence, in the writings of the great botanists,
many errors are found, which can only be corrected by observing plants in their native countries.

V. *Synonymy.*

250.

By Synonymes we understand the different names which a plant has received in botanical works, as also those which different nations assign to it in their native dialects. Both of these have undeniable uses. By the former not only learn we to understand the different views with which writers have described plants, but we find also references to plates, which are often very desirable, and we are thus able to give a complete history of the plant. The knowledge of vulgar and provincial names is often very useful for facilitating the finding of plants in their native seats, and also for acquiring a knowledge of their employments and uses.

251.

Scientific synonymes should be complete, sure, and free from superfluity. They are complete when no work is passed over, in which a more exact description, or distinguished nomenclature, or a figure of the plant, is contained. It is easy to perceive, that the use of a botanical library, as complete as possible, is important for this purpose.

Synonymes are sure, when the cited places really treat of the plants in question, and not of others. Innumerable errors have crept into the science by the statement of false synonymes; which errors can only be avoided by the most careful sifting of these statements. It is of the first importance for obtaining certainty, that no book should be cited, without having been compared at the time when the plant was examined. Nothing is more ruinous than to make a parade of borrowed citations. We draw upon ourselves by this means the guilt of propagating errors. We also proceed with certainty in respect to synonymes, when we minutely compare the plant in question with the description or figure.

Synonymes are redundant, when insignificant writings, or such as throw no particular light upon the plant, are quoted. Linnæus gave a beginning to this evil custom, by raising

L

books destitute of merit into authority; and the more recent
editors of his works have not only exactly transcribed these
citations, but have added a great many others which are to
the last degree unimportant,—by which means space has been
occupied to no purpose.   It is also quite superfluous to transcribe the specific character, or the description of the plant,
from the cited works; but it is of the utmost importance for
completeness, that we should admit, without abbreviation, the
occasionally diffuse nomenclature of Caspar Bauhin, Plukenet, and other older writers, because otherwise we cannot be
perfectly certain what is the plant in question.

252.

The order in which citations are made, is the chronological; we must therefore be sufficiently acquainted with the
history of the science to know the successive times in which
writers appeared.   Some indeed reverse the chronological
order, by putting the most recent works first, and the oldest
last.   But it is much more suitable, and is attended with essential advantages, to begin with the oldest writers, and proceed to the most recent.   We thus avoid repetition, and best
become acquainted with the earliest discoverers of plants.

It may be asked, with what writers we should commence.
Linnæus used to cite, out of the sixteenth century, only Clusius, Dodonæus, and, although more rarely, Fox and Dale-
champ.   He referred throughout to the *Pinax* of Caspar
Bauhin.   In later times it has been discovered, that those
who have been called the Fathers of Botany in the sixteenth
and seventeenth centuries, were acquainted with a much greater
number of plants than from Linnæus' citations could have
been believed.   Brunfels, Conrad Gesner, Tragus, and Tabernamontanus, have for some time been more industriously
consulted than formerly.   But to go beyond the time of
Brunfels, and to extend synonymes to the books of herbaceous
plants used during the middle ages,—to the writings of the
Arabians, Romans, and Greeks, perhaps even of the Jews,—
is as troublesome as it is useless and superfluous.   We
justly leave inquiries of this kind to the scholar, who studies

the history of his science, and avail ourselves in some instances of the results of his investigations.

253.

The necessary saving of room demands, that in citations proper abbreviations should be employed. The names of writers, who might be confounded with one another, are given at greater length. The Gmelins can only be distinguished by adding their Christian names. The younger Linnæus is commonly denoted by the addition of *fil.* The cited works themselves are denoted by intelligible abbreviations, which should always be explained in a register. The number of the pages in which plants are mentioned, is always given without prefixing the superfluous letters, pag. or p. Plates are referred to according to their number, with the prefixed t. or tab.

The place of a description is marked by *. When we are doubtful whether a synonym suits, we add an interrogation (?)

254.

Vulgar names are of importance principally in the Floras of particular countries. To these they should be appropriated, and it is a reprehensible waste of space, when in general works, or even in Floras, all barbarous names are produced. This is an employment which ought to be left to the followers of Menzelius and Nemnich.

VI. *On the Form of Botanical Works.*

A. *Monographs.*

255.

By a *Monograph* we understand a complete account of any one family, tribe, or genus, nothing being neglected which is necessary for a perfect knowledge of it. Such accounts have in the highest degree promoted the progress of

L 2

the science, because attention, when limited to one object, observes much more and better than when it is divided. Yet this very limitation of attention may give occasion to a certain subtlety, or to too keen a penetration into particulars, and many examples might be cited to shew, that the founders of monographs are particularly easy to be induced to admit more species than nature warrants.

256.

Monographs should especially correct synonymes. It is also very useful, when they present us with plates of new or very difficult species. In this respect we can highly recommend the labours of Jacquin respecting the Oxalidæ and Stapeliæ, Schkuhr on Reeds, Brown on the Contortæ and Proteaceæ, De Candolle and Pallas on the Astragalæ, Lambert on Pines, Dunal on the Solaneæ and Anoneæ, Lehman on the Asperifoliæ, Humboldt on the Melastomæ, Cavanilles on the Malvaceæ, Biria on Ranunculi; Lyngbye, Turner and Dillwyn, on Algæ; Hedwig, Schwagrichen, and Hooker, on Mosses, and of the last author on Jungermanniæ.

VII.   *On Floras.*

257.

No branch of botanical literature is more useful, and at the same time more neglected than this. The Flora of a country or region should contain an exact account of all wild plants within the limits of that country or region. For a beginner, therefore, it is the first, and one of the most important aids for obtaining botanical knowledge. Confined to a certain circle, the compiler of a Flora can study the peculiarities of the plants of his region with diligence, distinguish true species from subspecies, point out transitions, correct in this way many errors, and lay the foundation for a more correct knowledge of plants. But there are a multitude of Floras which contain nothing but a catalogue of names of pretended native plants of their region, with transcribed specific characters, without in one instance correctly signifying the situations.

258.

If the first and most celebrated example of this kind is to be followed, as it was given by Linnæus in his Flora of Lapland, the qualifications for the undertaker of such a labour must be of a yet higher nature.

First of all, it is necessary that the compiler should be acquainted with the labours of his predecessors,—that he should study them, correct the synonymes, and point out the changes which vegetation has since experienced.

To this should succeed a general natural history of the region, an account of the soil, the mountains, and particularly of the mountain rocks which make their appearance there,— of the meadows, forests, seas, marshes, pools, and streams. The degree of latitude, and the height above the level of the sea, are supposed to be known. It is only when such a picture has been sketched, that the reader can form to himself a distinct idea of the nature of the region. The knowledge of the mountain rocks is especially important, in order that we may perceive how vegetation varies with them. For this purpose, it will be very useful to institute a comparison with the vegetation of neighbouring regions, or of such as lie in the same latitude with the described region.

259.

The order in which plants are treated of, may either follow the Linnæan System or the Natural Method. But, above all, the compiler of a Flora ought to admit no plant which he has not himself found in its place, because, otherwise, innumerable deceptions, and such as are scarcely credible, will be occasioned. He ought also never to transcribe the character of genera and species, but to develope it himself according to the examples lying before him. New species, which have not yet been described, must be defined in the most careful manner, and if possible, they should be figured. It will also be of much use to beginners, that the properties of plants, and the distinguishing marks of doubtful or difficult species, should be shortly and distinctly given.

260.

The synonomy of a Flora is properly confined to an account of the best plates, and to a reference to preceding authors. It is necessary to give the situations, especially of rare plants, with much exactness, and this is best done in the language of the country. Provincial names of plants are added, to enable us, when necessary, to obtain explanations respecting their situations from the inhabitants of the district. Lastly, it will not be superfluous to describe the uses to which plants are put. All these requisites have been fulfilled in the most perfect manner by Linnæus, in his Flora of Lapland. This, therefore, is the model for all future attempts.

VIII. *Descriptions of Gardens.*

261.

Another branch of botanical literature consists of catalogues and descriptions of plants which are reared in gardens. These catalogues are often merely registers, which have been printed to promote the commerce of one garden with others. In this case, nothing else can be required but that the plants should be correctly and exactly named. New species are either described in an appendix, as has been done by De Candolle, in his *Catalogus plantarum Horti Monspeliensis,* 1813; or they are merely cited, their description being left to future works.

262.

More circumstantial catalogues, like those which we have of the gardens at Kew, Copenhagen, and Berlin, contain much hat is superfluous indeed, because they frequently repeat well known specific characters; but they are useful, partly from giving more exact characters of new species, partly from a more careful sifting of synonymes, and partly from furnishing an account of the manner and time in which plants were first introduced into the gardens. The *Hortus Kewensis*, and Sweet's *Hortus suburbanus*, are particularly distinguished by the first of these excellencies, as is Linnæus's *Hortus Cliffortianus*, by carefully chosen synonymes, and Gouan's *Hortus Monspeliensis*, by uncommonly useful accounts of the structure and other external peculiarities of plants.

Plates of rare or new plants, which are reared in gardens, are expensive undertakings. We justly admire the workmanship and wealth displayed by the English, in the *Hortus Elthamensis*, in the *Botanist's Repository*, the *Paradisus Londinensis*, *Botanical Magazine*, *Botanical Register*, and such like works; and of the French, in the *Jardin de Cels*, and *de la Malmaison*. Even Germany may boast of its *Hortus Vindobonensis*, *Schönbrunnensis*, and *Berolinensis*, although all these works, on account of their high price, can be useful but to a few.

IX. *Plates of Plants.*

263.

Good plates of plants are among the best means of promoting the progress of botany. When they represent the form of the plant according to nature, and especially when they develope the characters of the genus and species, even to their minutest parts, they fulfil all that can be required, especially when no such expence is laid out on them as renders their price too high. The fathers of botany, in the sixteenth century, set an excellent example in this respect. Lobelius, Clusius, Fox, and the Bauhins, used wood cuts, mingled with the text, which gave very correct representations, at least of the general aspect of plants. Conrad Gesner and Fabius Columna first used copperplates, which often gave the characters of plants in a masterly manner. Morison and Plukenet, in very small room, gave an extraordinary number of figures of very rare plants ; and Dillenius reached the very summit of art in his incomparable representations of Mosses.

These examples of our predecessors should be imitated by us ; and we should recollect, that science ought not to be subservient to the luxury of the great, but should be communicated even to those who are destitute of wealth. Lehman, also, in his Primuleæ, and Hooker, in his Mosses, have followed the laudable practice of giving their figures in sketches merely, excellently shaded, by which means the price is very much lessened. As, on the other hand, copperplates are given frequently of well known plants, but in a style of excessive splendour, and at an expence which cannot be afforded by a private person, we are forced to lament that the science is rather retarded than promoted by this means. Among these expensive copperplates we reckon Sibthorp's *Flora Græca*, Count Hoffmanseg's *Flora of Portugal*, and the *Jardin de la Malmaison*.

We cannot advise the giving of figures of plants from stone, because we have not yet been able, in this way, to express the finer parts. The same objection may be made to impressions of plants in printing ink, of which Kniphoff has published a great many.

X. *General Works.*

264.

General works on the vegetable world contain either an enumeration of genera or of species. The former, which are called *Genera plantarum*, should exhibit the known genera, and explain their characters, either according to an artificial system, or according to the natural method. This has been done in a masterly manner by Tournefort in his *Institutiones rei herbariæ*, by Linnæus, Schreber, and Jussieu.

A complete enumeration of known species of plants, which we call *Species plantarum*, has hitherto only been offered by Linnæus, Richard, and Willdenow. An excellent account of *Species*, by Vahl, in his *Enumeratio*, was stopped in its commencement by his death. The latest attempts of this kind are by De Candolle and Schultes.

Excerpts from these *Species plantarum*, in which only the characters, the best plates, and the native country, are given, have been published by Murray and Persoon.

265.

Whoever undertakes a new work of the same kind, has duties to perform which few scholars are qualified to fulfil. For it is obvious, that labour alone will not do, nor the simple accumulation of the discoveries and remarks of others; but that, in the first place, an eye accustomed to the vegetable world for many years, an acute and incorruptible judgment, and, above all, that which I have already (245) called the botanical genius, is requisite for this purpose.

It is an indispensable qualification that as many plants as possible should have been actually seen and examined. Travels in foreign countries,—the use of great herbaries,—a very complete library,—a well stocked garden,—and a general intercourse with the first botanists of the age; these are the requisites to such an undertaking, without the possession of which the whole will be nothing but a work of mere compilation, and of little utility.

XI. *On Collections of Plants.*

Hedwig's Belehrung die Pflanzen zu trocknen und zu ordnen. Gotha, 1797, 8vo.

266.

The most exact descriptions and best plates leave something still to be desired by him who wishes to have a perfect knowledge of a plant. Hence the actual sight and examination of plants is the only mean of obtaining certain information. Now, in order to have this examination at all times in our power, we dry plants; and this may easily be accomplished with most of them,—some very juicy plants and sponges excepted. Such a collection of dried plants is called a *Herbarium*, and the necessity of herbaries is so generally felt, that beginners and accomplished botanists justly consider them as their most important treasures. These treasures, indeed, are subject, in certain circumstances, to waste and destruction. But, under proper management, and with careful attention, they last for centuries, as we still possess the collection of Caspar Bauhin, and in part that of Burserius, since the beginning of the seventeenth century; (*Linn. Amœn. Acad.* I. 143.)

267

The preparation of such a collection costs little labour, and occasions a trifling expence, provided we can obtain a number of folios for this purpose, and are acquainted with some expedients which must be employed in the work.

Of these the principal is, that plants must not be laid in while they are wet with rain and dew, but when they are completely dry, and that they be put down with all their necessary parts. Finer plants, which are not too juicy, and too much soiled, cannot be better dried than in folio sheets which are subjected to some degree of pressure. In this case it is not necessary that the paper should be inspected or changed, till the plants are perfectly stiff and dry. When there is a want of folios, or when soiled or very juicy plants are to be dried, they are placed between several sheets of blotting paper, and pressed down with stones. But in this case they must be frequently turned, and, in particular, must be defended from mouldiness. The employment of a press for plants, with a screw, cannot be recommended, because the pressure is too strong, and cannot be gradually increased. Very prickly plants, on the contrary, can only be subdued in this manner. Fleshy plants are placed for some time in boiling water, and then dried in blotting paper. But in this case the forms and colours are commonly lost. A dry heat is particularly desirable in this employment; on which account the drying of plants always succeeds best in hot summers, in airy dwellings, in heated rooms, and even beside ovens.

268.

When the plants are dried, they are put, according to the order of the system, or according to the natural method, in whole sheets of writing paper ; on the first side of which the name of the plant, its situation, and the time of its being laid down, are marked. Subspecies, and several large parts of the same plant, are placed in separate sheets. A hundred and fifty, or two hundred of such sheets are bound together between pasteboard covers, on which the genera are marked in the order in which they have been inserted, and an exact register of the whole is kept.

269.

The care with which such collections are kept is repaid in no common degree. This care requires, in the first place, that the plants should be most correctly determined,—that the authorities for their names should be given,—and that, in every case, it should be noticed from whom the plants have been obtained, and, where it can be got, that the handwriting of the sender should be subscribed. The collection must also be defended from insects and moisture. The former of these are not easily removed, especially in many families, as in the Cynareæ. But an industrious examination of the collection, in the course of which the insects are killed, and, at any rate, a solution of corrosive sublimate in spirit of wine, are the best means of protecting it from these causes of destruction.

PART IV.

PHYTONOMY,

OR

ON THE STRUCTURE AND NATURE OF PLANTS.

CHAP. I.

PHYTOTOMY, OR ANATOMY OF PLANTS.

GREW's Anatomy of Plants.
Malpighi, Anatome Plantarum.
Leeuwenhoek Opera.
Reichel de vasis plantarum spiralibus.
Böhmer de vegetabilium celluloso contextu.
Hill on the Construction of Timber.
Swagerman, in Verhandelingen van de Maatschappy. Te Harlem, Vol. XX.
Hedwig, Sammlung zerstreuter Abhandlungen, Th. 1. 2.
Krocker, Diss. de plantarum epidermide.
Comparetti, Prodromo di fisica vegetabile.
Sprengel, Anleitung zur Kenntniss der Gewachse.
Brisseau-Mirbel, Traité d' Anatomie et de Physiologie vegetales, Vol. I. 2.
Dessen, Exposition et defense de ma theorie de l'organization vegetale.
Link, Grundlehren der Anatomie und Physiologie der Pflanzen.
Rudolphi, Anatomie der Pflanzen.
Treviranus, vom inwendigen Bau der Gewächse.
Dessen, Beytrage zur Pflanzen-Physiologie.
Moldenhawer, Beytrage zur Anatomie der Pflanzen.
Kieser, Memoire sur l'organization des Plantes.
Dessen, Grundzuge der Anatomie der Pflanzen.

I. *On the Structure of Plants in General.*

270.

WE must endeavour to trace the structure of plants to
certain primitive forms, which we find as well in the rudest

beginnings of vegetables, as in all parts of perfect plants, and into which we can resolve all their organs.

These primitive forms may be reduced to three, the cell-form, the tube-form, and the spiral form. We discover these forms more or less in all vegetable bodies. But a closer examination shews us, that some forms of a simpler kind lie at the foundation of these, and that from them every organic part proceeds. We must begin, therefore, with these latter forms.

271.

Every organising fluid, when it is passing from the fluid into the solid state, shews small spheres or vesicles, and spiculæ or needle-shaped bodies of a diminutive size. The former we refer to the disengagement of hydrogen, which, as one of the constituents of water, is always the first to separate itself from it, because it is little soluble in water. Oxygen, on the other hand, remains longer dissolved in water, and accordingly the spicular and straight lined bodies which are produced by it are more slowly disengaged,—as, in an electrical process, negative electricity displays sparks and images of a spherical shape, whilst positive electricity produces those of a spicular appearance.

In the lowest organic bodies we find this simple spherical structure, and they may now therefore be considered as belonging to the animal or vegetable world. The simplest Coniomyci, as well as the simplest infusory animalculæ, have this vesicular or spherical structure. Afterwards the spiculæ, threads, and tubes, which we find in the Nematomyci, become associated with these spheres; (Tab. V. Fig. 5. 7.) Treviranus has lately exhibited these spherulæ and tubes in the spawn of frogs, in the cellular texture of the femoral muscles of the mammalia, in the spinal marrow of frogs, and in the nerves of the garden snail, (Vermischte Schriften, I. Tab. 14.) We find this same combination of spherulæ and spiculæ in every generative sap, as well as in every slimy fluid of plants. From these, therefore, are evolved the peculiar primitive forms of the vegetable world.

A. *On Cellular Texture.*

272.

What we call cellular texture in plants is, no doubt, when it is regular, somewhat similar to the cells of bees; but it is distinguished from them by the direction of the cells, and especially by this, that it seems to be as frequently void of all regularity, or to be fashioned in a quite different manner.

Where the cellular texture is present in a regular form, it consists of spaces, which, when cut in a longitudinal and cross direction, display six sides and six angles, and the entire circumference of which resembles a dodecahedron. These spaces are chiefly distinguished from the cells of bees, by being more drawn out in length. There is, however, another form of the cellular texture, which seems to be more primitive than this. That is, the vesicular or spherical, which seems to arise from the juxtaposition of the primitive spherulæ. It is easily conceivable that from this juxtaposition interstices must remain, which, indeed, we see distinctly enough, and which sometimes seem destined to serve important purposes in the future history of the plant.

The spherical cells become angular, when their sides touch and attract each other in several points. That a figure precisely hexangular should be formed from a circle, is partly a consequence of the effort to be regular, which is the more conspicuous in imperfect organic bodies, the nearer they approach to the productions of the unorganised world, whence regular crystals appear in the products of certain Fungi and Sponges; and partly it arises from this, that the hexangular form is, next to the circle, that which includes the greatest space, with the smallest extension of its circumference. We often see the spaces which remain between the cells, after their form is thus completely changed, filled with peculiar juices, and often these interstices supply the place of tubes, and conduct the unprepared sap upwards.

3

273.

The sides of the cellular texture are, for the most part, very thin, but yet completely impervious ; so much so, that the communication of the sap from one cell to another can only be explained by the supposition of an organic perspiration. Yet there are exceptions to this. The cells of the Epidermis are observed to have peculiar slits, (Tab. V. Fig. 2.), of which we shall speak more particularly when we treat of leaves ; and in Pines the extended cells are distinctly observed to have gaps, which are surrounded by a pretty high margin ; (Tab. V. Fig. 4.)

The function of the cellular texture is simply to contain and to prepare the sap. It is not destined to conduct upwards the unprepared sap, because in the bark and in the pith, both of which have a structure entirely cellular, the ascent of the sap is not perceived. There are, however, what have been called sap-vessels in the cellular texture, but these originally are nothing else but extended cells, which are often stretched to a considerable length.

B. *On the Sap-Tubes.*

274.

The second primitive form of all plants is the tube-form, appearing to the unassisted eye like straight-lined fibres. But by magnifying them we perceive that these apparent fibres have a real, though uncommonly small diameter ; that they are therefore real tubes, which proceed for a considerable length with a cylindrical shape, and are sharply pointed at both ends ; (Tab. V. Fig. 1.)

275.

Apparently these tubes are the perfect state of the second common primitive form of organic bodies, namely, the right lined. Because, although they more frequently occur than the third or spiral form, they are later in being produced, and are first observed, as we have said, in the Nematomyci,

2

In more perfect plants they are found, for the most part, in the neighbourhood of the spiral vessels : they constitute the basis of trees, and a great part of the young wood, and shew a toughness and a power of resisting violence, which, considering their fineness, is astonishing. That they arise from the first form, cannot be believed, because they proceed directly from the generative sap, like fine straight tubes, close to the spherulæ. But the stretched form of the cells is very like the tube-form. It is even undeniable that it constitutes, especially in the lower organic bodies, the transition-form from the cells to the tubes. In the fruit-stalk of the Musci Hepatici and Frondosi we have not yet discovered the proper tube-form, but only stretched cells, similar to tubes, which apparently answer the purpose of these latter bodies.

276.

The object of Nature, in the formation of tubes, seems simply to be, by means of them to lead upwards the unprepared sap. The similarity of the sap-tubes to hair-tubes leads us to consider them as a physical contrivance, by which the ascent of the sap is assisted, although the only principle upon which these last act cannot obtain a place in this structure, (376.)

The pointed extremities of these sap-tubes present some difficulty to this account. They lie with their ends obliquely placed to one another, and the ascent would seem to be interrupted by this position, if we did not here also admit the organic perspiration of the sap through partitions which in themselves are impervious.

C. *On the Spiral Vessels.*

277.

This is called the spiral form, because originally it consists of canals, the sides of which are entirely formed by spiral

M

fibres, of extreme fineness. But we must distinguish this form into the primitive and the derived.

278.

The primitive spiral-form consists of canals, the diameter of which is almost of the same size throughout, and is from the twelfth to the fiftieth part of a line, their sides being composed of those winding fibres, which can easily be un-rolled; (Tab. IV. Fig. 19.; Tab. V. Fig. 1.) We find an instance of this form in some Confervæ, in some of the Musci Hepatici, (Tab. III. Fig. 8.); and especially in the cellular texture which covers the surface of the *Sphagnum obtusifo-lium*; (Tab. III. Fig. 25.)

The fibres which, by their windings, form the sides of the spiral canals, have so uncommonly small a diameter, that we might suspect them to be any thing but hollow. Several of them, however, especially in the Scitamineæ, commonly stick together, and in this way they are formed into bands; (Tab. IV. Fig. 19.) They are also easily unrolled, so long as they are in their primitive state, because there is no connecting membrane, either external or internal, by which they are united; and this is the chief distinction between the air-vessels of insects and the spiral-vessels of plants, that the former have the winding fibres united by a peculiar membrane, and that a soft cellular texture always surrounds them.

279.

But a still more important circumstance essentially distinguishes the spiral-vessels of plants from the air-vessels of insects. The former never divide into branches; but where they separate, a new pair always places itself on the sides of the old ones, whilst the air-vessels of insects undergo every kind of ramification, from their origin to their finest branches. The primitive spiral-vessels are always in the company of the sap-vessels, and are chiefly found between the bark and pith, in the common plants, which are produced with two seed lobes. But they appear later than the sap-vessels, and are only first discovered when the young plant begins to shoot. They are found also in the root, as well as in the stalk : they partly compose the nerves and veins of the leaves and vessels of the corolla : they are found in the stamina, in the pistilla, in the fruit, and also in the funiculus umbilicalis of the seed.

280.

The interior canal of the spiral-vessels, in its natural state, is always found free from water. It is true, that if a piece of wood be dipped in water, this fluid penetrates into the canal. Also, when we permit coloured fluids to flow into the cut branches of plants, these fluids become apparent in the sides of the spiral canals; but they are also seen, and still more distinctly, in the neighbouring bundles of sap-vessels ; nay, they penetrate in considerable quantity, even into the cellular texture. We are not, therefore, entitled, from this entrance of coloured fluids, to conclude respecting the natural contents of these canals, because, in general, this penetration of coloured sap does not succeed in an uninjured root.

281.

In spiral canals, which grow rapidly, the fibres are often torn in such a manner, that they fall together in the shape of rings. These ring-shaped vessels, as they have been called, are, therefore, an entirely accidental variety of the primitive form of the spiral vessels ; and this is the more evident, because we find the same vessel in one situation as a spiral canal, and in another as a ring-shaped vessel. This change, besides, shews incontestibly, that the spiral vessels cannot conduct sap, since they are often nothing else but rings at a distance from one another, the circumferences of which are every where and extensively separated.

282.

But an important and essential change of the spiral canal is that presented by the *Vasa scalaria.* Under this name are included those canals with transverse openings, which do not at all shew the spiral winding of the fibres, and which

M 2

cannot be unrolled; (Tab. V. Fig. 3.) They are formed
by an original spiral vessel meeting with perpendicular fibres
in its sides, which fibres cross the winding lines longitudinally, and unite them together. These perpendicular fibres
belong to the original structure of the spiral fibres, and are
by no means part of the neighbouring cellular texture, because, from the first, we see them as peculiar fibres, and
not as partitions or membranes; and because, after a gentle
maceration, by which the cellular texture is destroyed, these
fibres last as long as the twisted fibres of the spiral canals
themselves.

But that this form is not accidental, but one which makes
part of the original vegetable structure, is evident from this,
that in certain families this appearance is so common, that in
the Ferns, in the Lycopodeæ, and in Grasses, we perceive
scarcely any other. In young wood, too, this form appears
very early, although, in the first shoots, the primitive spiral
vessels, having a great resemblance to the pith, long preserve
their unchanged shape.

283.

A remarkable variety of the spiral form is that in which it
appears porous, punctured, or surrounded by a reticulated
covering. This also is an original, and by no means an accidental form. It is most frequently observed in the roots,
and in the woody parts of plants; (Tab. III. Fig. 25.)

The origin of these vessels may be explained in the same
way as that of the *vasa scalaria*; that is to say, perpendicular fibres cross the winding fibres, and bind them together.
To this also is added the further circumstance, that, in the
instances we have mentioned, the spiral fibres often cling
more early together, and take the shape of bands, which being
crossed by the perpendicular fibres, form the net-work above
mentioned. Not unfrequently we observe particular parts
of these canals closely contracted, which gives to them the
appearance of bladders. We also sometimes remark oblique
fibres, especially in Sassafrass wood, which seem to be remains of the original twistings of the threads. To the older
tubes, a soft and vesicular cellular texture frequently attaches

itself; (Kieser Mem. Tab. IX. Fig. 40.; Tab: XIV. Fig. 67.)

But it is characteristic of these punctured vessels, that they are always larger in their diameter than the primitive spiral vessels or the *vasa scalaria*, so that in many kinds of wood, particularly in the Bamboo, and the common chair cane, we can see their sections with the naked eye. But not unfrequently the pores of the sides are so regular, and the bladder-form of this canal is so surprising, that we might be disposed to suspect a transition to the porous cell-form, especially as in our Pines, the latter varies so much, that, besides the pores, spiral windings also appear, as in the Yew and the Larch; (Kieser's Grundzuge der Anatomie der Pflanzen, Tab. V. Fig. 47. and 48.)

284.

As, then, the spiral vessels, and all their varieties, are uniformly found empty of fluids; as they shew themselves only in the higher plants, and constantly appear wherever a strong shoot is sent out; as they are always in the company of the sap-vessels; as, in fine, they maintain, by their constant diagonal direction, the middle situation between the perpendicular and the horizontal;—from all these considerations we must suspect that they are the instruments of the higher vital activity of plants, and that they are the organs by which the sap-tubes suffer an external excitement to the speedy propulsion of the sap.

II. *On the Structure of Roots.*

285.

The internal structure of the individual parts of plants is always composed of the three primitive forms which we have now described. With respect to roots, in particular, they consist, as was formerly (64.) stated, of the radix and radicle. The latter, as being the organ appropriated to the absorption of the sap, is furnished, for this purpose, in perfect plants,

with a multitude of very fine fibrils, or hairs, which are closed
at their extremities, (Keiser's Grundzuge der Anatomie der
Pflanzen, Tab. VI. Fig. 62.) These hairs in particular, with
their close and flaggon-shaped extremities, have a considerable
resemblance to the first appearances of the absorbent vessels
in the small intestines. As the former, like the latter, are
shut, we have in them a new proof of the organical perspiration we have mentioned, and which takes place, notwithstanding the peculiar impermeability of the partition. These
hairs are in immediate connection with the cellular texture of
the radicle; and as this first conveys the juice, that has been
absorbed, to the sap-vessels, it is evident that the unformed
fluids are already considerably changed, before they proceed
from the *radicle* into the *root*. The entire structure of the
radicle is protected by a fine cellular texture which surrounds
the sap-vessels in the centre.

On the ends of the radicle we often perceive drops of a
fluid, which is of a slimy consistence, and which, in all probability, has been derived from them.

286.

Some families of plants of the lower orders, the Ferns,
Palms, and Hydrocharidæ, as also the Naiadæ, instead of
these small hairs, have a spongy integument at the extremity
of the radicle. We observe it, in the form of a coif, or hood,
very distinctly in *Lemna* and *Callitriche*. This integument is not porous, but it consists of a very soft cellular
texture, which swells out in some small portions, but in other
respects is entirely closed. Here also, therefore, takes place
the absorbtion through the impervious partitions of the cellular texture.

287.

The radix, being the continuation of the stem into the earth,
has the same constituent parts with it, but with some differences, which are derived from the covering of the soil. Commonly the pith is wanting, and the centre of the root consists
of a woody kernel. Sometimes, however, it is hollow, and

pith is formed when the root is laid open to the air. The bark of the root is richly stocked with peculiar juices, which the more readily are collected in it, as the descent of the sap from the stem into the root is favoured by the size of this latter part.

This direction of the root towards the centre of the earth, is, without doubt, an effect of the common law of gravitation, to which plants are partly subject, as they are fixed by their lower extremity in the earth. This tendency, however, is considerably modified by other circumstances, which originate in the organization of the plant, so that in many trees we perceive fewer roots proceeding downwards into the soil, than those which we observe running horizontally.

288.

Tubercles, or *tuberculous* roots, are distinguished by their greater thickness, and by their fleshy appearance, (65.) They enclose within a cellular covering certain parts, in which the cellular texture is much crowded, and from which the higher forms, the sap-vessels and the spiral-vessels, take their origin, as being the beginnings of the future shoot. A root has, therefore, the greater means of production, the more tuberculous it is; and in many tubercles we can distinguish very accurately, at fixed periods, the harder kernel, from which the future shoots are to arise, from the surrounding soft cellular texture. Even in the higher parts of the stem similar thickened parts occur, in which the power of propagation reposes; since, universally, wherever the cellular texture is much crowded, new sap-vessels and spiral-vessels arise as the foundation of future shoots; whence, in the stem and branches, the transition from tubercles to buds is obvious.

289.

Bulbs appear above and upon the root, like more perfectly formed tubercles. They have, as their foundation, a solid substance, consisting of extremely compressed cellular texture. From this substance spring the germs of the leaves between scales, which are a continuation of that solid matter; and in the middle of these arises perpendicularly the future stalk, from the sap-vessels and spiral-vessels, formed by the compressed cellular texture. The solid fundamental body pushes towards the side, horizontally, the young shoot, which, being nourished by its parent substance, is not separated from it until it also has acquired a firm fundamental body, crowned with scales, and is able, consequently, to maintain itself.

Between this lateral impulse and the upright one, which produces the stalk, there is such an interchange, that the one of these impulses languishes when the other is most active. Hence it is usual, after the flowering of bulbs, to lay them dry, that the quiet lateral impulse may remain undisturbed. Bulbs, which have once completely blossomed and produced seed, usually die.

III. *On the Structure of the Stem.*

H. Colta, Naturbeobachtungen uber Bewegung und Function des Saftes in den Gewachsen. Weimar, 1806, 4to.

J. Chr. F. Meyer, Naturgetreue Darstellung der Entwickelung, Ausbildung und des Wachsthums der Pflanzen. Leipsig, 1808, 8vo.

C. Pollini, Saggio di Osservazioni e di Sperienze sulla Vegetazione degli Alberi. Veron. 1815, 8vo.

H. L. du Hamel de Monceau, La physique des arbres. A Paris, 1758, 4to. Vol. I. 2.

P. Keith, System of Physiological Botany, Vol. I. p. 284—362.

290.

The internal structure of the stem varies according to the great divisions of the vegetable kingdom, which we have noticed above, (171. and Part II. Chap. 6.)

In plants the seed of which contains an undeveloped and rich albuminous body, the bundles of woody fibres, consisting of sap-vessels and spiral-vessels, are spread through the whole stem, and are every where divided by cellular texture. This is most distinctly perceived in the trunks of Palms, in the stems of the Scitamineæ, Muscæ, Orchideæ, and Coronariæ. This dispersed situation proceeds from these plants having no cotyledons which embrace the young plant during its growth, and its consequence is the parallel progress of the nerves in the leaves, without veins, and without a reticular distribution.

In the Ferns alone we observe a construction different in this respect, that strong bundles of numerous *vasa scalaria* are intermingled with the sap-vessels; and, being surrounded by a peculiar brown cellular membrane, they stand in definite number and in fixed order between the rind and the pith. Here, also, the nerves of the leaves soon pass into veins, and into numerous ramifications.

291.

In all perfect plants there is formed, where the two seed-lobes embrace the rising plumula, and out of the knot which unites these, a connected circle of spiral-vessels and sap-vessels, which rises perpendicularly between the pith and the bark, and thus forms the concentrical layers of the parts of the stem. In the knots of the stem this circle is interrupted, at the same time that the cellular texture which is there crowded, affords an opportunity for the production of new spiral-vessels and sap-vessels. In the mean time a similar circle proceeds from the knots upwards. The first spiral-vessels which take their place are always the innermost: these maintain, for a long time, their original shape, and even their green colour. Those of later growth take their place more towards the outer parts of the stalk, have a greater disposition to become woody, and shew this by their speedy transition into *vasa scalaria* and punctured vessels.

292.

It follows from this, that the innermost and outermost layer of the stem is purely cellular, while the middle ring, on the contrary, is composed of the higher primitive forms. The outermost layer is called the Rind; it consists of the proper rind and the epidermis, which covers the former. The latter is probably formed by the deposition of sap, and by its hardening under the influence of the ingredients of the air. It has always a different colour from the proper rind: it is, for instance, white in the Birch and in the *Melaleuca leucadendron*; of a golden yellow in the *Aucuba Japonica*; in a more advanced age, it exhibits slight rents; by and by it thickens into a substance resembling cork, and, in consequence of the increasing thickness of the stem, it is thrown off, as is very distinctly seen in the *Platanus*. This latter phenomenon is a proof that the epidermis, in this condition, is no longer organised, nor is of any use to the tree. However, in a considerable number of trees, it has a certain permeability, by means of which the ingredients of the air can have an influence through it upon the interior layers.

293.

The proper rind, which, in its earliest state, is of a green colour, assumes other hues at a later period. Its cells contain concentrated and peculiar juices, which, by being deposited upon its sides, make them impenetrable to the eye. Many of these cells are so extended by the sap, that they appear like proper sap-vessels; because they are stretched out in length, are surrounded by a very fine cellular texture, and are closed at both ends. In Pines, in Celadine, and also in the Rue species, these peculiar sap-passages can be most distinctly observed.

The rind cells proceed horizontally through the interior layers of the stem towards the inmost, and open by this means a very remarkable connection between all these layers, which is of consequence to the explanation of many phenomena. In most trees, however, this connection is interrupted at certain periods of their growth. The juices which rise in the inner bark, pass more readily the higher they ascend, into the generative sap, a kind of slimy organizing fluid. This sap, forcing its way from the outermost layers of the inner bark, tears the rind cells asunder, and fills the space which is thus formed between the rind and the inner bark. It is in this way that at the periods I have mentioned the rind is loosened, and a foreign branch, or a bud of another tree, can now much more easily be brought into this space, that it may derive its nourishment from the generative sap, and as it were take root

in it.   This is a short explanation of the artificial production
of trees, to which we shall again return, (306.)

294.

The bark cannot arise directly from any of the layers that
lie beneath it.   It owes its origin entirely to the generative
sap, and may be reproduced when this flows outwards.

As the bark forms the place of deposition for the peculiar
juices of the plant, and as these are frequently of a thick,
hard, balsamic or oily nature, it becomes by these means a
sluggish conductor of heat, and protects the interior layers
from the cold as from other external causes of injury.   It
maintains also the connection of all the interior layers with
each other, by means of the radiated cellular vessels, which
proceed from it to the pith.

Great as these benefits are which the bark affords especially to the shrubby plants, its removal and the wounding of the
rind are not however attended with immediate danger to the
life of the tree.   On the contrary, when the peeling of the
bark is effected cautiously, the young layers of inner bark
and of wood, which lie under it, arrive much earlier, though
rather violently, at the state of hard and perfect wood, from
the influence of the air.   Even the fruitfulness of the tree
may be encreased by this peeling of the bark, because the inner bark and the alburnum, laid open to the direct influence
of the atmosphere, are more powerfully excited, and the juices
become more concentrated.   Nevertheless, a peeled tree of
this kind must necessarily die sooner, unless its superabun-
dant vital power affords an opportunity for the production of
new bark from the wood.

295.

The layer which lies under the bark, and which is called
the Inner Bark (*liber*), is readily distinguished by its whitish colour, and by its distinctly fibrous and often mesh-
formed structure, as well as by its great flexibility, its toughness, durability, and power of resisting the ordinary causes of
destruction, especially putrefaction.   A closer examination

shews, that its apparent fibres are really tubes, entirely of
the same construction with the sap-vessels which have been
already described.   The bundles of these tubes are bent
from one another, where the horizontal and radiated cells of
the bark intersect them.   It is hence that the appearance of
meshes is formed; (Tab. V. Fig. 3.)   No trace of spiral vessels is found in the inner bark.

296.

This is peculiarly that part of the stem in which the juices
ascend, as may be distinctly seen by a horizontal cut into the
trunk during spring.   The higher these juices ascend, the
more are they changed into the organic slime, which is denominated the Generative Sap, and, in this sense, the inner
bark may be considered as the organ from which all the other
parts are produced.

297

The peculiar woody circle, which lies beneath the inner bark,
is composed of all the three primitive forms.   In its earliest
state, when it most resembles pith, it consists only of the primitive spiral vessels, together with the sap-vessels which constantly accompany them, and which are intersected by the radiated and converging rind-cells, sent towards them through
the inner bark.   It is evident that these latter vessels must always become closer set, the nearer they approach the pith.   The
more recently deposited layers contain for the most part *vasa
scalaria* and punctured vessels, but more seldom a primitive
spiral vessel between the individual layers.

298.

In the older branches and stems of many trees, the distinction
between the younger and older woody layers, is easily observed.   The former, which are usually called the *alburnum*
(*alburnum*), are known by their white colour, spongy texture,
and inferior durability.   Many trees, which either grow rapidly, or whose organisation is peculiar, deposit nothing but
alburnum; in many trees, what is called debility of the al-

burnum (*splint swache*) proceeds from their bad situation, and from other causes, which hinder the quiet lateral impulse, (416.)

As the growth of most trees takes place in determinate periods, it is from this circumstance that the annual rings, which we observe in the wood, take their origin. The earliest spring growth is commonly the richest; hence the greatest number of new layers are deposited by it, but, on account of the continued ascent of the sap during the summer, these layers do not experience the gentle lateral pressure to such a degree, as is necessary for the thickening of the sides of the cells and of the sap-vessels, and for the consequent production of wood. The second growth proceeds somewhat more softly; fewer new layers are deposited; but the subsequent lateral pressure from the bark-cells assists the thickening and hardening of the wood more powerfully, on which account the outermost layers of any annual circle are always the firmest and the richest in resinous and oily juices. In many tropical trees, the same distinction of the annual rings is perceived, because in these also there is a periodical change of vegetation during the dry and wet season. On the other hand, the wood of many trees even of our climate shews no annual circles, because they either undergo no double pressure from the sap, or because their organisation resists the alternate thickening we have mentioned. Even what is called the Silver-grain (*Quer-gefuge*), is not equally distinct in all woods,—although it is present in them all, since in all cases the bark pushes the silvery horizontal processes towards the pith. The firmest woods have commonly the most distinct silver-grain, as may be seen in the Oak, the Beech, and the Elm.

299.

The innermost part of the stem, namely the Pith, is as completely cellular as the bark. In young shoots it is full of sap, and is closely united with the woody circles. At a later period the juices are dissipated; the pith becomes dry and white, and seems no longer to be so closely connected with the wood. The quicker the plant grows, the more is the pith detached from the wood, until at last it entirely disappears, and leaves the stem hollow, as we commonly observe in the umbelliferous plants; or vacant spaces occur, and the pith is found only about the joints. Sometimes these vacancies are divided by regular partitions, as we observe in *Juncus glaucus*, in *Cicuta virosa*, in trees of the Walnut kind, and in Rose bushes; in which last, from the regular construction of these spaces, a connected cellular texture may be remarked.

The pith vanishes in the hardest woods, because these press ever more and more towards the centre, and, by uniting with the cells of the pith, render them at last completely indistinguishable.

300.

As the layers of the stem become united in the joints, and the primitive vessels are there crowded together and take a different direction, the pith cannot be supposed to proceed through the joints unchanged. There are indeed no peculiar partitions, which intersect the cavity of the pith in the joints; but it is so intermingled with the other parts, that its continuation is evidently interrupted. But the stronger branches are excepted from this remark; for they are knotted and push out new shoots, without our being able to observe the interruption of the pith. The forming sap here pushes into the space between the bark and the inner-bark, to form a reservoir from which new shoots may be unfolded.

301.

From the interruption of the pith in the joints, it follows, that this substance is by no means so essential to the production of the fruit, as some naturalists have believed. And this idea is still further opposed by the fact, that in the *Syngenesia necessaria* we observe the perfect seed only in the circumference; whilst in the middle, where the pith might have had some effect, we see the seed either imperfectly formed, or entirely wanting. There are also a great many trees, which, without having any peculiar pith-cavity, yet produce rich fruits. Finally, the nature and structure of many fruits are quite inconsistent with this origin, since the soft, spongy, and entirely cellular pith cannot possibly generate organs, which are often as hard as bones, and therefore contain a crowd of spiral-vessels and sap-vessels, which are entirely wanting in the pith.

Besides, the use of the pith is evidently altogether confined to the time when the young shoots are sent out, when the connection between its cells and the radiated cell-vessels of the wood seems to be subservient to the deposition and preparation of the sap. In more advanced age, when the formation of wood takes place, the radiated vessels themselves afford the means of this deposition, and render the pith cells superfluous. Hence we often enough observe, that in hollow trees, while not only the pith, but the whole of the wood is destroyed, they yet continue to grow, provided only the inner bark remains.

IV. *On the Structure of Buds.*

F. C. Medicus's Beytrage zur Pflanzen-Anatomie. Manheim, 1799, 1801.

Dessen's Pflanzen-physiologische Abhandlungen. Leipsig, 1803.

Darwin's Phytonomia.

Aubert du Petit-Thouars' Essay on the Organization of Plants. Paris, 1806.

302.

We must first establish a general idea respecting germs, before we proceed to a more particular consideration of the formation of buds, because these latter are only germs unfolded, and matured into a variety of shapes, although in Latin, and its kindred dialects, the word *Gemmæ* is used as well for germs as for buds. By germs, we understand every condensation of the peculiar juices, or of the particular matters from which new individuals of the same kind can be produced. In their simplest form these germs may be observed in the small granular or spherical bodies, which are produced in

the vessels of Confervæ, as also in those which ooze out upon
the epidermis of Lichens. These, when they have come to
their mature state, are separated from their parent body, and
constitute new individuals, which retain not only the essential,
but the accidental nature of the parent plants; for it is a leading
character of propagation by germs, that these are properly to
be regarded as a continuation of the parent plant, by the condensation of its substance, on which account even accidental
peculiarities and diseases propagate themselves in this way,
and from this cause the shades of colour in the Lichens and
Sponges are so constant, that we are forced to assume them
also into the characteristic description. But this very circumstance renders the determination of species, in these lower organic bodies, a matter of doubt, for propagation by seed is,
in their case, out of the question.

303.

In trees and woody plants, the structure of the germ is of
a more complex nature. The parts of the stem and branches
become crowded in particular positions, pass into the substance of each other, and, in this manner, reservoirs and
joints are produced, which we observe both in the leaf-
stalk of trees of the citron kind, and in the fleshy leaves
of other plants. These reservoirs consist of a compact cellular texture, and of the congregated rudiments of new sap-
vessels and spiral-vessels, and may be artificially produced,
namely, by making an incision in a branch, and thereby promoting the appulse of the sap. In every attempt to produce buds artificially, it is a necessary condition of success
that these reservoirs should first be formed. They are indeed produced in all woody plants, even when no proper
buds appear, as, in tropical trees especially, we observe them
occupying the place of buds.

304.

These last mentioned bodies, in the case of our fruit and
forest trees, commonly appear, during the time of the second
growth, in the axis of the leaves, or at the extremity of the

branches, and, towards harvest, increase gradually in circumference and size. They always arise out of those reservoirs,
are outwardly surrounded by variously coloured scales, which
are often bound together by a substance of a resinous nature,
and contain within them leaves or leafy scales, which are
placed upon one another and bound together in a peculiar
manner. They are found, for instance, in the Ash, (Tab.
IV. Fig. 5.), mutually riding, as it were, on each other; and
they have a similar construction in the Alder; (Tab. 1V
Fig. 2.) In the Salisburia they stand clenching each other;
(Tab. IV. Fig. 7.) In the Horse-Chesnut, and in the Medlar,
they are folded into each other; (Tab. IV. Fig. 6.) In some
trees these scaly coverings are in a very small number, as in the
Guelder-rose bush, (Tab. IV. Fig. 8.); and especially in the
Tulip-tree; (Tab. IV. Fig. 3. 4.) These scales have either
the future leaves lying between them, as in the Alder, (Tab.
IV. Fig. 2.); or the leaves are found only in the centre of
the bud. In most instances the future leaves are folded and
bent in a variety of ways. A woolly sort of substance also is
often found between them, which evidently serves to keep off
the cold, to prevent the influx of superfluous moisture, and
to be a defence against other external injuries. The bud is often so protected by the closely shut and firmly agglutinated
scales, that no external power can have effect upon it, unless
it is inflicted by a very powerful cause. Until these coverings unfold themselves, the bud can only be nourished by the
reservoir from which it arises.

305.

In many instances buds contain only leaves, but in other
cases they inclose also the rudiments of the future blossom.
Hence, in fruit trees, we commonly divide them into Wood
and Fruit buds. The former, which are of a smaller size and
more pointed, contain only the future leaf; the latter of
greater roundness, discover, when they are cut through, the
germ of the coming flower. However, they pass into one
another; for the active and more perpendicular movement of
the unformed sap is their producing cause; but the fruit-buds

N

require a slower lateral movement and the co-operation of the
rind-cells, on which account stimuli and wounding of the rind
often force the tree to put forth fruit-buds. The same purpose
is promoted by the constrained, horizontal, and bent position
of the branches. It is on this account that gardeners so ma-
nage their fruit-trees, as to remove the branches which rise
directly upwards, and lead out into a fan-shape only those
that rise obliquely, that in this way the tree may be urged to
put out more fruit-buds.

306.

The idea is altogether false that wood-buds are produced
by the wood-circle, and fruit-buds by the pith or bark.
Every bud, as we have already said, arises from a reservoir,
which owes its origin to the generative sap. As this is the
product of the inner bark, all buds originate properly in
the *inner-bark*, and in so far as the wood also contains
inner-bark vessels, these may be considered as contributing
to the formation of buds. But we must more especially attend to the connection of the buds with the inner-bark itself,
to understand particularly the success of the insertion of
buds, at the time when the rind is loosened from the inner-
bark, and the intervening space is full of generative sap, (293.)

307.

Finally, the position of buds on the stem is worthy of
notice. Generally we find them either placed opposite to
each other, or alternating. If, however, we attend to the
series of their positions on the stem or branches, we frequently, at least, perceive a spiral line on which they are set. We
see in this case, again, the continually equalised and ever
renewed contest between the perpendicular and horizontal
direction.

Buds plant their roots, which are properly continuations of
the sap-vessels, between the inner-bark and the rind, and
every bud ought to be considered as a new individual, which,
separated from its parent body, has the power of being propagated. Hence the art of the multiplication of trees. In

imperfect plants, propagation is commonly by means of buds. The forms and colours still remain in this way, after the parent plant has been completely divided

Those plants must be related whose buds are fitted to be propagated together ; but to what extent this relationship must exist is not quite clear. It is certain that evergreens may be propagated on plants which are deciduous, if they belong to the same genus, (Hopkirk, Flor. Anom. p. 19.) *

It is certain that the usual mode of propagation by layers, grafts, and shoots, diminishes the power of the plants to produce seed. It is hence that *Salisburia adiantifolia, Saccharum officinarum*, and *Bambusa arundinacea*, very seldom produce blossoms or seed with us.

V. *On the Structure of the Leaves.*

C. Bonnet, Recherches sur l'usage des Feuilles dans les Plantes. Geneve, 1754.

308.

The leaves are an expansion into a surface of those primitive forms which in the stem stand near each other, or were inclosed within one another. We hence find the leaves to be altogether of a cellular structure, in those plants whose stem contains no other form, as in the Mosses, among which, however, the *Sphagnum obtusifolium* shews the same fine spiral fibres in its cells, which we find on the surface of the stem ; (Tab. III. Fig. 25.)

---

* Virgil speaks as a poet, not as a natural historian, when he sings, (Georg. II. 69.)

Inseritur vero et fœtu nucis arbutus horrida,
Et steriles platani malos gessere valentes :
Castaneæ fagus, ornusque incanuit albo
Flore pyri ; glandemque sues fregere sub ulmis.

The best introduction to the artificial propagation of trees is in Munchausen's Hausvater, Book V. vid. 675—758.

The leaves of those plants, whose stems contain scattered and parallel bundles of spiral vessels and sap-vessels, with intervening cellular texture, exhibit only parallel nerves, without the appropriated veins, as in the Grasses, the Palms, the Coronariæ, the Irideæ, and the Scitamineæ. In Ferns we observe a peculiar distribution of the nerves and veins. These seldom anastomose with one another, but more commonly issue in clear and pellucid points, which, by a more careful dissection, shew the extremities of the spiral vessels in a vermicular form. Something similar is observed in the *Hypericum dubium* of Smith, and in some species of *Crassula*. The distribution of the nerves and veins is carried peculiarly far in the leaves of the Aroideæ, and the Melastomeæ, for they are united on the margin of the leaf by large anastomoses, which run parallel with the margin.

309.

The nerves and veins of leaves are a continuation of the bundles of spiral and sap vessels, and therefore they remain uninjured during the maceration of the cellular texture. They thus exhibit, frequently, a beautiful skeleton, the delicacy and almost endless ramifications of which are astonishing ; (Seligman's Nahrungsgefasse in der Blattern der Baume, Nurnberg, 1748, folio.) But the cellular texture, which fills the interstices of this net-work, and which, on account of its juicy consistence, is distinguished by the name of Parenchyma, is of as much importance as the net-work itself.

The cells of leaves have a different structure, according as they are nearer the upper or lower surface. Near the upper surface, they are more extended in length, and take the form of prisms or cylinders, in preference to any other shapes. Besides, the upper surface of leaves is frequently covered by an apparently impermeable, and somewhat brilliant epidermis, in which we observe no further remarkable organization. On the lower surface, again, the cells are more extended in width ; here and there also spaces void of juice occur, or some of the cells seem to have lost their sap, and to have become filled with air. Their partitions undergo a change, since they often, instead of the right-lined direction, assume a crooked, folded, or winding form; and the cells or spaces of the cellular texture, which want their sap, are thus placed, by means of peculiarly constructed slits, in immediate communication with the external atmosphere.

310.

These slits are commonly oval-shaped, and pointed at the extremities, being encompassed by a border, which consists of a granular or glandular mass, and can frequently be torn off; (Tab. V. Fig. 2.) The partitions of the cellular texture either unite with this border, or they pass round it without touching it.

The size of these organs is as various as their number. In the Coronariæ, where they are largest, their longitudinal diameter is from the twelfth to the twentieth part of a geometrical line, and their diameter, in the cross direction, is from the twenty-fourth to the fortieth part. They are exceedingly fine in the most perfect plants, as the Myrteæ, Rosaceæ, Leguminosæ, and Caryophylleæ. Two hundred of them, at least, might lie upon a geometrical line.

Their number is as various. The smaller they are, the more numerous they usually are. In general, we can count from fifty to two hundred of these slits upon a square line.

311.

These organs have some resemblance to the air-vessels of insects, especially when we compare them with the raised pores of the chrysalis of the Sphinx populi; (Mein. Comment. de part. quibus insect. spirit. ducunt, Tab. II. Fig. 16.) But there is this difference between these two kinds of organs, that the pores of insects always contain the stem of the airpipes, whilst the slits of plants are in no immediate contact with the spiral vessels. Yet it is very worthy of notice that these two sets of vessels are produced together; and as the Ferns first shew the spiral vessels, the first symptoms of slits are also seen in them.

312.

The appearance of these slits, in certain families, has, how-ever, some other remarkable circumstances, which at least somewhat limit the relation of the spiral vessels to these organs. Among plants of an entirely cellular structure, slits have been actually observed, although they are but rare, in the Marchan-tia, and in some species of Splachnum.   Those plants of the higher orders, which have no green leaves, are destitute also of slits.   Although, commonly, they appear only on the un-der surface, they are yet observed on both surfaces in the Coronariæ, the Grasses, the Palms, and even in Pines.   But they are found only on the upper surface of water plants, whose leaves are spread out flat upon the water, and in such land plants as have their leaves lying flat upon the ground.

But these slits are found on every leafy integument, pro-vided it be not too much set with hairs. They are according-ly observable on the exterior surface of the calyx, and serve, when the calyx takes the place of the corolla, or is united with it, as an excellent mark of distinction between these two coverings of the sexual parts.   They are as invariably want-ing in the proper corollar integument, as in the sexual parts themselves, (175.) Yet in one instance they have been ob-served in the epidermis of the Cherry; (Vom Bau und der Natur der Gewachse, Tab. IX. Fig. 43.)

313.

The use of these organs is by no means confined to one function ; but, as in plants, and even in the lower animals, the same organ can perform two apparently opposite func-tions ; so these slits appear to be destined as well for the re-ception and preparation of gaseous matters, as for exhalation. The former of these functions seems to be established by the facts, that leaves absorb more powerfully with their under than with their upper surface, and that the slits are more nu-merous in juicy plants, which are nourished more by the sur-faces of the leaves than by the roots.   And that the slits ex-hale, and even serve for evaporation, we learn from the expe-riments of Treviranus,. in which plates of glass. fixed to the

2

under surface of leaves, were covered, after some time, with drops of dew, while they were little or not at all stained when they were fastened to the upper surface.

314.

Generally the leaves, and leafy integuments, are the organs, which, by exhalation and absorption, like the breathing organs of animals, maintain the proper composition of the plant, and contribute essentially to nourishment and propagation.

Experiments and observations have instructed us, that healthy and green leaves, during sunshine, take in carbonic acid, and give out oxygen in the state of gas. In the shade, and at night, as also when they are sickly and take a different colour from the green, they take in oxygen gas, and give out carbonic acid. By these two functions, however, the condition of the atmosphere is not to any great extent altered, either with respect to its quantity of oxygen, or of carbonic acid. Let us then enclose the green parts of vegetables in a definite portion of air. In this case the quantity of oxygen, during sunshine, is increased to such an extent, that from twelve square inches of green blades, ten cubic inches of oxygen are produced in a few minutes. The increase of carbonic acid from plants, which are confined in the shade, may also be observed from the resolution of lime-water placed in a similar confined space. In the open air, on the contrary, scarcely more than the usual quantity of oxygen is imparted, by this means, to the atmosphere, because it is impossible that all the leaves can be at the same time illuminated by the sun ; for in bushes, forests, and gardens, a greater proportion of leaves is always in shade, and, consequently, by the production of carbonic acid, the quantity of oxygen gas is balanced. It thus happens that the produced oxygen is constantly consumed, as well by the shaded leaves as by animals, and also by the soil, which is incessantly taking it up. It maintains an equality with the carbonic acid given out in the shade. This carbonic acid is constantly deposited during the night, along with the dew : and during sunshine, plants consume as much as they give out when they are in the shade,

It has been observed, that a smaller quantity of carbonic acid is always given out in the shade, than the quantity of oxygen absorbed ; and, also, that the quantity of the latter increases when branches or leaves are cut off. Both circumstances seem to shew, that oxygen is not merely concerned in the formation of carbonic acid, but is also appropriated as a part of the plant: and the more so, because, according to the latest experiments, juicy plants, and those with fleshy leaves, consume the greatest quantity of oxygen, and form the smallest product of carbonic acid.

315.

Which of the two surfaces of the leaves performs these functions, or how they are divided between the two, is not yet completely ascertained.   Most experiments favour the idea that the upper surface of the leaves is especially employed in exhalation.   This surface is also the better adapted for this purpose, as it is better exposed to the light of the sun, and can give out the oxygen through the closed partitions of its cells, just as easily as, in perfect animals, this same substance forms a communication for itself through the shut vesicles of the lungs, and the equally impervious partitions of the other vessels.

316.

The exhalation of oxygen gas, during sunshine, is an effect of several conspiring circumstances.   One necessary internal condition is the vital activity of the plant, which being excited by the light of the sun, decomposes the carbonic acid water in such a way, that while the oxygen is given out, carbon and hydrogen are fixed and become appropriated.   The exhaled oxygen gas is by no means derived from the decomposition of water into its constituent parts, because, sometimes, the quantity of the oxygen corresponds exactly with the quantity of carbonic acid which has been consumed ; and because, sometimes, no oxygen gas is produced, when water,

I

deprived of its carbonic acid, is exposed to the light of the sun. Universally this effect ceases when the leaves begin to fade, to become discoloured, and to fall. It is most powerful in leaves which fall periodically, because their irritability is considerably greater than that of evergreen and fleshy leaves. This effect is, in the last place, most actively produced, when the electrical excitation of the atmosphere is greatest, on which account the greatest quantity of oxygen is exhaled from leaves during Spring, after a storm, and in the morning.

317.

But the exhalation of oxygen gas is closely connected with a remarkable property of leaves, namely, their green colour. As this colour in the rainbow stands exactly in the middle, between the two outermost tints, the red and the violet,—as it is bounded on the one side by the yellow, and on the other by the blue,—as all experiments further shew that the red and yellow tints are more of an oxygenous, and the blue and violet more of a hydrogenous nature,—it is extremely probable, that the green colour is the effect of a neutralization between the two extreme colours, or that it arises, when the light has attracted exactly as much oxygen as was required by the hydrogen and carbon which remained. And this theory seems to be confirmed by the following observations. All plants, so long as they are withdrawn from the light of the sun, are of a pale yellow colour, and regain this same hue when, as in the instance of the Endive and Cardoon, they have been covered with earth and blanched. In this condition, they are rich in oxydized juice, as their sweet taste, and the tenderness of their parts, shew. Besides, these blanched plants give out nothing but carbonic acid water, saccharine matter, and mucilage. As soon, however, as the light of the sun has called forth the proper activity of the plant, it empties itself of its superfluous oxygen, and forms those partly resinous, partly oily substances, which we find connected with the green colouring matter. The green colouring matter evinces its resinous nature by this circumstance, that it dissolves completely in spirit of wine; but, as it is not deposited by water from this solution, but continuing mixed with it, gives out a very nauseous hepatic smell, it is likely that azote goes also to the composition of this green colouring matter. We shall come back again to this subject, and only notice farther here, that a higher degree of vital activity in leaves, awakened by the light of the sun, often produces a blue colour from the green. In this case, there is probably an excess of hydrogen above carbon, as the putrefactive fermentation of woad and indigo, which is encouraged for the production of the blue colour, seems to shew. The blue colour of woad and indigo passes again, with mineral acids, into green, and lastly into yellow.

Decayed and falling leaves are yellow and red, because the oxygen remains in them after the vital activity is gone.

318.

This peculiar breathing of plants through the leaves has the most important influence upon their whole economy. By this only the proper mixture of the juices, and the production of fruit, is accomplished. It is hence that the leafing of plants is so necessary to the setting and ripening of the fruit, and that an attack of lightning, by which the leaves are destroyed, is injurious to hops and to all other plants. The reception of gaseous matters is equally important to the formation of the proper juices, and to the perfection of every other vital function.

319.

But we must also treat of the absorption of fluids in the shape of drops or of vapours, and of their evaporation, as being important functions of the leaves.

That the vapours and rain-drops are absorbed by leaves, is evident to sight. This also is confirmed by the fact, that a multitude of plants which have insignificant roots, yet grow very freely by absorbing with their green surface the nourishing fluidity of the atmosphere. In the parched deserts of Africa, where the quantity of rain in a century rises scarcely

to the height of an inch, the most juicy plants are often
found to grow to an astonishing height. They can only be
nourished by means of their green surfaces. In hot-houses,
too, we never attain a brisk growth so much by watering the
roots of the plants, as by an artificial wetting and sprinkling
of the plants from above. Evident as all this is, it is still
a difficult matter to explain this absorption, upon common principles, through the closed sides of the cells. We
might indeed ascribe this effect to the under surface of the
leaves, on which principally the slits are seen; but as dew
and rain much more frequently fall than ascend, we cannot
avoid confining this absorption of the vapours and fluid drops,
to the upper surface, on which supposition, we are again
forced to betake ourselves to an organic perspiration.

320.

The evaporation of leaves is one of the most obvious and
important of their functions. No person can deny it, who
has noticed the drops of clear moisture on the points of leaves,
even in hot-houses, where they cannot be affected by the dew;
or who has traced the movement of a mist in a still evening,
as it raises itself from fields planted with vegetables; or who
has seen the rising of clouds from forests, and the ascent of
vapoury columns from the same places before the formation
of a storm. In fact, plants lose, by evaporation from their
leaves, the greatest part of the moisture which they take in
by their roots; the proportion of the water absorbed, to that
lost by evaporation, is as 15 to 13, seldom as 4 to 1. It is
hence that a branch without leaves, when it has been placed
in water, becomes heavier than one in a state of frondescence,
because it wants the organs through which it may relieve itself of its superfluous nourishment. The organs which are
chiefly employed in evaporation are the slits, and also the
hairs, which latter organs are therefore more abundant in
young shoots, and in those parts whose evaporation is most
active.

321.

Evaporation has an essential influence on the economy of the plants themselves, and on the whole economy of nature. The activity by which the plant empties itself of its superfluous matters, operates as an incitement to the other functions, and a plant is, in truth, the more healthy, the more freely it eva-porates. Yet there may be an excess in this also, especially when not only unformed juice, but the prepared and proper sap, is given off. The sudden and powerful operation of the sun-beams after a passing drizzling rain, favours not un-frequently the perspiration of oxidized slime and of sweet drops, which are known by the name of " honey-dew." Swarms of insects are thus invited, whose young brood over-spread the surface of the leaves as a fine powder, and ren-der them incapable of performing their functions. This is the simple explanation of the blight, or of the mildew, as every person may convince himself by observation, (426.)

322.

The evaporotion of leaves has a great influence on the ge-neral economy of nature. As in the transition from the form of drops to that of vapour, a greater portion of heat is con-sumed, the quicker this transition takes place ; we find in this fact a principal cause of the low temperature which the juices of living plants exhibit even during the greatest sum-mer heat. Nay, the shade of a leafy tree will always afford a greater coolness to sentient animals, than the shade of life-less objects.

The influence which the evaporation of leaves has upon the whole atmosphere, as well as upon the earth and its waters, produces very extensive effects. Forest regions are not only cooler, but also more productive of rain, than steppés and sandy deserts, where vegetation is entirely wanting. All the streams of the world have their sources in mountain chains covered with woods; and although the melted snow is their immediate cause, they would neither continue to be poured along, nor grow to a river, unless forests and woods, by their evaporation, incessantly afforded the necessary stores of water. The largest rivers in the world flow in South America, in Upper India, and in Northern Asia, through forests of immeasurable extent.

323.

We perceive in leaves a remarkable difference in point of duration. Some of them, the evergreens, remain very long, and fall off at least without any regularity. These are commonly of a firmer and more tough consistence, as they are very small and needle-shaped, and contain a number of peculiar, resinous, or oily juices. This characteristic leads us to the conclusion, that the irritability of such leaves is not adequate to their complete exhaustion.

With respect again to other leaves, which have a fixed periodical change of budding and falling, we can find no other cause of this but their more perfect irritability, which, having been exposed for a length of time to stimuli, is at last exhausted, as in all the higher organised beings the vital activity acts periodically. External accidents have indeed an influence upon these phenomena, but the weathering of the seasons cannot be the only cause of this change, since in hot-houses and green-houses, we see that tropical plants, which enjoy the same heat and the same nourishment during the whole year, yet undergo this periodical change.

VI. *On the Structure of Blossoms.*

324.

We refer to the distinction which we have stated above, (89, 90. 101.), between the calyx, the corolla, and the nectary, whilst we now apply ourselves to a more careful examination of the structure of these parts. We have already stated, (185.), that the calyx has the same structure with the leaves, as it springs from them.

It is the Corolla, therefore, which must chiefly occupy us at present. When it is not united with the calyx, it is distinguished by a surface, which, in general, is of an extremely fine cellular structure, and the superficial cells of which rise up into fine prominences or pyramidal shaped knobs, on which we often perceive fine drops of a fluid. This construction occasions the brilliant appearance which many blossoms assume in the light of day, and especially during sunshine. We call this fine surface the Corollar Integument. When the calyx and corolla are united, when, as in the Rosaceæ, the *filaments* seem to be fixed in the calyx, it is, however, from this corollar integument that they arise. The parts of the corolla, also, frequently alternate with those of the calyx, as in the Ribes, the Rhamneæ, the Salicariæ, and the Melastomeæ, because both these parts spring from one base, and shew the corollar integument sometimes only on the inner surface and sometimes on both surfaces. In the Polygoneæ we find this same transition of the parts of the corolla and calyx into each other, evincing that these two organs, notwithstanding their usual separation, are yet very nearly related, and can pass into each other.

This corollar integument covers the proper *parenchyma*, which is the seat of the colouring matters, and which displays a great variety of tints in different blossoms. The cellular texture is by no means regular in blossoms; yet the sides of the cells are not so bent, as they are seen to be on the epidermis of the under surface of leaves. When this cellular texture, the seat of the coloured fluid, is taken away, we then perceive the spiral vessels, and less frequently the sap-tubes: they appear in bundles on the basis of the petals. But most frequently we observe them single towards the circumference, apparently ramified, and anastomosing with each other in great arched lines, until at the margin they gradually pass away, so that the most powerful magnifiers are scarcely able to shew us distinctly their extremities.

As the spiral vessels resist maceration, very fine skeletons may be prepared and kept, as well from the blossoms as from the leaves.

325.

The structure of nectaries, and the apparatus belonging to them, corresponds, in this respect, with the structure of the corolla, that the nectarilymata and the nectarothecæ are often parts of the corolla, or such likenesses of the parts of the corolla, that we sometimes hesitate to which of the organs they should be assigned. In order to be convinced of what has now been stated, we may examine the nectarilymata of Phylica, (Tab. II. Fig. 15.); of Agathosma, (Tab. II. Fig. 22.); and of Büttnera, (Tab. IV. Fig. 18.) The proper nectaries, in the strict sense of the word, are cellular or glandular organs, which we find on the receptacle, or at the base of the filaments.

326.

If we attend to the appearance of the corolla in the different families of plants, we observe, in those of the simplest and lowest organization, that there is either no trace at all of this organ, or a feeble one ; but where it does shew itself, it is nothing but a colourless scale, or it consists of a fine and very pellucid membrane. Coloured coverings for the sexual parts, first appear in the Musci frondosi, which, beside these, have also a permanent cap, by which the fruit is covered till it be perfectly ripe. In the Pipereæ, and most of the Naidæ, we find scarcely any thing but some fine scales below the sexual parts, which it is difficult to consider as representatives of the corolla ; (Tab. III. Fig. 4, 5.) The Aroideæ supply the want of the corolla by sheaths, which are often of a beautiful colour, and from which the spadix projects, or here and there some white hairs arise, which may be considered as the representatives of the corolla ; (Tab. II. Fig. 11.) In the Cyperoidæ there are also some scales only ; whilst, in the Grasses, there are fine pellucid membranes, which we must regard as the corolla. But the outermost valves are sometimes coloured, as in *Triodia, Avena versicolor*, and *Sesleria cœrulea.* The Restiaceæ and Junceæ begin to exhibit a regular corolla, which, in some genera, is beautifully coloured. The colouring of the corolla proceeds through the

Palms, Sarmentaceæ, and Coronariæ, onwards to the Irideæ, Scitamineæ, and Orchideæ, where it is evolved in the greatest magnificence. Although these, in some respects, are families of a lower rank, we yet thus see, that in the progress of nature towards a more complete evolution of forms, it is seldom that a harmonious construction of all the parts takes place, but that commonly one organ is exquisitely fashioned, while others remain imperfect ; since the Pine tribe, the Armentaceæ and Urticeæ, which, in many respects, stand higher than the before named families, yet want, with some exceptions, a proper corolla, their sexual parts being commonly protected merely by scales.

The Polygoneæ and Chenopodeæ also shew only a corollar integument of the calyx. In the Santaleæ, Thymelææ, Proteaceæ, Laurinæ, and Tricoccæ, the calyx is also formed with the same integument, and might be mistaken for a true corolla. It is in the Nyctaginæ and Primuleæ that the corolla first begins to be distinctly separated from the calyx, and to take the place of a peculiar organ.

327.

If we attend more particularly to the colour of blossoms, as their most striking characteristic, it is evident that the operation of the light of the sun upon the exceedingly fine structure and on the juices of the delicate parenchyma of the corolla, is the principal cause of the evolution of these colours. This is evident from the fact, that not only tropical plants have the utmost magnificence of colour in their blossoms, but that also in the polar regions some very warm tints appear upon the flowers ; because every person knows, that where the sun does not set for several weeks, he must exert an uncommonly powerful influence on vegetation,—as is also apparent from the rapid ripening of the summer crop in the polar countries.

As an internal cause of the colours of blossoms, we must attend also to the green colouring matter of the leaves. The tints of the blossoms arise from this, by a change in the proportion of oxydation, as we perceive in the corollar integu-

ment of the calyx, and as we also see it in the colouring of
the bracteæ : since most of these pass again, when treated
with alkalies, into the green hue. The red juice of many
blossoms becomes, by means of alkalies, first blue, then green,
and, lastly, yellow. The iron in the soil has also a considerable influence in changing the red colour of the *Hydrangea*
into a blue. When, from all these facts, we conclude, that
the green colouring matter, as it passes into the corolla, frees
itself from its superfluous hydrogen and azote, and, in that
way, becomes more oxydized, we are supported in this conclusion by a variety of considerations ; not only by the before-
noticed change of colours by alkalies, but also by the solubility
of the colours of blossoms in water ; and more than all this,
by the frequent exhalation of azotic and hydrogen gas from
blossoms.

328.

It cannot be denied, however, that a multitude of difficulties still remain, and that many hypotheses must yet be
adopted and rejected, before we can flatter ourselves that we
have come near the truth.

Of the utmost consequence, in particular, is the great variation of colour, of which the *Hibiscus mutabilis*, and *Gladiolus versicolor*, afford the most striking examples, although
the fact is seen, in an inferior degree, in many other blossoms,
which, when they are first unfolded, are coloured by tints
different from those which they afterwards assume. It seems
that this variation of colour passes most frequently into the
red, because many white and blue flowers take this colour
in their later stages ; nay, in a few cases, the yellow colour,
as in the *Medicago media* Pers., passes into the violet. It
cannot well be denied, that a variation in the proportion of
oxygen lies at the foundation of this fact.

329.

The smell of blossoms is another remarkable property, the
explanation of which will be facilitated, at least, in some degree, by what we have already said respecting colours. It

O

cannot be denied, that in all odorous matters hydrogen predominates. Along with this, the finest parts of the peculiar juices are drawn out, and occasion the manifold smell of flowers. That hydrogen is given out by flowers, may also be concluded from the powerful evaporation of blossoms, which, according to some observations on the *Arum cordifolium* Bory, can even generate drops of water. To the same purpose, also, are the experiments which have been made respecting the inflammability of the atmosphere of White Dittany, by lighted bodies, as well as respecting the flashes given out by many flowers on sultry summer evenings.

That azotic gas is produced from flowers, and that they even regularly exhale it along with carbonic acid, at the same time that they inhale oxygen, has been shewn by Saussure, (Recherches Chimiques sur la Vegetation, p. 127.), and by Grischow, (Untersuchungen über die Athmungen der Gewachse, sect. 154.); but we shall return again to this important observation. Saussure, indeed, has denied that hydrogen is exhaled from flowers, and he attributes the inflammation of the atmosphere of dittany to the burning of essential oils; but these also consist, for the most part, of hydrogen.

330

Every thing seems to shew, that the corolla is not only a covering of the sexual parts, but an organ by which the polarised primitive matters are directed to their evolution, and to their different attractions. The return of the sap to a more oxydized condition, and the evident evacuation of hydrogen and azote, appear to have as essential an influence on fructification, as the deposition in the honey-juice of flowers of oxydized mucilage, during the evolution of hydrogen,—of which we are about to speak.

331.

The situation of the nectaries, at the basis of the sexual organs, shews us, that the oxydized sap must be deposited in these organs, before the more volatile matters can ascend into the parts of fructification.

It is hence that the nectaries have commonly such a position, that the evacuation of the pollen from the antheræ is directed towards them. This is so evident in the Irideæ, that it is impossible to deny the connection between the nectaries and the organs of fructification. This relation is still more striking, when we observe an inclination of the pistilla, with their stigmata, towards the nectaries, at the period when the former have attained their perfect state. Finally, the evolution of the sexual organs at different times, or what is called the Dichogamy, is a very obvious proof that in many cases fructification is accomplished by the nectaries. When we thus observe, that, in the same flower, the antheræ are much sooner ripe than the stigmata, or the reverse, it is evident that these latter organs cannot be impregnated by.the former, in so far as they belong to the same plant. It hence happens that the first blossoms always fall off, and the fruit fails, when the dichogamy is gynandrous, because the early unfolded stigma finds no antheræ to impregnate it; and when these become capable of this office, the stigma of these first blossoms has already lost its susceptibility. When the dichogamy is androgynous, the last blossoms suffer the same failure, because, when the stigma of the last blossoms has come to perfection, there are no antheræ remaining to impregnate it.

332.

From all these considerations it is evident, that there must be some other helps to impregnation ; and this becomes the more obvious, when we observe that the sexual organs are often so placed, that, according to any usual and mechanical plan, the fructification cannot happen. We must, no doubt, ascribe something to the efficacy of the winds, in transporting pollen from a distance ; and it is certain, that in the Grasses, and some other plants which have no nectaries, this cause may operate, especially as the antheræ of these plants are large and pendant. Insects, also, which suck the honey from the blossoms, are most excellent assistants in impregnation. They wipe off the pollen from the bent antheræ, and bring it to other blossoms of the same species. Nor is there much danger of a mixture of species, and a production of hybrids in this way, because it is proved respecting bees at least, that in each excursion they gather only from flowers of the same species; (C. K. Sprengel, Entdecktes Geheimniss der Natur im Bau, und in der Befruchtung der Blumen; Berlin, 1793, 4to. Smith's Introduction to Botany, ed. 3., p. 256, 257.)

VII. *On the Structure of the Sexual Organs.*

333.

The analogy between the filaments and the petals, which was formerly stated, (181. and 187.), as well as their frequent union, lead us to conclude that the former have a similar structure with the latter. Where the filaments are so fine that they cannot be dissected, their structure, of course, remains hidden from us. But in some of the greater flowers among the Liliaceæ, we meet with filaments of considerable diameter; and the filaments of the Malvaceæ can also be easily examined. In these instances, we observe that very fine bundles of spiral vessels proceed through the whole length of the filaments to their summits, and are so lost at their points, that we cannot exactly point out their mode of connection with the antheræ. The filaments likewise contain a very fine cellular texture, and have sometimes also a corollar integument.

This simple structure, however, varies according to the manifold variations of the external form of the filaments. Particularly, we observe in the Urticeæ an articulated structure of the filaments, by means of which, in the *Parietaria*, *Forskolea*, and *Antiaris*, they are inclosed in the lobes of the calyx, previous to their being fully ripened, and afterwards spring forward with great elasticity to scatter the pollen from the antheræ. More minute experiments have not yet been made respecting the internal structure of these joints in the filaments. Meanwhile the contracted portions seem to occasion a considerable swelling of the parts which lie between them, so that, upon the slightest touch, the elasticity of the filaments is set in motion, and their tendency to the upright position is favoured. In a similar manner the filaments of the *Erica aggregata*, (Tab. III. Fig. 13.) are bent, before they are fully ripe, into a large curve, and it is by their elasticity that they afterwards become erect. The same thing is observed in the *Hirtella*, (Tab. VII. Fig. 4.), the filaments of which, after the antheræ are ripe, become of considerable length.

In the Euphorbiæ, the filaments appear to be geniculated, or as if they had a joint, (Tab. VI. Fig. 5.) But probably each filament consists of two parts, the lower of which is a similar stalk for the monandrous male blossom, as that which supports the germen. According to this idea, which was first advanced by R. Brown, (*General Remarks on the Botany of Terra Australis*, page 24.), what is commonly called the corolla in the Euphorbia, is properly a common covering of the flower, and incloses several male, and one female floret. This idea is further confirmed by the fact, that in some species of Euphorbia, we observe on the joint small laciniæ or lobes, which apparently are nothing else but the rudiments of the interior covering, or of the proper corolla.

The filaments of most of the syngenesious plants have also a peculiarly jointed structure, as they have been represented by Schkuhr, in the *Cacalia*, (Tab. 236.) ; and still more distinctly in *Baccharis*, (Tab. 244. Fig. h.) But, in this case, the joints seem rather to occasion a shortening and drawing back of the cylinder of the antheræ, because, in the more ripened state, the pistillum, with its stigma, which was before included in this cylinder, rises above it.

334.

With respect to the structure of the antheræ in general, we find, that for the most part they resemble an extended purse, surrounded by a very fine cellular texture, which is found throughout their whole substance, and contains, in each of its cells, a small pollenous body. It is worthy of remark, that the circumference of the antheræ is not increased by their ripening, but that they have the same size in blossoms that are not yet evolved as at a later period.

We are still destitute of any exact information respecting the manner in which the antheræ are connected with the filaments, or in what way the opening of the antheræ takes place. Meanwhile we observe in a great many plants, in the Irideæ, the Laurels, in *Asarum*, and *Stratiotes*, that the antheræ cling to the side of the filaments in such a manner, that these latter bodies, in some degree, overtop them. In most cases, the antheræ either lie horizontally, or swing on the points of the filaments, or these latter parts pass into the substance of the antheræ. Very fine sap-vessels extend from the points of the filaments into the antheræ, and conduct nourishment to them.

The ripening of the antheræ, like the ripening of the fruit, seems to be a kind of desiccation. Although sufficient sap be supplied, it ceases to be taken up. The partitions of the cells become thinner and drier. The cavities, in which the pollen is lodged, press it outwards by means of their elasticity, and when the extenuated and dried partitions of the cellular texture do not yield, they are rent by force.

This tearing open, or bursting of the antheræ, however, takes place according to fixed and secret laws of nature. The antheræ of *Solanum, Galanthus, Calectasia* (R. Brown), open at their summits. In *Galeopsis* this opening takes place by means of a fringed flap. The antheræ of the Syngenesious plants open longitudinally, each into two compartments. The antheræ of the Cucurbitaceæ open in winding lines. The antheræ of the Laurels burst from the lower to the upper surface, as also those of *Epimedium*, and of *Leontice*. The antheræ of *Triglochin* open around the circumference, and those of *Brosimum* open in a circle which surrounds the middle of the body. We have already noticed, that in flowers which have nectaries, this opening of the antheræ corresponds with the position of the nectaries, (331.)

335.

If we attend to the pollen, as it appears in most plants, we perceive it to have different forms in the different families.

In the Malvaceæ and the Compositæ, its form is the same, (Tab. IV. Fig. 15.) It appears, in these instances, to consist of regular spherical bodies, with bristly openings. In the Liliaceæ and the Geraniums, it consists of oval-shaped bodies, surrounded by elastic rings or hoops, (Tab. II. Fig. 24.) In the Onagræ we find obtuse triangular bodies connected by slimy threads. In the Proteaceæ the pollen is cylindrical, and somewhat bent; (Bauer's Illustr. Nov. Holl. Tab. III. Fig. H.) The two first forms seem to be very common. But the pollen varies its shape when it is put into water or oil. In the former it becomes inflated, takes a more or less perfectly globular form, and gives out, often with great elasticity, its contents, like a small cloud, which does not mix with the water. In mild oil it remains pretty much unchanged, only it becomes by degrees surrounded by a dark coloured ring, which by and by passes into the oil. In spirit of wine it contracts powerfully, and takes frequently, when it is treated with nitric acid, an obtuse triangular form, which originates in the contraction of the three rings. In nitric acid it gives out its contents in the shape of rays, which do not unite with the fluid.

336.

There are some remarkable variations from these forms of the antheræ and pollen, most of which we have already noticed, (107.) There is, however, still one variety in the structure of the pollen in some of the Naiadæ, particularly in *Chara* and *Zostera*, to which we must attend. In these instances, the antheræ seem to contain nothing but strings of confervæ, which do not unite with the water. These are as distinctly inclosed in particular reservoirs, as in the Fuci; and in some Ferns they stand single beside the germen.

337.

So many varieties, which might easily be multiplied, in the structure of the pollen, naturally lead us to a difference of chemical contents. Hitherto, however, experiments of this kind have only been made with the pollen of common plants, which could be procured in abundance, and even in these the results of the experiments have been very various and even contradictory.

A variety of the albuminous matter, which is called by John, *pollenin*, seems, in fact, to be the chief ingredient in pollen.   This substance is uncommonly liable to decomposition, readily becomes corrupted, gives out a great quantity of ammonia, and communicates to the pollen the naseous animal flavour which we frequently perceive in it.   It is insoluble in the ordinary menstrua.   Beside this, the pollen, in all probability, contains wax, although not in the same state of mixture in which we have it from the cells of bees, on which account some chemists entirely deny the existence of wax in the pollen.   Extractive matters, of a gummy or resinous nature, constitute the other component parts of the pollen, (John in Schweigger's Neuen Journal, B. II. Heft. 3. s. 247.   Grotthuss in the same Journal, B. II. Heft. 3. Stolze, in Berlin Jahrb. der Pharmacie, B. VII. s. 159.) Although these results are not universally admitted, they yet entitle us to conclude, that an animal matter predominates in the pollen, and that it is made up of azote and hydrogen, in union with albumen and gluten.   After the deposition in the nectaries, and in the coloured portions of the corolla, of the superfluous oxygen, hydrogen and azote make their appearance   as the chief product of vegetation ; and of this product we are speedily made sensible by the before-noticed exhalation of azotic and hydrogen gases from the flower, by which means the substance of the plant assumes a resemblance to animal matter, whilst the plant is preparing to give, by its fructification, the highest proof of its vital activity.

338.

We naturally begin the consideration of the female parts of the flower, with that of the Germen.   This, in its unimpregnated state, is principally a cellular organ, in which the ovula, or rudiments of the future seed, appear like small vesicles filled with pure water, and can scarcely be distinguished from the cells themselves.   From the fruit-stalk, or from the receptacle, bundles of spiral and sap-vessels proceed into the germen, disperse themselves through it when it is furnished with partitions, and become united in the central column or placenta. From this central column arise, by the mere act of vegetation, those vesicles I have mentioned; and at a later period, when impregnation takes place, a new activity, awakened by this peculiar stimulus, is called into action. In some instances, the exterior covering of the germen secretes nectar, because at that time there is an evident overflowing of the mild oxydized mucilage into the germen.

339.

The Pistil is commonly a solid column, but sometimes it is observed to have a hollow space throughout its length, although this is always shut at the point in which it passes into the germen. On the other hand, the communication between the pistillum and the germen is maintained by means of sap and spiral vessels, which pass into the dissepimentum, into the central column or placenta, and constitute the passage to the ovula. We have already endeavoured to shew, that the number of the pistilla corresponds with the loculi of the germen, and that where the pistillum is single, it has in all probability become so from the union of several pistilla, (188.)

340.

The structure of the Stigma is as wonderful as it is simple. Universally, wherever we have examined it, we have found its surface moist, and studded with very fine warts or hairs, which are always closed as in the roots. Whatever, therefore, passes into the interior of the stigma, or pistillum, must in this case also make its way through the impervious and shut extremities of these organs. In many plants, particularly in the Lobelias, nature has been careful to protect the tender structure of the stigma from external injuries, by a particular contrivance. This is a peculiar veil (Tab. II. Fig. 23.) which covers the stigma; and which, in some genera allied to Lobelia, consists evidently of two valves.

341.

The position of the stigma, as well with respect to the male parts as to the pistillum in particular, presents remarkable differences. We know that in the Orchideæ, a common columna genitalium supports both the stigma and the two antheræ. The same union takes place in the Stylideæ, in *Cleone*, in *Podostemon*, and in *Andrachne*, as well as in Aristolochia. In some of the Proteaceæ and Scitamineæ, the two sorts of sexual organs become united at their base, and thus furnish a proof of the kindred nature of their structure.

A remarkable difference of arrangement takes place in the stigmata of the Syngenesious plants. The proper surface of the stigmata appears to be less intended for the collecting of the pollen, than what are called by Cassini the Collectors, or hairy surfaces which are placed opposite to the proper stigmata. In many of the Caryophylleæ, the greater part of the pistillum is of the same consistence with the stigma, whilst in the Saxifrages and Oxalidæ, as also in the Ericæ, only the extreme point of the pistil can be called the Stigma.

CHAP. II.

PHYTOCHEMY, OR DOCTRINE OF THE COMPOSITION OF PLANTS.

Sennebier, Physiologie vegetale, tom. ii. p. 298.

Keith, Physiological Botany, vol. i. p. 375, vol. ii.

K. Sprengel, Von der Natur und dem Bau der Gewachse.

J. Ingenhousz, Versuche uber die Ernahrung der Pflantzen.

Treviranus, Biologie. B. 4.

A. V. Humboldt, Aphorismen aus der Chemischen-Physiologie der Pflantzen.

Einhof in Hermbstadt's Archiv fur Agricultur-Chemie.

H. Davy, Elements of Agricultural Chemistry.

J. F. John, Ueber die Ernahrung der Pflantzen.

G. Wahlenberg, de sedibus materiarum immediatarum in Plantis.

H. Steffens, Beytrage zur innern Naturgeschichte der Erde.

Rauch, Regeneration de la Nature vegetale.

I. *General Remarks.*

342.

The composition of organic bodies is distinguished by more than one circumstance from the union of the elementary matters in unorganised substances; and, on these accounts, the examination of them becomes as difficult as it is instructive.

The first peculiarity in the composition of organic bodies, is their great liability to change, and their constant tendency to decomposition; while, at the same time, as long as life remains, this tendency never perfectly succeeds. As soon, however, as organic juices are withdrawn from the dominion of life, they undergo a change of their elements, and a decomposition of their constitution, which are attended with remarkable consequences.

Since we can thus only examine the composition of organic bodies, after they have ceased to live, we cannot be always certain that the results of our chemical examination really explain the way and manner in which the juices are mixed in living substances. Indeed, a highly etherial matter, which, as it were, gave life to the sap, seems often to make its escape, at the moment when the fluid or organic matter loses its individual character, and is subjected to examination. We are thus often reminded of the *spiritus rector* of the blood, which our predecessors admitted, and which we have no reason to treat with too much contempt.

The change which organic matters undergo when they cease to live, is of so peculiar a kind, that it cannot take place in inorganic bodies, unless they are mixed with organic matter. It is an internal change, which, in the juices of plants and in other vegetable matters, commonly begins with the evolution of carbonic acid, and ends with the plentiful production of acetic acid. It is called Fermentation. In animal matters, again, and in those juices of plants which approach to an animal nature, another decomposition takes place, during which hydrogen and azote principally are disengaged. This change is called Putrefaction. Both of these changes afford products, which did not previously exist as such in the organic body, and from which, therefore, we cannot conclude, with any certainty, respecting the component parts of the living substance.

343.

The second peculiarity of the composition of organic bodies, is, that it is more or less independent on the inorganic bodies that may happen to be present. It cannot be denied, indeed, that the composition of the earth and water, by which plants are nourished, has a considerable influence on their ingredients, and that at all times plants, and especially the lower orders of them, partake more than animals of the composition of the substances by which they are surrounded. In general, however, it is a law of nature, that every organic body forms its own ingredients out of the elementary materials which surround it; and that the quantity of lime contained in plants which grow in pure sand, or which spring amidst granite, is not less than the quantity of the same material contained in those plants which grow on a calcareous soil. In a few instances, some compound matters seem to pass unchanged into plants and the presence of common salt and of soda in plants which grow on the sea-shore or on calcareous soils, is as undeniable as the passage of metallic substances, and of many odorous and colouring matters, into the blood and secreted juices of the animal body. On the other hand, it is established, that neither odorous nor colouring matters pass unchanged into the unwounded roots of plants; and that we cannot accomplish the absorption of coloured fluids by these parts in any other way, but by cutting the twigs, so as to bring these fluids into direct contact with the sap-tubes.

344.

The third peculiarity of the composition of organic bodies, respects the absence, commonly, in such bodies, of any of the ordinary chemical ingredients in an entirely disengaged state. Completely disengaged acids are as seldom to be met with in the vegetable as in the animal juices. For the most part they are united to a base, and are first disengaged during fermentation, by the influence of mineral acids, or by some other operations. It is equally seldom that we find, in a free state, in the vegetable kingdom, any of those ingredients of which hydrogen is a part. These, like alcohol, are first disengaged, in consequence of a change which takes place during the saccharine fermentation. It is true, that, in a few instances, free acids are found in plants, and that volatile oils are given out by them. But these matters are for the most part to be considered as excrementitious, as the oxalic acid which is exhaled from the Chick-pea; besides, the hydrogen, in volatile oils, is too closely united to other matters, to be regarded as an entirely free body.

345.

The fourth peculiarity in the composition of organic bodies, consists in a kind of circulation, which the simple connections of the elementary bodies in the sustaining juices undergo. As in the higher animals, the chyle by degrees becomes freed from its oxygen, while it is mingling with the gall and passing through the glands of the intestines, that at last, in the thoracic duct it may pass into the state of blood by the union of azote, by the evolution of phosphorated iron, and of the colouring materials; as this neutral fluid frees itself in the secreting organs from its hydrogen, azote, oxygen, and carbon, that it may suffer a new oxydation in the lungs, and be prepared for undergoing again the same changes;—in the same manner plants attract carbonic acid water saturated with azote; mix it with their own substance, and sometimes add more evolved hydrogen and carbon to the oxygen; and at other times free themselves from their superfluous oxygen and carbonic acid by exhalation from the

1

leaves. But, in the same manner, they repair their loss of oxygen, by which means new affinities take place, until at last the elementary substances in the blossom separate from one another, from which separation the decay and partial death of this organ take place.

In this ceaseless circulation, we cannot consider deoxydation as the ultimate object of the chemistry of the vegetable kingdom, especially as a manifest progress to a still more powerful oxydation may often be remarked. It is in this manner, that, by means of the more complete predominance of oxygen, resin is produced from oil, gum from mucilage, and fixed substances from those that are volatile.

346.

These general considerations shew us the difference between the composition of animals and vegetables. The difference evidently consists in this, that azote and hydrogen are the prevailing matters in animal substances, oxygen and carbon in the vegetable world ; on which account, animal juices commonly putrefy, but vegetable saps pass into a state of fermentation. Not as if these matters were confined exclusively to each of the two organic kingdoms, since not only do albumen and gluten pass into a state of putrefaction, and disengage ammonia ; but we have also shewn the evolution of azote and hydrogen from blossoms, and the predominance of both these matters in the pollen. Our concern at present is chiefly with the general difference of composition, which always observes the assigned relations in the two great kingdoms of nature. It is hence that in transition forms, and in organised bodies of the lower orders, we usually regard it as a common proof of an animal or of a vegetable nature, if, when the body is burnt, it gives out an animal odour, which arises from a pecular union of azote and hydrogen with carbon.

II. *On the Common Sap.*

347.

The matters which plants attract from the soil, are not
for the most part its compound ingredients ; they do not
attract either earth, or metals, or salts, or extractive matter ; but, according to all observations and experiments,
they take up only carbonic acid water, united with azote ;
and all improvements of the soil, all manuring of it, have
no other object than that of increasing this sap, for the
purpose of evolving more powerfully, and in due proportion,
its proper products. From the time of Helmont, it has been
understood, that water is the only source of all the nourishment of vegetables. Plants have long been reared amidst
circumstances of such a nature, that no earthly ingredients at
least could be taken up by them. The experiments of Bonnet, Kraft, and Duhamel, are the most decisive on this
point. In later times, it has been completely established by
Ingenhousz, Percival, Schrader, and Braconnot, that plants
thrive amidst substances that are altogether insoluble, provided they are supplied with carbonic acid water, and that
they even present the same constituent parts, as when they
are reared in the earthy soil.

And this account is strengthened by the necessity which is
known to exist, for exposing any soil, the chief ingredient of
which is carbon, to the influence of the air, that oxygen may
be attracted, and carbonic acid be thereby formed. The
careful and complete turning up of Clover and Lucern fields
in spring, not only loosens the soil, but promotes also the production of carbonic acid. Hence the repeated ploughings,
by means of which Peter Kretschmar, seventy years ago,
wished to render manure needless, must have turned out very
unprofitable.

If we reflect still farther on the recent experiments which
have shewn, that there is a considerable consumption of
carbonic acid from absorption by the leaves ; and if we consider that all those circumstances are favourable to the growth

of plants, which increase the quantity of carbonic acid in the
moisture of the soil, as well as in the atmosphere, we shall
find ourselves obliged to admit that this substance is the proper nourishment of vegetables. To these circumstances belong
dew and rain, which convey more carbonic acid to plants,
than that which is supplied by spring water ; on which account,
this latter fluid is always inferior to rain-water for the sprinkling of plants. From the same cause arises the uncommon
fertility of volcanic countries, which, according to Gagliardo,
give out a peculiarly great quantity of carbonic acid. Our
own black mould is also so productive for the same reason,
namely, because the extractive matter which it contains is incessantly attracting oxygen from the atmosphere, to form carbonic acid. Hence also arises the advantage in horticulture
of screens for fruit trees, because the surface of the earth,
when overshadowed by great leaves, attracts carbonic acid
vapours more strongly, and these also can less readily escape
than they must do from a bare field. On the same account,
Vetch-Oats and Summer-Rye succeed better when sowed
among Peas. Beans also are sowed among Oats, and this
mixture is called Rough-cast. Clover is commonly sowed
with some other crop, in particular with Rape, Flax, Peas,
and even with winter-crops.

It is hence that heath is so little injurious to forests, that
it rather affords to the young shoots the advantage of shade,
and a more powerful attraction of carbonic acid.

But every just theory of the effects of manure, is in the highest degree favourable to this assertion. Manure consists commonly of intermingled animal and vegetable matter ; in the
excrements of swine in particular, we find this latter substance, and especially seeds, so unchanged, that they readily
vegetate. But the excrements of animals are extremely apt
to become putrid, and it is necessary, in order to subject
them to a slower fermentation, during which carbonic acid
is produced, to mix them with straw, and other refuse of
vegetables. The manure of sheep, when mixed with
straw, is therefore much more lasting than the excrement
by itself , and the refuse of the fold, unless it is very

powerful, produces little effect upon the second crop. When the manure, however, has passed through this fermentation, it will for ever communicate extractive matter and carbon to the soil; but before these can serve for the nourishment of plants, they must again be combined with the oxygen of the atmosphere.

The burning of the soil, too, operates by the production of oxygenised carbon, because every burning ends with the attraction of oxygen.   Hence also arises the advantage of smoking the ground, as it is practised in Italy ; (Bertuch Garten Magazin, b. 3. sec. 239.)   Ashes must be considered as carbon half oxidated during the burning, and united with alkali. Accordingly, when they are laid out, they have a powerful influence in the melioration of the poorest soils, as is proved by the excellent example of the improved agriculture around Dankerode, south from Harzgerode ; (Georg. in den Mögelin-schen Annalen, b. 3. s. 419, 448.)

The green manuring (*sovescio* of the Italians) proves the same fact; but the slow fermentation of green vegetables should only be employed in fertile and warm soils.   Potato plants, Zostera marina, and the common sea-ware, serve as manure, in the same manner, by the fermentation which they undergo in the soil.

Lastly, Nothing proves the truth of this account more clearly, than the distinguished advantage which the soil derives from calcareous matter; since the attraction of this substance for carbonic acid is, as every person knows, very powerful.   The most luxuriant vegetation arises on calcareous soils; the strongest wood in the world grows in the calcareous islands, which are formed by the Coral Reefs of the South Sea, as also upon the volcanic territory of Mascaren's Island. By a mixture of lime, we prepare the best compost, when it is placed in layers between animal manure, clay, and old mud.   The powerful effect of *Chara vulgaris* in fertilising land, depends on the calcareous slime with which it is covered.

348.

Not only carbonic acid water, but also the azote which is mixed with it, constitutes a principal part of the peculiar nourish-

P

ing sap of vegetables. Azote is produced in the earth, partly from the corrupted refuse of animals, partly from the clay or loam which in a greater or less quantity are mixed with the soil or mould, and partly, in the last place, from the atmosphere, which, in three out of four parts, is composed of azote. The advantage of a certain portion of clay in the soil, is not only evident from this circumstance, but also from the further consideration, that it retains the moisture for a longer time, and supplies it to the roots.

Azote appears also to be taken up into plants from manures. Tobacco derives its pungent taste and unpleasant flavour from animal manure. Grain produced on land highly manured with sheep dung, contains too much gluten to be employed in he brewing of ale. Manure composed of the refuse of animals, as pieces of horns and claws, produces abundance of straw fit for litter, but little meal, and too much gluten.

349.

It having been shewn, that the moisture drawn up by plants is only carbonic acid-water impregnated with azote, the crude ascending sap of plants must have in a great degree the same composition; and the only difference will be, that the living principle has operated upon the fluid, and has mixed it, the higher it has ascended, with the greater number of peculiar vegetable matters.

In fact, the crude ascending sap exhibits properties which completely confirm this account. As it mounts in the inner bark of trees, it appears as a clear fluid, of a pleasant taste, and producing a titillating effect upon the tongue. It easily ferments, and evolves during this process carbonic acid gas, which, when united with hydrogen, is known as an intoxicating drink. If the fermentation is continued, azote is disengaged, which swims upon the surface, and the liquor itself becomes of the taste of vinegar. As this composition of the ascending sap is the same in almost all plants, it is evident that it is directly derived from the moisture of the earth, and that it is distinguished from it only by the oxygenised slimy matter which it has attracted.

350.

The attraction of sap from the earth, differs in the same plants in the different periods of their growth. In the first periods, before the plant has blossomed, a greater quantity of nourishment is taken up, the more the plant is covered with leaves; because we have already seen (321.) that the evaporation promoted by the contracting power and irritability which the plant maintains, is a very important function, and one which must therefore strengthen its absorbing power. But we must take into account, that, by means of the more powerful frondescence, the ground is more shaded, and carbonic acid is deposited in greater quantity. Perhaps we ought also to take into account the evacuation of superfluous sap from the roots, (285.) It is certain that any species of grain succeeds so much the better among the stubble of a previous crop suitable to it, as, for instance, Wheat after Clover or Pulse, the more luxuriant that previous crop had been, which must be ascribed to the shadowing of the soil; so that clover in general promotes the fertility of the soil, the more luxuriantly it springs. It is possible that this proceeds from the more lively attraction of sap from the soil, by means of which exactly such a quantity of the latter crop is called up, as is best for the crop of grain which is to succeed it. We must also take into account the extractive matter which a plant imparts to the soil by its decay; in comparison with which, the portion of matter which the plant takes from the soil is scarcely worth reckoning. At least a strong and rank fallow may always be regarded as a manure. The thick grass sward, which is produced on the soil after a long rest, increases its fertility in an uncommon degree.

It is an undeniable fact, however, that a plant draws the greatest quantity of nourishment from the soil, at the time when it blossoms and forms its fruit. On a superficial consideration, we might have supposed the reverse. We might have thought, that as the period of growth ends with the blossom, a more powerful absorption could not take place than formerly; but when we reflect, that the blossoming and setting of the fruit are entirely new operations which increase

P 2

the contraction in the whole body, and raise its entire irritability, we must be sensible, that this is the time when the expenditure of sap must be the greatest, although we should not take into consideration the fresh exhalation from the blossoms, or the evolution of azote and hydrogen from them.

In fact, all our experience, both on a great and on a small scale, confirms this remark. In horticulture, it is known, that a plant requires the most powerful irrigation, when it puts on its blossoms and fruit. In agriculture, it is known, that it makes an important difference whether the juicy or the ripened stalk remains in the ground. The latter exhausts it so much the more powerfully, as the dead roots are no longer able to give out again the superfluous sap. It is certain, that from this cause Flax in particular exhausts the soil so much, that no winter crop at least can be obtained after it. It even seems to be true, that the richer the crop of grain is, and the more nutritious its parts, the more is the soil exhausted by it ; not that the moisture of the earth passes directly into the fruit, but because a greater expense of power, and a more lively elasticity in all the parts, is required for the formation of the nutritious parts of the fruit, so that the absorption of fluid sap must also be brisker. Einhoff's calculations teach us, that among common plants, French Beans contain the greatest quantity of nutritious matter, namely 85 *per cent.* ; Wheat contains 78, Pease $75\frac{1}{2}$, Lentils 74, Rye 70, Barley 65, and Oats 58. From this we may judge of the exhausting power which these plants exert upon the soil.

At the same time, it cannot be denied, that plants with strong and deep-seated roots, or with juicy tubercles, take a great quantity of nourishment from the ground, even when they are not in flower. Among these we may reckon Beet and Potatoes. Recent experiments have taught us, how much the former of these roots contributes to the manuring of the land, when it is suffered to remain in it, and when, in the end of harvest, the sheep are enclosed in a field of this plant after its leaves are shed, and time is allowed them to gnaw its roots, and to manure the field. These putrefying roots replace completely to the soil, whatever it had lost by nourishing the plants. Potatoes, when planted in a field which had previously been fallow, keep back the winter crop. For this reason, they are rather reared in a field which had before been under crop; dung is applied after them, and then pease are sown.

III. *More intimate Constituents of Vegetables.*

351.

It is not possible to state precisely how every one of the more intimate constituents of plants is formed from the common sap; but in a great many instances we can give a distinct explanation, and in others, where the component parts are still unknown to us, we must satisfy ourselves with probable conjectures.

The first matter formed from the common sap, is the organizing mucilaginous matter, or the forming juice, respecting which we remarked above (271), that it contains the two primitive forms, namely, spherules or vesicles, and ray-shaped or straight-lined bodies. The mucilaginous matter is a tasteless and inodorous fluid, which, without undergoing the acetous fermentation, passes, after a considerable lapse of time, into a kind of putrefactive decomposition. When treated with mineral acids, it forms oxalic and saccho-lactic acid. By dry distillation, mucous acid, with a burnt smell, is disengaged from it, and from this ammonia is formed by fixed alkali. We thus see that the common sap, when it passes into mucilage, changes its constituent principles, and that azote in particular is disengaged during this process. With respect to the acids which mucilage exhibits, they can scarcely be considered as present in the living plants, because they are first disengaged by means of mineral acids and distillation. However, as carbonic acid consists of 72 parts of oxygen, and 28 of carbon, the oxalic acid, which we find as a constituent part of some plants, may easily be derived from it; for this acid consists of 70 parts of oxygen, 26 of carbon, and 2 of hydrogen.

Mucilage passes into gum and starch. Into the former of these it passes only after it has escaped from the plant, when it is thickened by the influence of the oxygen of the atmosphere. It then allows the azote to escape, which forms ammonia, or the azote becomes united to lime, of which we find a considerable quantity in gum. Gum has no longer the same tendency to become putrid as mucilage.

In the same manner starch is produced, probably from a similar thickening of the mucilage, only this takes place in the living plant; because by the microscope we distinctly see the grains of starch, like a deposit in the cellular texture. Starch gradually passes, when mixed with water and exposed to the warm air, into the acetous fermentation, during which it becomes covered with mouldiness.

352.

Mucilage undergoes a further change when it passes into saccharine matter. As, according to pretty good authorities, saccharine matter consists of 74 parts of oxygen, 17 of carbon, and 8 of hydrogen, it is evidently formed by a higher oxydation of mucilage, and, at the same time, by a change in the proportions and quantity of hydrogen. The azote seems in this case to have entirely vanished, since, during the dry distillation, not a trace of it appears. From starch and mucilage is formed sugar by oxydation. The tendency of mucilage to oxydation favours the production of sugar. When Beet, therefore, is exposed to the powerful action of the sun's light, it affords less sugar, and it is accordingly covered with earth to withdraw it from the light. When sugar is treated with spirit of wine, it loses its sweet taste, and passes into the nature of gum. It is worthy of remark, that saccharine matter, dissolved in water, undergoes the vinous fermentation, which must arise from the disengagement of hydrogen, and from a change in its proportions. Sugar is produced in a dry and crystallised state, as a secretion of some plants, and in union with quartz; and in innumerable plants, it appears as a honey juice or nectar, but, in these cases, it is often united with the peculiar and even with the poison-

ous ingredients of the plants, as is shewn by the hurtful properties of the honey produced by bees from the *Azalea pon-
tica* and *Kalmia latifolia.*

353.

Azote being thus a constituent part of mucilage, it is
also, in an especial manner, evolved in albumen and gluten, so that these are considered as properly the ingredients
of animal matter. In the leguminous fruits, these are the ingredients which render the plants so peculiarly nutritious.
In the juice of the *Agave Americana*, which is the favourite
drink in Mexico, there is so much of these animal matters,
that this liquor smells strongly of putrid flesh. Albumen, in
moderate warmth, passes into a putrid decomposition, during
which sulphurated hydrogen gas and azotic gas are disengaged, and sulphur and phosphorus are generated. With
weak nitric acid, it disengages azotic gas, and in a stronger
heat it gives out cyanogen : by distillation we obtain carbo-
nated hydrogen, ammonia, and empyreumatic oil. It consists, according to the French analysis, of 52 parts of carbon,
23 of oxygen, 15 of azote and of hydrogen.

Gluten is only distinguished from albumen by its greater
consistence, and by this further circumstance, that it is insoluble in water, but moderately so in spirit of wine.

354.

We further find extractive matter, tannin, and colouring
matter, all existing as ingredients of plants. The extractive
matters are indeed very various, but they seem to be formed
from mucilage by a difference in the proportion of their
oxydation, and by a mixture of the peculiar matter of plants.
Extractive matters are insoluble in water, and they also so
far partake of the nature of mucilage, that they enable oil to
mix with water, and to be used as soap. They contain also so
much oxygen, that they redden the blue juices of vegetables,
and the bitter taste of most of these substances proceeds from
carbon oxydated to a certain pitch, and which has been de-

rived from hydrogen ; (Braconnot, in Berlin Jahrbuch der Pharmacie, b. 20. s. 205.)

Tanning matter is produced by such a change in extractive matter as renders it fit to coagulate animal jelly, and to cause a black deposition of salts of iron. There is also a dark green deposition of salts of iron, by means of the common extractive matter ; and tanning matter may be almost completely changed into extract, by repeatedly evaporating a solution of it, by means of heat. Caustic alkali also takes from tannin its astringent taste, and it is hence also probable, that a difference in the proportion of oxygen distinguishes those two substances from each other. This is also in part evident from the disengagement of the gallic acid during the mouldiness which takes place on a solution of the gall-nut ; because, during the production of this matter, the characters of tannin are lost, and that acid probably does not exist in the matter previously, but is produced during the change.

We have already stated, when speaking of the green colour of leaves, (317.), and of the tints of blossoms, (327.), that the colouring matters of plants are of very different kinds. They can sometimes be removed by water, and sometimes by spirit of wine ; very frequently they are united with a residuary powder, or a substance containing azote, which gives the putrid smell to the colours of many plants, and with a weak nitric acid even produces azotic gas.

355.

An important ingredient of plants is the mild oil, which we procure in its purest state from the fruit of the Olive and Beech, but which we also very often obtain from the cotyledons, and albuminous bodies in the seed. As, according to corresponding analyses, mild oil consists of seventy-five parts of carbon, and twenty-five of hydrogen ; as nature also prepares the same matter from the saccharine juices of fruits, and art can again change this oil into saccharine matter, by treating it with the oxide of lead, it is extremely probable that it owes its origin to a complete deoxidation of the saccharine matter, and to an increased proportion of hydrogen.

356.

Mild oil passes into a kind of wax, when it is slightly oxidated, and attains, by that means, a certain degree of firmness; for art can produce wax from oil by nitric acid, and can change it into oil again, by taking away the acid. Wax is produced by Nature herself, in the fruit of the *Stillingia sebifera*, *Rhus succedanea*, *Myrica cerifera*, and in the bark of the Wax-palm (*Ceroxylon Andicola*).

357.

Volatile oils, which are commonly united with extractive matters, are distinguished from mild oils, by containing a greater proportion of hydrogen than of carbon. Of the former substance, they seem to contain, in many cases, two-thirds, or even three-fourth parts. It is remarkable that they are chiefly found in the covers of seeds, and, in the case of the Scitamineæ, in the vitellus even, but almost never in the albuminous bodies, and in the cotyledons of the embryons. When volatile oils are oxydated, they run into a hard substance, which is called Camphor, and in which it has been supposed that a peculiar acid resides, although good chemists consider it as the same with the benzoic acid. This last substance is found in most of the balsams, or in those bodies which, beside volatile oil, contain also resinous extract. The strong smell of this acid, and its inflammability, as well as its weak affinity to the alkaline base, point it out as rich in hydrogen.

The volatile oils pass through the balsams into resin; because, by evaporation and the application of acids, we can procure these oils from resinous substances.

358.

We must now consider those bases in plants which can be united with acids,—the origin of which is involved in much obscurity.

With respect to the alkalies, it is certain that potassa is one of the most common substances to which the vegetable acids are united; but it is undoubtedly not drawn from the soil in the state in which it is known to us, but is prepared, by the organizing powers, from original materials, the nature of which we only conjecture, but cannot state with certainty. As ammonia is undeniably composed of azote and hydrogen, and is found especially in gluten and albumen, we may, without incurring the reproach of being hasty in our conclusions, assume the same ingredients for potassa and soda, only in quite different proportions, and probably with a certain share of oxygen, by which they are put in the state of oxydes. The latter substance, soda, is either drawn by plants from the soil, or it is first decomposed by them into its constituent parts, and with these the plants form salts similar to those which are found in the soil, (343.)

The increase of the quantity of potash in woods which have been subjected to putrefaction, and which was observed by Schreber, Wiegleb, and John, (John, Uber die Ernährung der Pflantzen, s. 156. 157.), favours the opinion, that alkalies are produced by the disengagement of hydrogen and azote.

359.

We must form the same opinion respecting the earths which plants contain. Lime is found not only in the ashes of plants, but in albumen and in gum : it cannot, however, be derived from the soil, because, as we remarked formerly (343.), it is found even in those plants which have grown in pure sand, or amidst granite, or which have been reared in porcelain vessels, and nourished only by carbonic acid water. If we should suppose, that still the pure carbonic acid water might contain dissolved lime, this is merely a supposition, which rests on no foundation of proof.

Nor is the production of siliceous matter in plants more easily accounted for. This substance is not only found in considerable quantity in the larger Grasses, which grow in boggy places, (*Arundo phragmites, Poa aquatica, Festuca fluitans*); it forms also the chief constituent of the saccharine matter which exudes from the joints of the Bamboo. But this siliceous matter cannot come from the soil, because it cannot be dissolved but by a red heat with alkalies. We shall not, therefore, be too hasty, if we suspect that plants form siliceous earth, from carbon and hydrogen, especially as the diamond, which is the most perfect of the gems, is almost entirely composed of carbon and hydrogen. Lime, also, as it is a principal product of the animal kingdom, and is found almost every where along with azote, is probably a compound of this latter body with carbon.

360.

With respect to the metals which we find in the ashes of plants, they may be reduced, so far as is known to us, to the following, namely, iron, manganese and copper. The last of these is found in the tubercles of the Scitamineæ. Iron is found in the ashes of almost all plants, and frequently united with oxide of manganese.

Although we know not the elementary bodies from which metals are formed, it is very probable according to Dobereiner's experiments, that carbon, in union with hydrogen and azote, suffers a change during the process of vegetation, which assimilates it to a metallic nature; perhaps this change has an influence upon the colours of blossoms, and by the power of heat, the metal in the organic body is transformed to a metallic oxide.

361.

When we proceed to the consideration of the more remote constituents, a wide field is opened to conjecture. Sulphur, phosphorus, the elementary matter of cyanogen, the fundamental principles of halogen, and iodine, are more or less related substances, which, by their great inflammability, and tendency to assume the gaseous form, attract oxygen from air and water, and thus form peculiar acids.

Sulphur exists in albumen, and in other vegetable juices; and sulphuric acid salts are found in the ashes of almost all plants. Albumen also contains phosphorus, and phosphorated lime, and salts compounded with the same acid are produced in the juices of many plants. The phosphoric light of putrid wood is a proof of the existence of this substance, and of its gradual oxydation by the admission of air. A. F. Mornay has observed a phosphorescent plant, of the family of the Contortæ, in Brazil, (Philosophical Transactions, 1818.) The nauseous smell which phosphorus gives out, and the existence of the same in the salts of cyanogen, favour the suspicion, that phosphorus is produced by a certain combination of hydrogen and azote. We can scarcely doubt of the existence of this combination in the base of cyanogen. We find this matter commonly in a slight state of oxydation, as cyanogen, there being scarcely 2 *per cent.* of oxygen in it, in the Bitter Almond, in the Cherry Laurel, in the Black Alder, and in various other narcotic plants, (Gay Lussac in Schweigger's neuem Journal.) Halogen or chlorine has the same disposition to become volatile, and the same tendency to assume the gaseous form. Its strong attraction to oxygen is the cause why we commonly meet with this substance as hydrochlorous acid, and why we find it in the ashes of plants, in combination with alkalies.

Iodine is very nearly related to this substance: it is commonly found as an oxyde in the ashes of sea-weed; it smells like chlorine, and evaporates, when exposed to heat, with a purple colour, and a smell of chlorine. It does not affect the colours of plants either as an acid or an alkali : it is most soluble in alcohol and ether, but it takes from water, as do also chlorine and azote, nearly two parts and a half more of oxygen than its own weight. Even the highly poisonous nature of iodine, which has been lately remarked, favours the idea of its being a compound probably of hydrogen and azote.

362.

There are some other newly discovered matters in plants, which may be arranged more or less with those. At least, there are some bases of acids, which are probably similar combinations of the elementary matters, as those which form the alkalies. Morphium, which is united in opium with the meconic acid, proves its affinity to alkalies by this circumstance, that it restores to their former colour the blue vegetable juices, after they had been reddened by acids. Besides, it shoots out into fine needle-formed, or prismatic crystals, and is scarcely soluble in hot water, but very much so in alcohol. Of precisely the same nature is Strychnin, which Pelletier has found, by means of corrosive alkali, in the crow's eye, and in Ignatius' beans. Many strong-tasted matters owe their origin, probably, to the same element; (Annales de Chimie, tom. x., 1819.)

Other peculiar substances have quite different relations, although they stand in the same class with that we have now considered, as being the bases of acids. Cinchonin, which is found in the chinarinde, combined with the acid of cinchona, polychroit in saffron, and inulin in the Alant, are but individual examples of such matters, being formed by different proportions of the same elements of plants.

363.

From all these facts we conclude, that the simple elementary bodies which plants take up from the soil, are united and separated again in the most varied and multiplied proportions, and that it is in this manner that all the more intimate constituents of plants are produced. If, with the view of illustrating this conclusion, we attend to the relations of the vegetable acids, we perceive that the further vegetation has proceeded, there is less oxygen, and more hydrogen and azote in the acids. We remarked formerly (351.), that the oxalic acid, next to carbonic acid, is the richest in oxygen. To these succeeds the tartaric acid, which we find in a free state in the pith of the Tamarind, and in other plants, combined with a base. It is distinguished from the oxalic acid only by 4 *per cent.* additional of hydrogen, and 1 *per cent.* less of oxygen. The citric acid, which is found in great quantity in orchard fruits and berries, contains as much hydrogen, 9 *per cent.* more of carbon, and 10 *per cent.* less of oxygen, than the tartaric acid. The malic acid is still richer in carbon, but less plentifully stored with oxygen, on which account it is possible, by means of nitric acid, to change it into oxalic acid.

———

CHAP. III.

PROPER PHYTONOMY, OR ON THE LIFE OF PLANTS.

I. *Effects of Stimuli.*

Mustel, Traité Theorique et Pratique sur la Vegetation, Vol. I. 4. Rouen, 1781.

Sennebier, Physiologie Vegetale, Vol, I. 5. A Geneve, 1800.

P. Keith's System of Physiological Botany.

G. R. Treviranus's Biologie.

R. Sprengel, Von dem Bau und der Natur der Gewachse.

J. Sennebier, Experiences sur l'Action de la Lumiere Solaire dans les Vegetaux.

Linné, De Somno Plantarum, in Amœn. Acad. Vol. IV.

J. Hill, The Sleep of Plants.

Hunter, Versuche uber das Vermögen der Pflantzen und Thiere, Warme zu erzeugen.

Al. Wilson, Beobachtungen uber der Einfluss der Klima's auf Pflantzen und Thiere.

J. Ingenhousz, Vermischte Schriften.

Bertholon de S. Lazare, Uber die Electrizitat, in Beziehung auf die Pflantzen.

Saussure, Recherches Chimiques sur la Vegetation.

Plaudus Heinrich, in Hermbstadt's Archiv fur Agricultur-Chemie.

Balde, in Laurop's Annalen der Forstkunde.

Salome, in Hermstadt's Archiv.

V. Marum, in Rozier Journal de Physique.

364.

Plants live, not merely in the common sense of the word, which includes activity of every kind, but in that stricter sense, by which a higher and self-dependent activity is expressed.

If we compare natural bodies with one another, we find some which are produced by an attraction of primitive matters, and which are destroyed by a process of the same kind.

But others are reared by the internal powers of their own constituent parts,—occasion, by means of motions, which mechanics and chemistry have not yet explained, the ascent of the sap, its resolution, and an evident change of their own external proportions,—and even extend their agency beyond the bounds of their own existence, by propagating themselves. These three operations—the assimilation and maintenance of the sap,—activity arising from internal impulse,—and the production of new individuals, are the characteristics of the life of plants, which we have partly already considered (343.), and which we have partly yet to illustrate. At the same time, we must pay some attention to the slight traces of sensitive life, which is the highest degree of vital activity in plants.

365.

The vital activity of organized bodies is excited by stimuli, and receives the name of Irritability, since we cannot explain its effects either upon mechanical or chemical principles; for these last admit neither of extension nor contraction of parts, —of no remarkable attraction nor removal of matters,—and we consider the power of producing these changes, therefore, as one of a higher kind, than the lower powers with which mere matter is endowed. The explanations which we give of this peculiar power we term Dynamical: yet it is impossible to determine sometimes, whether a particular phenomenon is to be explained purely on dynamical principles, or whether it may also be illustrated by mechanical and chemical means. It cannot be denied, that the recent discoveries of chemistry, respecting many operations of Nature, have also given more light to the science of plants, than was formerly communicated by dynamical explanations. In particular, we are indebted to the introduction of the science of imponderable bodies, for a more connected view of many vegetable processes, than we formerly possessed. But still these imponderable substances act partly dynamically, at least contrary to the laws of gravity, and not always according to the laws of chemistry.

3

366.

Light is the principal stimulus for all living bodies, and for the vegetable world in particular : without its agency no plant grows, nor continues to exist. We have already considered (317.), the chemical attraction of light for the oxygen of plants, as a very important phenomenon, and have also remarked, that the consequences of this attraction of light for oxygen do not take place without the exertion of the internal activity of the plant, and we must therefore admit a real contest between light and the power of the plant. The light of the sun, however, is not a material substance, but a pure principle of activity, because, without any perceptible loss of time, it traverses the immeasurable extent of heaven,—because it penetrates even through space that is deprived of air,—and because it every where acts in opposition to the laws of gravity. It operates, therefore, on plants, simply as a vital stimulant, and excites all the functions of the plant to such a degree, that all its secretions are more powerfully performed ; on which account even corn is not so nutritious in wet seasons, when there have been many tempestuous days. It is hence that branches and leaves push towards the light. We hence also remark, that plants, which have long been withdrawn from the light, suffer an exhaustion of their powers, become white, and wither, when light has suddenly been admitted to them. Nay, in one instance, (Cludius in der Garten Zeitung, b. i. s. 386.), after a long removal of light, an evident quaking of the plant was observed, as if it had been agitated by the shivering of a fever, and even a leaf of the *Dionæa muscipula*, which has been torn off, makes repeated attempts to open itself to the light. These attempts are seen in undulating movements of the marginal cilia, during which the surface of the leaf is successively opened and shut, until it entirely opens, and thereby loses its excitability ; (Nuttal, Genera of North American Plants). It seems even to be established by ancient and modern observations, that the light of the moon has some influence on plants ; (Wilson, Uber der Einfluss der Klima's auf Pflantzen und Thiere, s. 16.)

367.

The periodical change in the direction of leaves, which has been called the Sleep of Plants, is undeniably connected with this stimulating operation of light. It is established, that during the clear light of the sun, the leaves become erect, and move their upper surface to the light, whilst, on the contrary, during the absence of light, they either hang downwards, and turn to the horizon, or they take an upright position, so that the under surface of the leaves is turned more outward. On account of this particular position of what have been called Sleeping Plants, we cannot properly ascribe this direction to sleep, because the leaves do sometimes even raise themselves during this state with greater energy, and press upon the stem or leaf-stalk, for the purpose of turning their lower surface outwards. This change is much rather, therefore, the consequence of the contest we have already mentioned, between the activity of the plant, and the great activity of nature. This change is the more evident, and the sleep of leaves the more striking, the finer and more compounded the organization of the leaves is. We hence most frequently observe it in the pinnated leaves of leguminous plants, although also in some others, as in Atriplex.

That an internal and self-dependent activity is to be taken into account in this sleep of plants, is plain from the fact that this sleep does not equally follow from a short withdrawing of the light, but only from its complete and long continued removal; as also from this other circumstance, that leaves fall asleep or awake at fixed hours, whether the sky be serene or troubled, exactly as happens with regard to animals. Other stimuli, too, and especially heat, have a great influence upon this phenomenon, because, in the cold, leaves awaken later, and fall more easily asleep, notwithstanding the influence of light.

368.

Blossoms also experience a similar change from the influence of light. Most of them unfold themselves in the morning at stated hours, and again close in the evening, (*Blumen-Uhr*). A considerable number, especially the Oxa-

Q

lidæ of the Cape, and the Mesembryanthema, require a very
warm sunshine, and the full light of day, to unfold themselves. Bory St. Vincent forced them to open themselves,
by means of a powerful light artificially produced by burning
glasses, (Ann. Gener. des Scien. Phys.) Many again only open
themselves during the hours of mid-day, as the *Mesembryanthemum noctiflorum, pomeridianum, dolabriforme,* and *stra-
mineum.* Some others seem to be too tender for the strong
stimulus of the clear light of the sun: they hence open only
during the evening hours, and shut again when the sun has
attained a certain elevation. Of this description are many
Silenæ, Œnotheræ, the *Cactus grandiflorus* and *triangularis.*

369.

Light displays its stimulating influence also upon plants
that are beginning to germinate. But necessary as it is for the
springing plant, it is equally injurious when it is permitted
to exercise too powerful an influence, especially upon very
fine seeds and unusually tender plants. The seed of Ferns, of
the Ericæ, and some others of a similar nature, only germinate,
when, with sufficient warmth and moisture, they receive as
little of the sun's light as possible. For larger seeds, and more
hardy plants, the clear light of the sun is not too powerful a
stimulant.

370.

The second common stimulus which the life of plants requires is Heat, which is as little a material substance as light,
since, like it, it operates contrary to the laws of gravity, dif-
fuses itself in rays on all sides, and passes through the Torri-
cellian vacuum. Without a certain proportion of warmth,
no plant can spring or thrive: this principle exalts the activity
of plants, and when it is very powerful it exhausts their
powers, as it does also those of animals. Its long absence renders plants as susceptible of a new excitement as animals. It
operates most advantageously upon the vegetable world, when
there is a regulated change of temperature, and when thus

cool days are interchanged with warm weather, whence we
account for the luxurious vegetation of alpine regions and of
tropical climates, where a considerable coolness during the
night succeeds the heat of the day.

371.

It is worthy of remark, that seeds and bulbs most perfectly
withstand the cold, because they suffer no considerable increase of susceptibility, by the withdrawing of the vital stimulus during sleep. Generally, the effects of heat and cold
in plants must be considered in connection with the action of
their internal powers. As the powers of plants are connected
with a certain organization, no plant will accommodate itself
to a certain temperature, unless its organization permits this
change. We can transplant vegetables from foreign countries
into our own; but we cannot, by centuries of cultivation,
force plants from warmer regions, to become perennial in
colder climates, as the common examples of Cucumbers,
French Beans, and Potatoes, shew. It is believed respecting
many plants, that by custom they become more hardy and
agree with our climate ; but this is a deception which proceeds
from our want of information respecting the nature of many
regions, which, on account of their latitude, we presume to be
very hot. It is true that we have partly accustomed to our winter some Japanese plants from the fortieth degree of north latitude ; but Japan contains mountains, and lies east from Asia,
where the temperature is always lower than in western regions. The Japanese plants, too, are not more hardy than
those from the mountainous districts of America, which lie
under the same degree of latitude, (391.)

372.

Another very important stimulus for the whole vegetable
world is presented by electricity. Obscure as the science of
atmospherical electricity still is, we yet know with certainty that
the influence of electrical excitation, and the discharge of positive electricity, is one of the most powerful stimulants to
the vital action of plants. We perceive this from storms.

Even the blighting of corn in low lands, by means of power-ful storms of lightning, shews the operation of the exhausting stimulus, which the discharge of atmospherical electricity af-fords. In the same manner Buck-wheat fails to be produc-tive when it has been exposed to lightning ; (Thaer, Grund-satze der Rationellen Landwirthschaft.)

We cannot state, in opposition to this, that artificial elec-tricity from a machine produces little effect on very fine leaves, when they are disposed to sleep, because this change of state is one of those vital phenomena which depends principally upon the effects of light. It is certain that seeds germinate more actively, and that branches put forth buds more early, when they are electrified.

373.

The kind of electricity which is excited by two bodies of different susceptibilities of oxydation, with intervening moist-ened conductors, and which we call Galvanism, or the Gal-vanic Activity, has a powerful influence upon the whole of na-ture, particularly upon organized nature, and, therefore, upon the vegetable world. Indeed, neither simple galvanism, nor the voltaic pile, has any remarkable influence on the motions of sensitive plants ; but it has been proved that still the germina-tion of seeds is very much promoted by the positive pole of the voltaic pile, when no extraordinary discharge takes place. On the contrary, the negative pole of the voltaic pile exhausts in every case the vital activity of plants.

We are more generally concerned, however, with the pro-cess of natural galvanism, which incessantly takes place in every organized body, where opposite principles are evolved, where an excitation of the elementary matters takes place, and where layers of different capacities of oxydation are con-nected by moistened conductors. In all these cases, that ac-tion takes place which we call Galvanic, and produces effects, which chemistry can then only in some degree imitate, when she avails herself of artificial galvanic batteries.

The organic perspiration through impervious coverings, which we have often spoken of, is a vital function, which Wol-laston has successfully imitated at the pole of the voltaic pile.

The wonderful combination of primary matters, and the changes which are produced in ingredients derived from the soil and from water, can only be explained by this influence of the galvanic power, and it is in the same manner that we daily perceive more distinctly that wherever heat assists in the solution of bodies, electricity is produced, as in every chemical process the electrical relations are changed. Art has already imitated, by means of the voltaic pile, many of these operations of nature, by proving that carbon, colouring matters, alkalies, and earths, are but oxides of metals, which distinctly resume their metallic nature at the negative pole of the voltaic pile. Finally, even in the operation of impregnation, there is undeniably an electrical or galvanic process, which is evolved by the elementary contraction of opposite substances, as ancient observers of the physiology of vegetables had observed.

374.

Oxygen is one of the most important and common stimulants in the vegetable world. We have already (314. 329.), seen that plants attract oxygen, and it is certain, that in the germination of seeds the influence of this body is of considerable importance. It has, indeed, been long known, that salts, lime, and weak acids, powerfully promote germination,—that seeds, which have long been laid by, are forced to unfold themselves by hydrochlorous acid,—and even that an excessive excitement is produced by too powerful an application of these salts and acids. For experiments have made it obvious to sight, that plants, treated with these acids, spring up speedily indeed, but on account of their too speedy and active growth, are destitute of any proper power of increase, and that they hence easily decay without bearing fruit. This experiment, which has been confirmed by Lampadius in particular, shews, that oxygen is not properly a nourishing substance, nor becomes assimilated to the plant, but that it only increases the vital activity in the capacity of a stimulus; (Hermstadt's Archiv der Agricultur-Chemie). This, also, is further evident from other experiments, in which we see that almost as much oxygen is either exhaled or employed in the composition of the carbonic acid which is exhaled, as had been taken into the plant. It is also evident from the fact, that lime, in combination with sulphuric acid, as we find it in gypsum and marl, properly communicates no fertility to the soil, but only a stimulating power; on which account, marl, as a manure, must always be interchanged by animal manure, if we would not render the soil altogether barren; (Mogelinsche Annalen). Free acids in the soil, especially when the latter stands commonly under water, are rather hurtful to vegetation, since the extractive matter of such soils is not dissolved; (Thaer, Grundsatze der rationellen Landwirthschaft). On this account, the mud which is taken from streams and pools, must first be mixed with lime, if we would assist by it the fertility of the soil.

375.

Experiments have shewn that the other stimulants act upon plants as well as on animal bodies. Arsenic destroys the irritability of plants and germinating power of seeds. Opium, too, acts on plants, although in a different manner from the poisonous body we have last mentioned, which exalts the susceptibility to an extreme degree, but entirely destroys the power of action. Opium, on the contrary, exhausts the susceptibility, by increasing the active power. Cherry-laurel water, and Ticunas poison, bring plants speedily to a state of decay, and take from them their irritability.

Hydrogen gas seems, on the contrary, to act less injuriously on plants than on animals, as it has also been found, that plants with juicy and green leaves live entirely uninjured in azotic gas, and exhale carbonic acid gas. But they die when we take from them, by means of lime water, the carbonic acid gas, which seems to be constantly changed by them into a source of nourishment.

II. *Other Proofs of the Higher Life of Plants.*

Brugmans et Coulon, de mutata humorum in regno organico indole a vi vitali vasorum derivanda.

Carradori Sulla Vitalita delle Piante.

C. Bell, in Memoirs of the Society of Manchester.

V. Maurum, Diss. qua disquiritur, quousque motus fluidorum et ceteræ quædam animalium plantarumque functiones consentiant.

Medicus in Act. Theodoro-palat.

Calp. Fr. Wolf, Theoria Generationis.

Patr. Blair, Botanical Essays.

Seb. Vaillant, Discours sur la Structure des Fleurs.

Lefebure, Experiences sur la Germination des Plantes.

C. Richard. Analyse du Fruit, consideré en general.

L. C. Treviranus, Von der Entwickelung des Embryo.

376.

We find other proofs of the higher life of plants in the ascent of the sap, and in the distribution of it through all the organs.

It is impossible to overlook the fact, that the sap-vessels, on account of he similarity of their structure to that of hairtubes, are chiefly appropriated to the purpose of raising the sap to a certain height, (276.) But were the sap-vessels simply hair-tubes, it were impossible to explain why the sap in them rises constantly to a greater height, since the hairtube retains the fluid which it has once taken up, at one fixed height. It would also be impossible to explain why light has so powerful an influence upon the ascent of the sap, since fluids, in hair-tubes, stand as high in the dark as in the light of the sun.

Neither heat, nor the pretended hygroscopical nature of the vessels of plants, can altogether explain this phenomenon; because, were the ascending sap attenuated by heat, and were the upper parts of the sap-vessels rendered, by the same means, more fitted for receiving the attenuated sap, the heat of the upper aerial strata must always be greater than that of the earth around the roots. But in hot-beds, and in hothouses, we only bring plants to a lively growth, when we render the earth about the roots much warmer than the strata of air above the earth; and it is not possible, therefore, to explain the ascent of the sap from this attenuation by heat. The hygroscopic nature of the vessels of plants is scarcely worthy of a refutation, since the moisture, which constantly surrounds them on all sides, directly contradicts this explanation.

There remains nothing further, therefore, than to assume a higher power in the sap-vessels, which being excited by the sap and strengthened by light, heat, and electricity, and being also probably favoured by the elastic air, or by the exhalations which are produced in the spiral vessels, forces the fluids upwards. The periodical change of the ascent of the sap, without any stated change of the external influences, must also lead to this idea, since the repeated motion of the sap in our trees, during, perhaps, a similar temperature of the atmosphere, and the similar repeated ascent of the sap in tropical trees, and in those which are reared in hot-houses, completely establish this opinion, (298.)

377.

To confirm our conclusion respecting the higher life of plants, we must further remark, that they are capable of resisting the external influences of heat and cold,—although many of the phenomena of this kind can partly be explained by chemical effects. It is certain that plants in warm baths, the temperature of which is from 150° to 180° of Fahr., and that others on the brink of the margin of volcanoes, grow briskly, where the air is warmed above the boiling point. On the other hand, we see a great many trees in the polar zone resist a cold which is from 20° to 25° of Fahr. We see that the internal temperature of trees in winter is always higher than the temperature of the atmosphere,—so much so, that this internal temperature of some trees seems never to fall below 52°, and never to rise above 75°.

If we reflect on the manner in which plants and other living bodies withstand external cold and heat, we shall find that this is a necessary consequence of their internal activity. By means of this, evaporation must go on incessantly so long as the plants are in leaf. In consequence of this function, the heat that previously existed in a free state is employed in the maintenance of the evaporation, and a diminution of the degree of temperature must thus of necessity arise. To explain the power by which plants resist cold, we might refer to the production of an internal heat from the transition of fluid into solid parts, by which means heat must always be disengaged; but this would be to avail ourselves of a process of nourishment, which is speedily accomplished, and of which we can scarcely take advantage in winter. We must therefore recollect, that the roots of trees stand in a layer of earth, which is not easily freezed, especially when it is covered with snow; that the sleeping life of plants in winter indures the removal of stimuli as easily as that of animals, who, during their winter sleep, are surrounded by very cold air or by ice. We must recollect, that when the thermometer, held in a hole bored in the stem of a tree, rises during winter, this is a natural consequence of the slow transmission of heat through the rind filled with resinous sap, and that, besides, snow around living trees does not dissolve sooner than around dry stakes.

The production of an internal heat in plants is therefore at least very doubtful; and we must as yet limit it to the rare cases in which a considerable heat is generated, namely, in the spadix of the *Arum* species, and, according to one report, also in the *Pandanus* when it flowers; although this phenomenon proceeds less from any preponderance of vital activity, than from a process which is truly chemical, and which relates to the evolution of elementary bodies in the blossoms; (Sennebier, Phys. Veget.; Hall in Bradley's and Adams' Med. and Phys. Jour.; Bory St Vincent in Ann. Gener. des Sciences Phys.)

378.

We come now to another vital phenomenon in plants, namely, visible motions. As, however, there are a multitude of brisk and often of wonderful movements, which may be explained by mechanical or physical arrangements, we are led to inquire to what class we must refer these peculiar vital motions. Now, when we attend, in the first place, to the absence of mechanical arrangements,—next to the fixed order which occurs in these movements,—and, lastly, to the exciting cause, which operates by irritation,—we shall find it difficult to consider this kind of motion as of any other than a dynamical nature.

When we see the antheræ, which previously were included within the concave parts of the corolla or calyx, suddenly and powerfully raise themselves up and scatter their pollen, we can only attribute this to the relaxation of the contraction of the parts. When we see fruits or capsules spring up with yet greater elasticity, or scatter their seeds, we find upon examination that this also is a consequence of the preceding contraction, which in the Acantheæ, for example, is occasioned by peculiar hooks on the dissepimentum ; (Tab. I. Fig. 37.)

To the same class we refer other phenomena, which, in our opinion, are falsely considered as effects of higher powers. The springing back of the sexual parts of the *Medicago* species on the vexillum, by the contact of the carina, is as certainly a consequence of the excited elasticity of the parts, as the elevation of the fruit-stalk of the Stylidiæ, which was previously pressed into double curvature, and was held in a state of contraction by a peculiar pointed petal, until, after the complete evolution of the blossom, this petal relaxes, and thus leaves the fruit-stalk at liberty, which then stretches and raises itself up, although it still continues curved ; (Fred. Bauer, Illustr. Nov. Holl.) The irritability of the arched margin of the corolla of the Leeuwenhoekia, R. B. is to be explained in the same manner. An irritability has also been ascribed to the parts of the glume or the *Leersia lenticularis*, Mich., which is known in North America under the name of the Fly-catching Grass ; but it has been lately shewn, that the notched cilia of the valves detain the proboscis of flies, when they stick in them, in a manner entirely mechanical ; (Nuttall, Genera of North American Plants.) In our *Drosera* also, as in the *Roridula* of the Cape, the catching of flies seems to be merely a consequence of the sticking of the insects to the glands of the leaves.

We cannot determine with certainty whether the opening and shutting of the lid of the water-bladder in the *Nepenthes distillatoria*, *Cephalotus follicularis*, and of the hollow leaves in the *Sarraceniæ*, is merely the consequence of a mechanical arrangement. In the *Cephalotus* at least there is a ring on the opening of the water-bladder, with numerous rib-shaped processes, to which we might ascribe a function similar to that of the elastic ring of the capsules of the Ferns.

Finally, the alternate approximation and separation of the cilia in the setting of the orifice of the Mosses, is a purely hygroscopic phenomenon, since every change in the quantity of vapour contained in the atmosphere sets this alternation a-going.

379.

But what, without doubt, are automatic motions, or such as we cannot account for by mechanical causes, but by higher powers, are, in the first place, the movements of fresh-water Oscillatoriæ during their growth. By the microscope, we observe these motions under the influence of sunshine so striking during the rapid increase of the bodies, that we can scarcely refrain from supposing, that we perceive a trace of animal life, or from considering these bodies as holding a middle place in the two organic kingdoms. Equally striking are the movements of the *Hedysarum gyrans* during its rapid growth. It is the smaller side leaves, however, which incessantly exhibit this circumvolving motion. Even the power of detaining flies which belongs to the *Dionæa muscipula*, may be ascribed, notwithstanding the wonderful mechanical means for producing it, to the irritability which, from other proofs, we know to belong to the leaves of that plant.

The collapsing of the leaves of what are called Sensitive Plants, which belong to the species Mimosa, Schrankia, Æschynomene, Averrhoa, Oxalis, and Smithia, is, notwithstanding all earlier attempts to explain it mechanically, an indubitable consequence of irritability ; and this is the more evident that in one of them, the *Averrhoa Carambola*, it is not the leaves which were touched, but those placed opposite to them, which collapse and fall down.

380.

But the most important proof of the higher, and even of the sensitive life of plants, is derived from the events that occur during fructification. The regular order in which the antheræ free themselves by degrees, and indeed one after another, from their pollen, is a phenomenon which we cannot well account for but by the admission of higher powers. This regular order is seen most distinctly in the Garden Rue and the *Parnassia palustris.*

The splitting of the two-lipped stigmata in the *Mimulus* and *Gloxinia* before fructification, and the quick shutting of the two lips, the moment that a particle of pollen is placed upon the inner surface of the stigma, also announces the existence of a susceptibility which does not require to be much exalted, in order to pass into the feeling of an animal body. But we must also take into account the consequences of the increase of the plant,—of the gradual unfolding of its parts, and of dichogamy ; because in the Malvaceæ, as well as in the Syngenesious plants, the late appearance of the stigma, and its prolongation above the filaments, are to be ascribed to the latter causes.

381.

Finally, we think we shall be able to prove, that the whole process of fructification is truly a dynamical operation, in which every thing depends on the excitement of a new life in the germen.

The times are past when previously formed embrya were supposed to exist in the ovula of the unimpregnated germen. This previous existence of embrya is contradicted in the strongest manner, not only by the production of hybrid plants, when two different kinds are forced together for the purpose of artificial impregnation, by which means the produced plants exhibit the combined characters of both the parents, (Jos. Gottl. Kœlreuter, Vorlaufige Nachricht von einigen das Geschlecht der Pflantzen betreffenden Versuchen. Gallesio, Theorie der vegetabilischen Reproduction) ; but also by the changes of form which a plant undergoes by reproduction, and still farther by the difference in the number and insertion of the unimpregnated ovula, from that of the impregnated seeds, and by the constant failure of the former, when they are not awakened by the pollen into new life.

We might indeed suppose, with the natural historians of the early part of the eighteenth century, that there is a direct passage of the pollen into the ovary, and a material union of this substance with the ovula; by which supposition, we should explain the germination of the seed entirely on atomical principles, (Sam. Moreland, Phil. Trans. vol. xxiii.) ; but the following reasons lead us to reject this supposed direct passage. In the first place, there are no visible canals through which this passage could take place, except the sap-vessels of the pistillum and the spiral vessels, both of which connect the germen with the stigma. But sap-vessels and spiral vessels perform quite different functions, as we have already seen.

In the next place, the surface of stigmata, as we have already remarked (340.) is covered with shut warts and hairs, which equally prevent this direct communication.

It must also be noticed, that in the Labiatæ, Asperifoliæ, some of Urticæ, and Rosaceæ, the pistillum only directs itself sideways towards the ovula, without appearing to stand in direct communication with them; and that in the Contortæ the pistilla are covered by a shield-shaped surface resembling the stigma, without being directly united to it. Finally, we must allow its privilege to analogy, which informs us, that in the lower animals, partieularly in the Molluscæ, impregnation is accomplished without the passage of any material substance from the impregnating organ into the receptacle of the seed. In this case it is purely a galvanic process, in which, by means of a chain of different organs lying together, the vital activity is unfolded to the degree necessary for the production of new individuals. A process exactly similar takes place in the vegetable kingdom. The pollen, brought into contact with the stigma, awakens in it, as well as in the germen, a new life. New secretions and depositions of organising matters take place, by means of which new forms are produced corresponding to the pattern which properly belongs to each plant.

That electricity performs an important part in impregnation, has long been suspected ; and the contraction of electrical matters in the blossom, and in the parts of fructification, seems to favour this idea.

382.

As we thus consider the stimulus of the pollen to be a necessary condition of the evolution of seed from the ovula, we must at the same time defend ourselves against both the ancient and later objections to this doctrine, and must be prepared to encounter the reproach of having taken a partial view.

It is true that diclinous plants often seem to bear fruit without having been impregnated by pollen. Not only ancient pretended observations respecting the setting of the fruit in the female hemp plants, after the male plants had been taken away, but more recent experiments, which have the appearance of having been carefully made, have bred a doubt respecting the necessity of impregnation by pollen to the evolution of the seed (Spallanzani, Fisica animale e vegetabile.) Supposing these experiments to have been always made with indisputable exactness, fidelity, and care ; supposing also that perfect seed has been obtained from female plants of Spinage, Hemp, and Mercury, which were completely isolated, the reason of this is to be sought in the frequently androgynous nature of diclinous plants. For Gleichen has shewn respecting Spinage, Schkuhr respecting Mercury, and Kastner respecting Willows, that female plants often bear hermaphrodite blossoms, at least that they are androgynous. Something similar must take place with respect to Hops ; for when only female plants are set, male plants are found to have risen among them ; (Thaer, Grundsatze der rationellen Landwirthshaft.)

383.

If we examine still further the changes which take place in the germen after impregnation, we find this idea of a new life being awakened, confirmed, in the first place, by the swelling which this organ exhibits at the expense of the other parts of fructification, and their coverings.   These last decay and fall off.   The ovary and the receptacle, and in *Hovenia dulcis* the fruit-stalk, begin to swell ; and the ovula, which before were only simple vesicles or spherical cells, being now filled with pure water, begin to undergo wonderful changes.

The skin of the seed begins to thicken by depositions from the organising fluid.   After some time, which cannot exactly be defined, we are able to distinguish a double covering of the seed ; the exterior, which from analogy is called the *Chorion*, and the interior, which is denominated the *Amnios*. The latter, which is for the most part full of a sweet and slimy fluid (*liquor amnii*), shews after some time a small point, either swimmimg or fixed to the side of the vessel, which is the first trace of the embryon.   It seldom occurs that we find more than one embryon in the same ovulum, although this is observed in the Agrumæ and in Fuchsia.   It still more seldom happens that no embryon is found in a proper seed, but that it first makes its appearance in the shoot.

The farther evolution of the seed is different according to the type of its structure; that is to say, the embryon either does not increase, but remains unevolved, and continues to be like a point, a line, or a fungus.   In that case, it does not consume the moisture appropriated to the nourishment of the germ, but this matter becomes thickened by the absorption of of its more volatile parts, and passes at last into the nature of albumen.   When this occurs, as in the Grasses and Scitamineæ, we can discover the scutellum or the vitellus appearing as an instrument of evolution.   In other instances, the embryon becomes thickened at one end, by which also it attaches itself to a substance, which holds the place of the cotyledons, and is called the Cotyledonous Body.   In some of the Naiadæ and Zamiæ, as well as in the Pine tribe, this body incloses the germ within itself.   By the germ it is itself divided, and

2

thereby its transition to the true cotyledons becomes evident. According to the observations of L. C. Treviranus, the thickened cotyledonous termination of the *Ruppia* has the structure, and supplies the place of the albuminous substance; (Von der Entwickelung des Embryo.)

In most of these albuminous plants, the cotyledonous body is first formed along with the germ. The Palms, for instance, the Liliaceæ, Junceæ, and Sarmentaceæ, send out from the seed, along with the germ, a horizontal thread, which becomes thickened into a tubercle, and from which the root proceeds downwards, and the plant upwards. A great many families of plants retain the thickened fluid of the germ, as albuminous substance, and yet the embryon is evolved with its cotyledons. We observe this in the Umbellatæ, Polygoneæ, Nyctaginæ, as well as in the Caryophylleæ, in which the albuminous substance remains in the middle, and the embryon is placed around it. In plants of the higher orders, however, it commonly happens that the evolution of the embryon with its cotyledons is performed so much at the expense of the albuminous substance, that this substance is either entirely consumed, and becomes one with the chorion, or there remains only a small trace of it.

384.

Although the chemical changes in the germ are of very high moment, yet this process cannot otherwise be fundamentally explained but in a dynamical way.

The object of first importance in germs is their vital activity; and this in many plants dies so speedily, that acorns and coffee-beans cannot be preserved above a few months, without losing their power of germinating. On the other hand, this power is retained in many other seeds for an amazing length of time, especially when light and air are excluded; and it is from this cause that we must account for the otherwise incomprehensible phenomenon, that the bottom of dried pools, or earth which has been stirred to a great depth, produces plants, which do not make their appearance for many miles round; (Thaer, Grundsaze der rationellen

Landwirthschaft.) The peculiar power of plants seems even to be increased by this long keeping of their seeds, apparently because, by the drying up of the albuminous substance and of the cotyledons, the constituent parts become more concentrated and powerful. We hence choose old seeds of melons, for the purpose of obtaining a rich crop of the best flavoured fruit. Old Lint seed also gives commonly the finest and best Flax.

It is a further consequence of the different varieties which the internal power of seeds exhibits, that many seeds germinate very early, and others very late; for we cannot find a sufficient foundation for this difference, either in the structure or in the apparent composition of the parts of the seed. While the Umbellatæ, Rosaceæ, and Proteaceæ, often do not appear till after two years, most of the Grasses, Cruciform, and leguminous plants, on the contrary, germinate in a few days.

385.

There is also a series of chemical changes, connected with the germ, by means of which, as a sort of conditional causes, the vital power of the germ is awakened. In the first place, the seed attracts carbonic acid water through its umbilicus, by which means a swelling of the albuminous substance or of the cotyledons takes place, and an evident effervescence is begun. During this latter operation, carbonic acid gas is disengaged, and hydrogen is in part set free. The external heat, the regulated influence of the light of the sun, and the oxygen of the atmosphere, are the stimuli which now awaken the life of the germ, and enable it to make use of its proper nourishment, the carbonic acid water impregnated with azote, for its full evolution.

If the embryon remains still undeveloped in the seed, more arrangements and preparations must be had recourse to, before it will completely unfold itself. In many of the lower plants, the necessity of these preparations consists in this, that the awakening life, directed by no fixed original type, produces fluctuating forms, which have often no resemblance

R

to the type of the family or species. The germinating mosses thus produce conferva forms, which often remain for a long time after the young Moss has attained its full size ; (Nees, de Muscorum propagatione Dissertatio.) Germinating Ferns and Algæ produce lobed cellular forms, which remind us of the structure of the Musci hepatici, and which are improperly named Cotyledons, since the existence of these cannot be suspected or proved in the seeds of these lower organic bodies. We have formerly shewn, that in the Naiadæ and in some other families, the thickened cotyledonous end of the germ, and, in the Palmæ and Junceæ, the lateral tubercles, furnish the apparatus, by the help of which the further evolution of the embryon takes place. In the Scitamineæ and in the Grasses, the vitellus or the scutellum are the organs by which the sap, when prepared, is conducted to the embryon, and by which its evolution is favoured.

We must here also, as in most of the lower families, take into account the sheath of the root, or the warty prominences from which the radicle first proceeds, and which are equally instrumental in preparing the evolution of the germ.

These warts, or club-shaped appendages, attract also, in higher plants, the moisture of the earth, and push the radicle from the umbilicus. This is always the first external appearance which indicates the germ. The distinction which Claud Richard makes between Endorrhizes and Exorrhizes, is in so far correct, that in the former, the entire embryon not being developed, the roots are first formed from the warty substances ; while the latter, on the contrary, possess a root already formed ; but this also urges on the roots and warts, by which it is succeeded ; (Link, Anatomie der Pflantzen.)

Plumulæ and radicles are separated by what is called the Knot, which constitutes the partition between the descending and ascending motion. From the knot sap-vessels pass into the cotyledons. These take up the sap from the radicle, and prepare it, by means of the processes for respiration, with which they are furnished by the slits in their epidermis. From them the sap proceeds, apparently through the same sap-tubes, but at different times, back again to the upper part of the knot, and rises from thence into the plant. We thus perceive, that in plants of the higher kinds, the cotyledons are a necessary instrument of germination and growth ; on which account, when both the cotyledons are cut off, the plant of necessity dies.

386.

The distinction which has been made, chiefly since the time of Ray, between the Acotyledonous, Monocotyledonous, Dicotyledonous, and Polycotyledonous plants,—a distinction which still prevails in the system of Jussieu,—entirely vanishes upon a more exact and more general observation of nature.

Among the Acotyledonous plants were mentioned the Fungi, Algæ, Lichens, Homallophyllæ, Musci hepatici and frondosi, Ferns, and Naiadæ. But it has been already shewn (302, 307.), that Fungi, Algæ, and Lichens, are propagated by germs only ; and that Homallophyllæ, Musci hepatici and frondosi, and Ferns, during germination, produce fluctuating forms, which might be regarded as representatives of the cotyledons. In the Naiadæ, the thickened cotyledonous end of the root renders, in most instances, the cotyledons superfluous, and in some cases the albuminous substance, (*Zostera, Ruppia, Zannichellia, Potamogeton.*) In other Naiadæ (*Lemna, Hippuris*), the embryon is entirely unevolved in the middle of the albuminous substance ; and in others (*Callitriche, Ceratophyllum, Myriophyllum*), the embryon evidently divides itself into two cotyledons.

With respect to what are called Monocotyledonous plants, Jussieu has reckoned them in this order, the Aroidæ, Cyperoidæ, Grasses, Palmæ, Restiaceæ, Junceæ, Sarmentaceæ, Coronariæ, Irideæ, Hydrocharidæ, Scitamineæ, Museæ, Orchideæ, and Pipereæ. But in none of these plants can we admit the existence of a proper cotyledon. Indeed, Mirbel (Annal. du Mus.), Fischer (Ueber die Existenz der Mono- und Polycotyledonen), Smith (Linn. Trans.), and Treviranus (Entwickelung des Eyes), consider the vitellus in the Scitamineæ, or the bodies resembling scutella in the Grasses, as cotyledons ; because there is a distinct transition from

R 2

these bodies into the stem. But, without entirely denying this connection, which, however, cannot be shewn in all cases, there is a complete objection to this account in the structure of the vitellus, and in the composition of its juices. In the vitellus those small granular bodies are wanting, which are deposited by the mucilage and albumen, and which we observe in the cotyledons; while, on the other hand, it contains in the Scitamineæ peculiar ingredients, namely, volatile oil and aromatic matter. Other authors, as Claud Richard (Analys. du Fruit, p. 27.; Ann. de Mus.) consider the sheath of the first leaves (called *blastus*) as the cotyledon, though with still less propriety, because these are no way distinguished from the leaves that come after them. In fact, the Grasses and Cyperoidæ have no proper cotyledon. As little have the Aroidæ, in which the embryon for the most part lies, with the end of its root thickened, in the middle of the albuminous substance: the Naiadæ also are destitute of this structure. In the Junceæ and Palmæ, a tubercle proceeds during germination out of the first horizontal shoot, and from this tubercle the plant and radicle are first evolved, (385.) Intermediate forms, as Zamia and Cycas, have a distinctly divided cotyledonous body, which also shews itself in the Pipereæ. The same organ is found in the trees of the Pine tribe, which have been falsely ranked among the polycotyledonous plants; because their first leaves are different from those that follow in number and form. But, notwithstanding this, they are still leaves, and not cotyledons. In the Sarmentaceæ and Coronariæ, the embryon is unevolved, and stands either in the centre of the albuminous substance or surrounds it. It has sometimes a thickened cotyledonous end (*Hemerocallis, Hæmanthus*); in the *Gloriosa* it is even divided into cotyledons. In the Hydrocharidæ the thickened end of the embryon seems to hold the place of the albuminous substance; and in the Orchideæ, where, from the fineness of the seed no parts can be distinguished, the young plants unfold themselves in the same manner as Ferns, that is to say, first a soft, cellular, lobed body, and after it a tubercle arises, out of which the young plant springs.

We thus perceive, that what are called the Monocotyledonous plants have no proper cotyledons. As little do we find polycotyledonous plants, from the number of which the Pine tribe, as we have remarked above, must be struck out. If, in the case of dicotyledonous plants, the seed-lobes are divided or cleft, as in *Erodium, Canarium,* and *Lepidium,* they are fundamentally still but two, and all perfect plants, reckoned upwards from the Polygoneæ, must thus be considered as dicotyledonous.

387.

In the study of nature, we must accustom ourselves to suspect that in all bodies there are transition forms, and never to believe that one and the same type remains without change. There are thus transitions from cotyledons to leaves (in the Liliaceæ); transitions from albuminous substance to cotyledons, and even to the root of the embryo. There are transitions from leaf-stalks, and even from branches, to leaves, as in the Cereæ and Acaciæ, (181.); transitions from leafy appendages to leaves in the Cistæ; from leaves to bracteæ and to the calyx; from the calyx to the corolla in the Liliaceæ; from petals to filaments (*Pancratium*), and to nectaries (*Contortæ.*) Nay, in the Canneæ and Orchideæ, the opposite parts of fructification are so run into each other, that even here transitions must be admitted.

388.

When we are examining the evolution of the parts of plants, we must further attend to the law of nature, which was before stated (183.), namely, that simple forms always precede those that are complex and subdivided. In imperfect plants, where the embryon resembles a line or a point, and is unevolved, the expansion of it takes place in parallel surfaces, when the plant unfolds itself, (290.) A more complex expansion takes place when this complexity was prefigured in the manifold subdivisions of the parts of the seed, and when a more powerful crowding towards the first knot is evident. From this law (183.) that the more early forms are always

simpler than the later, as, for example, that the root leaves
are simpler than the stem leaves, there are, however, many
apparent exceptions in the Acaciæ of New Holland, the first
leaves of which are separated into many parts, while those
that come later appear to be simple. These later leaves,
however, are rather intermediate forms between the leaf-stalk
and the leaves. These last have not arrived at their evolution ; they have thus become abortive, and the leaf-stalks
supply their place, (181.)

389.

Another law to which the consideration of the evolution of
plants leads us, is that of numerical proportion. As all division and unfolding of parts proceeds from the spiral vessels,
and where these fail, from the sap-vessels, no other division,
according to the rule formerly given (279), can be fundamental, but that which proceeds from one to three ; because two
new vessels always place themselves on the sides of the original spiral vessels. Hence, the number three prevails in all
the lower organic bodies, as far as the Amarantheæ. By
doubling this number we have six, and by tripling it we have
nine. Hence *Butomus* and *Hydrocharis* belong to the same family. In more perfect plants two new vessels place themselves
on both sides of the original spiral vessels, and in them, therefore, the number five must prevail, the double of which gives
ten. When we perceive a fourfold and eightfold subdivision,
we may assume an abortion or union of parts, as the law of
nature in these cases, (178.)

2

———

CHAP. IV.

ON THE DISTRIBUTION OF PLANTS UPON THE EARTH.

Linné, Stationes plantarum, in Amœn. Acad. vol. iv.

Giraud Soulavie, Géographie physique de regne végétal.

F. Stromeyer, Historiæ vegetabilium geographicæ Specimen. ; Dissertatio.

G. R. Treviranus, Biologie, b. ii. s. 44, 137.

Humboldt et Bonpland, Essai sur la Geographie des Plantes.

Willdenow, im Magazin der Berlin, Gesellshaft naturforschender Freunde.

Wahlenberg, Flora Lapponica.

Dessen, Flora Carpathorum principalium.

Dessen, de Vegetatione et Climate Helvetiæ septentrionalis.

Brown's General Remarks, geographical and systematical, on the Botany of Terra australis.

Brown's Observations, systematical and geographical, on the Herbarium collected by Professor Smith in the vicinity of Congo.

Humboldt, Prolegomena ad nova genera plantarum.

Schouw, de sedibus plantarum originariis.

Jahrbucher der Gewachskunde.

Ritter's Sechs-Karten von Europa, mit erklarendem Text.

Titford's Sketches towards a Hortus botanicus Americanus.   Table of climates and habitats of plants.

390.

The geography of plants makes us acquainted with the present distribution of plants upon the earth and in the waters, and endeavours to refer their growth to external causes.   It is thus a part of the *Physiology of Plants*, since it investigates the laws according to which climate, temperature, soil, elevation above the surface of the sea, and distance from the equator, as also accidental external circumstances, operate upon the production of plants.   It is connected in some measure with the *History of Plants*, or with investigations respecting the origin, diffusion, and gradual distribution of plants. Yet it must be distinguished from this; and when a sure foundation of facts is laid and arranged, it exerts an essential influence on the science of the cultivation of gardens and fields, on the rearing of forests and other civil occupations.

391.

We may investigate the laws of the distribution of the families and tribes of plants in different climates by two methods.

In the first place, we divide the surface of the Earth into certain zones, in which we seek for the plants that are produced, and thence draw general results. This method is indeed a laborious one, and is especially difficult on this account, that we are not yet completely acquainted with all the parts of every zone of the earth; while the lower families of plants have commonly been neglected by most travellers. Yet we can draw conclusions, with some probability, respecting unknown plants from those that are known; at least, this method leads to greater certainty than the following, on which only, however, most of our labours have been conducted. In this second method, we place the Floras of countries of different climates before us; we compare the plants which they contain, and in this way form conclusions respecting their distribution. But, as we are not in possession of complete Floras of all countries and their individual regions, it cannot but happen that false inferences and contradictions will arise, while we do not take into account the productions of neighbouring countries, or of those that lie between the districts which have been examined. Besides, we can only make use of the Floras of particular degrees of latitude and longitude, but not those of the whole zone; because most of the compilers of Floras have been acquainted with the products of vegetation only within a certain circle.

392.

Without entering, in this place, into the history of plants, we may state it as the fundamental law of the geography of plants, that the lower the organization of the body is, the more generally is it distributed. As infusory animalculæ are produced in all zones, when the same conditions exist; we find, in the same manner, that Fungi, Sponges, Algæ, and Lichens, and even Musci frondosi and hepatici, are distributed every where upon the earth, in the sea, and in the waters, when the same circumstances propitious to their production occur. We have seen that the idea of genera and species can be applied with so much less strictness, the less perfect the vegetable is; and hence, although the same or similar forms of Conyomici, Nematomici, Gastromici, and Sponges, are produced in all zones, we cannot pronounce in all cases respecting the identity of the species. If travellers had not so much neglected the imperfect plants of foreign countries, this assertion might easily have been proved by innumerable testimonies. But we must receive with caution and limitation, even what they have told us respecting the growth of common European cryptogamous plants in the most distant regions and waters of the earth, because many of these travellers had no exact knowledge of the cryptogamous plants. The most distant countries of the earth, Europe and New Holland, the inhabitants of which are antipodes to each other, have, according to the testimony of Brown, a witness of the best information and highest credit, a considerable number of Lichens, almost indeed two-thirds of those that have hitherto been discovered in New Holland, of the same species with those that exist in Europe. Of the Musci hepatici and frondosi, nearly one-third belong equally to New Holland and to Europe. And, with respect to the Algæ, not only Confervæ, but Fuci, are common to the most distant seas. *Laminaria Agarum*, Lam., for instance, is found in Greenland, in Hudson's Bay, in Kamtschatka, and in the Indian Ocean. *Halidrys siliquosa*, Lyngb., *Sphærococcus ciliatus*, Ag., and many others, have a distribution equally extensive.

The Naiadæ and Rhizospermæ also are found in the same manner almost in all waters, as the *Marsilea quadrifolia*, *Zostera marina*, and the native Potamogetons and Lemnæ shew, which Brown found also in New Holland. Even the Grasses and Cyperoidæ share in this general distribution. A great many native members of these families, as the *Carex cæspitosa, Scirpus lacustris, Glyceria fluitans, Arundo phragmites, Panicum Crus Galli,* and so forth, grow also in New Holland. Humboldt has confirmed these observations, in respect to the growth of European Mosses, Grasses, and Cyperoidæ, in South America.

Higher and more perfect plants, on the contrary, are less generally distributed by nature, although, by cultivation, they also can be forced to vegetate in the most distant countries, provided favourable circumstances occur. Of these circumstances, a considerable number must always co-operate for the perfect evolution of plants of the higher orders. Yet there are some exceptions to this rule. *Verbena officinalis, Prunella vulgaris, Sonchus oleraceus, Hydrocotyle vulgaris, Potentilla anserina,* and some other common European plants, grow also, according to Brown, in New Holland. Almost the seventh part of the phanerogamous plants that vegetate in North America are found in Europe; yet we cannot deny that many of them have been transplanted hither, (401.) On Mascaren's Island, Bory St. Vincent found *Cladium Germanicum,* Schrad., *Cyperus fuscus, Potamogeton natans, Hydrocotyle vulgaris,* and some other European plants.

393.

The same distance from the equator, or the same degree of latitude, produces rather a resemblance in the forms,—an agreement in the families and genera,—than the same species, chiefly because, besides this geographical latitude, the height above the surface of the sea, the temperature during the growing season, the soil and constitution of the mountains, even the degree of longitude, and several other circumstances, have an influence on vegetation.

There are a great many perfect plants which exclusively belong to the tropics, which never pass beyond them, and which are found equally in Asia and Africa, in America and the South Sea Islands, and even in New Holland. Although, as we have said, these are rather families, as **Palmæ**, Scitamineæ, Museæ, Myrteæ, Sapindeæ, and Anoneæ; or genera, as *Epidendrum, Santalum, Olax, Cymbidium,* and so forth : yet there are particular species, which grow in all parts of the world only between the tropics, as, for instance, *Heliotropium Indicum, Ageratum conyzoides, Pistia stratiotes, Scoparia dulcis, Guilandina Bonduc, Sphenoclea zeylanica, Abrus precatorius, Boerhavia mutabilis,* and so forth.

But most commonly there are other species, which, under the same degree of latitude, supply in the new world the place of related species in the old. *Dryas octopetala,* indeed, grows equally upon the mountains of Canada, and in Europe ; but *Dryas tenella* of Pursh, which is very like the former, grows only in Greenland and Labrador. Instead of the *Platanus orientalis,* there grows in North America the *Platanus occidentalis ;* instead of *Thuia orientalis,* Thuia occidentalis ; instead of *Pinus Cembra,* in Europe and Asia, there grows in North America, *Pinus Strobus ;* instead of *Prunus Laurocerasus,* in Asia Minor, there grows under the same latitude in North America, the *Prunus Caroliniana.*

394.

There are many exceptions to this rule, however, depending on circumstances that have been already noticed. In the first place, countries are wont to share their Floras with neighbouring regions, especially islands lying under the same latitude, as the Azores possess the Floras of Europe and of Northern Africa, rather than those of America, because they are scarcely ten degrees of longitude from the coast of Portugal. Sicily, and still more Malta, possesses a Flora made up of those of the south of Europe and the north of Africa. The Aleutian Islands share their Flora with the north-west coast of America and the north-east of Asia. But the most distant countries, lying under the same latitude, may have the same, or a similar vegetation, while countries, or islands which lie between them, have not the least share in this particular Flora. The island of St Helena, which is scarcely eighteen degrees of longitude from the west coast of Africa,

and which lies a little further south than Congo, has yet no
plants, which are found in those last named regions; (Roxburgh's List of Plants seen in the Island of St Helena, Ap-
pend. to Beatson's Island of St Helena.) Japan has a great
many plants common to southern Europe, which, however,
are not found in these regions of Asia that lie under the same
latitude.

395.

We must further remark, that the eastern countries of the
old world, and the eastern shores of America, as far as the
Alleghany Mountains, have a much lower temperature than
the western regions; or that it is always colder in Siberia
and the north-east of Asia, than under the same latitude in
Europe; and that even Petersburgh is colder than Upsal,
and Upsal than Christiania, although they all three lie in the
60th degree of north latitude. In North America the difference is still greater, and there are commonly fifteen degrees of
Fahrenheit's thermometer between the temperature of the east
and west coast. It hence happens that many plants, which in
Norway grow under the polar circle, scarcely reach the 60th
degree on the limits between Asia and Europe. To this class
belong the Silver-fir, Mountain-ash, Trembling Poplar,
Black Alder, and Juniper. Even in the temperate zone,
the vegetation of many trees ceases sooner in the east than in
the west. In Lithuania and Prussia, under the 53d degree,
neither Vines, nor Peaches, nor Apricots thrive; at least
their fruit does not ripen, as also happens in the middle of
England. The most remarkable example of this great difference of temperature is furnished by the *Mespilus Japonica*,
which grows at Nanga sacki and Jeddo, under the 33d and
36th degrees of N. Lat., and which also grows in the open air
in England, under the 52d degree of N. Lat., when it is
planted against a wall, (Botanical Register, vol. v.)

396.

The same degrees of latitude, in the southern and northern
hemisphere, are connected with very different temperatures,

and produce a completely different vegetation. This, however, must be understood rather of the temperate and frigid
zones, than of the tropical climates, which, as we have already noticed, are pretty much the same over all the earth.
But the summer is shorter in the southern hemisphere, because the motion of the earth in her perigee is more rapid.
The summer is there also colder, because the great quantity
of ice over the vast extent of sea requires more heat for dis-
solving it than can be obtained ; as also, because the sun
beams are not reflected in such quantity from the clear surface of the sea water, as to afford the proper degree of heat.
It hence happens, that in the southern hemisphere, the Flora
of the pole extends nearer the equator, than in the northern.
Under the 53d and 54th degrees of south latitude, we meet
with plants which correspond with the Arctic Flora. In Magellan's Land, and in Terra del Fuego, *Betula antarctica* corresponds with *Betula nana*, in Lapland ;—*Empetrum rubrum*
with *Empetrum nigrum*,—*Arnica oporina* with *Arnica montana*,—*Geum Magellanicum* with *Geum rivale*, in England,—
Saxifraga Magellanica with *Saxifraga rivularis*, in Fin-
mark. Instead of *Andromeda tetragona* and *hypnoides*, of
Lapland, Terra del Fuego produces *Andromeda myrsinites*:
in place of *Arbutus alpina* and *Uva Ursi* of the Arctic polar circle, Terra del Fuego produces *Arbutus mucronata*,
microphylla, and *pumila*. Aira antarctica reminds us of the
Holcus alpina of Wahlenberg; and *Pinguicula antartica*
recalls to our recollection *Pinguicula alpina*.

We must recollect, however, that in South America, the
great mountain chains of the Andes stretch from the tropical
regions, almost without interruption, to the Straits of Magellan, (from the 52d to the 53d degree of S. Lat.) ; and that, on
this account, tropical forms are seen in that frigid southern
zone, because, as we shall have occasion to remark more fully
afterwards, the tract of mountains every where determines
vegetation. It is hence that the Straits of Magellan are pro-
lific of *Coronariæ*, *Onagræ*, *Dorsteniæ*, and *Heliotropiæ*,
which in other parts of the world grow only within the tropics, or in their neighbourhood.

In general, the vegetation of the southern hemisphere is very different from that of the northern, and there is a certain correspondence between the Floras of Southern Africa, America, and New Holland. Most of the trees are woody, with stiff leaves, blossoms sometimes magnificent, but fruit of little flavour. In Southern Africa, as well as in New Holland, it is the form of the Proteæ which prevails as if appropriated to these regions. Instead of the South American *Ericæ*, we find the *Epacridæ* of New Holland. *Lobeliæ*, *Diosmeæ*, and a great number of rare forms of compound blossoms, and of *Umbellatæ*, are common to all these southern regions.

397.

For understanding the growth and distribution of plants, we must also attend to the soil. Similar plants are found in similar soils, completely separated from each other, and with respect to which no supposition of interchange can be entertained, provided only the climate be not too different. Salt soils produce almost every where particular Chenopodeæ, species of Chenopodium, Atriplex, Salsola, Salicornia, and Anabasis. Calcareous soils produce always the most numerous and distinguished forms of plants. Volcanic mountains, particularly basalt, produce few forms, but those of a distinguished kind and very variable. Alluvial mountains, particularly in the neighbourhood of streams, usually, like marshes, produce forms that are always the same. The primitive mountains, on the other hand, almost every where separate the Floras of countries. Thus the Pyrenees,—the Alps which divide Italy from France and Switzerland,—those which separate Upper Italy from the Tyrol and Carinthia,— and the Carpathian mountains,—divide the Floras of the southern from those of the northern countries.

It is hence of so much importance, along with Floras, to describe the mountain rocks and the different soils, (258.) The first example of a map of this kind was given by De Candolle, in the second volume of his *Flore Française*, in which, however, the mountains only, the southern shores, and the alluvial land, are distinguished, and the heights above the sea are given. Wahlenberg's Map of Lapland, in his *Flora Lapponica*, is much more correct. Maclure has given a Map of the constitution of the Mountains in the Free States of North America, (Geographische Ephemeriden). I know not that any person has given a more pleasant and instructive account of the soils and mountains of his country, in relation to their Floras, than the excellent Villar, respecting the Alps which divide Italy from Switzerland, in the Preface to the *Histoire des Plantes de Dauphiné.*

398.

This brings us to a very important circumstance, which must be taken into account in every examination of the causes of the growth of plants, namely, the height of their station above the level of the sea. As, upon the whole, the temperature on the highest mountain tops seems to be the same with the temperature at the polar circle, it is commonly believed, that under the snow-line, and near to it, the same vegetation is found as in the polar regions. The limit of perpetual snow, under the Equator, is at the height of 15,000 feet; in the 35th degree of N. Lat. it is at 10,800; in the 45th degree, at 8400; in the 50th degree, at 6000; in the 60th degree, at 3000; in the 70th degree, at from 1200 to 2000 feet above the level of the sea; and, at the 75th degree of N. Lat., the snow-line lies almost upon the ground In general, it has been ascertained by observations, that the same vegetation is produced at the same distance from the snow-line. It must be considered, however, that towards the pole, the summer is shorter, but hotter, than under the snow-line upon the tropical mountains, where winter and summer cause no change of temperature. On this account, a better vegetation must be produced in the polar regions during summer, especially as the plants are there exposed to the uninterrupted light of the sun : while, on the other hand, from the uniform temperature on the highest tropical mountains, throughout the year, a very different Flora is there produced from that which springs towards the pole. As yet we know only, from Humboldt's immortal labours, the vegetation upon the highest chains of the Andes, in South America; for the Floras of the much higher mountains in Northern India, which are called the Himalaya Mountains, and of the Mountains of the Moon, in Africa, are entirely unknown to us. The Flora of the Andes, at the height of 14,760 feet, is almost entirely of a peculiar kind ; and if a pair of Ranunculi, a *Gentiana*, and a *Ribes*, remind us of the Flora of the poles, the remaining productions are completely peculiar, and prove that the height above the level of the sea is very far from producing universally the same forms. Yet we must add, that in Europe, at least, many northern and even polar forms appear under the snow-line of the Pyrenean and Helvetian Alps. Of this the Dwarf Willow, Dwarf Birch, Saxifrages, Ranunculi, Cerastia, and other genera, are striking proofs.

Of perfect plants, the *Daphne Gneorum* seems in Europe to hold the most elevated station, since, on Mont Blanc, it stands at 10,680 feet ; and, on Mont Perdu, at 9036 feet high. The growth of woody plants ceases, on the Alps of central Europe, at the height of 5000 feet ; and, on the Riesengebirge, at 3800. Oats grow on the Southern Alps at 3300, and on the Northern scarcely at 1800 feet. The Fir grows on Sulitelma, in Lapland (68 degrees N. Lat.), scarcely at the height of 600, and the Birch scarcely at the height of 1200 feet. On the other hand, upon the Alps which divide Italy from France and Switzerland, Oaks and Birches grow at 3600, Firs at 4800 ; and the same plants grow on the Pyrenees above the height of 600 feet.

In Mexico, the mountain chains, and, in particular, the Nevado of Toluca, are covered, above 12,000 feet high, with the occidental Pine (*Pinus occidentalis*) ; and, above 9000 feet, with the Mexican Oak (*Quercus Mexicana, spicata*) ; as also with the Alder of Jorullo (*Alnus Jorullensis.*)

On the Andes, Palms grow at the height of 3000 feet. The woody Ferns, (*Cyathea speciosa, Meniscium arborescens, Aspidium rostratum*), are found as high as 6600 feet ; as are also the pepper species, Melastomeæ, Cinchonæ, Dorsteniæ, and

some Scitamineæ, rise to the same elevation. At the height
of 14,760 feet, we still find the Wax Palms, some Cinchonæ,
Winteræ, Escalloniæ, Espelettiæ, Culcitia, Joanneæ, *Vallea
stipularis, Bolax aretioides*, and some others.

399.

The growth of plants in society, or as individuals, is very
interesting. Many forms are so appropriated to certain regions, that amidst constant changes they still take in a great
tract of land, and are produced in exuberant abundance.
Others, on the contrary, stand quite insulated, and seem as if
they would utterly disappear, did not Nature, in a manner
which is often inexplicable, provide for their continuance.

While with us, *Polygonum aviculare, Erica vulgaris,
Poa annua, Aira canescens, Vaccinium Myrtillus*, grow always in society, and cover great tracts of country, we observe, on the contrary, that *Marrubium peregrinum, Carduus cyanoides, Stellera Passerina, Carex Buxbaumii, Cirsium eriophorum, Lathyrus Nissolia, Hypericum Kohlianum,
Schœnus ferrugineus*, and *Helianthemum Fumana*, are confined, in an insulated state, within a very narrow space, beyond which they never pass. The Cedar of Lebanon, *Fors-
tera sedifolia* of New Zealand, *Melastoma setosum* on the
Volcano of Guadaloupe, and *Disa longicornis* on some spots
of the Table Mountain of the Cape, are examples of this
completely insulated growth, which renders the idea of the
migration of plants at least very doubtful.

400.

If we proceed through the separate families, we shall find
their geographical distribution pretty exactly ascertained, and
their increase or diminution determined according to the different zones. But, as many families consist but of single
groups, which are limited to certain zones or countries,—this
circumstance occasions always a variation in the account. If
we attend, for instance, to the Rubiaceæ, it is almost impossible to pronounce any general opinion respecting the geographical distribution of this family, because the first group of

S

this family, the Stellatæ, is almost peculiar to the temperate, and especially to the northern temperate zone, while the Spermacoceæ and Coffeaceæ are confined to the tropical zone. The Cinchoneæ, indeed, grow chiefly between the tropics, but always at a fixed height above the level of the sea, and they also pass beyond the tropics. Plants with compound flowers are indeed generally more abundant between the tropics, but the group, which I have named Perdicieæ (Labiatifloræ, De Cand.), is peculiar to South America, and descends even into the southern frigid zone.

Respecting Ferns, it is understood that, in the temperate zone, they constitute the sixtieth part of the whole vegetable kingdom. But how little certainty attends such conclusions, may be understood from this, that in New Zealand, the number of Ferns is to the number of other plants as 1 to 6; in Norfolk Island, and Tristan d'Acunha, as 1 to 3; in Otaheite as 1 to 4; in Mascaren's Island as 1 to 8; in Jamaica as 1 to 10; in St Helena as 1 to 2; and in Egypt we have as yet found but one species. The Grasses seem in all zones to maintain nearly the same proportion: They constitute the tenth or fifteenth part of the whole Flora. The Umbellatæ are evidently in greatest number in the temperate zone. They constitute about the thirtieth part of other plants. Towards the pole they diminish in number; and in the torrid zone there are scarcely any other Umbellatæ but some intermediate forms, which only appear at a very considerable height upon the mountains. The Cruciform plants exhibit a similar proportion. In the temperate zone, they are to the remaining plants perhaps as 1 to 20. Towards the pole they decrease in number, and between the tropics we find scarcely a trace of them. The reverse is the case with the Malvaceæ. Whilst these constitute, between the tropics, the fiftieth part of the other plants; in the temperate zone they bear to them the proportion of 1 to 200, and in the polar zone they fail entirely. The Leguminous plants are in greatest quantity between the tropics, where they form the twelfth part of the whole Flora. In the temperate regions they fail considerably, and in the polar zone they are to the other plants as 1 to 35.

The Primuleæ are almost the only plants that are common to the frigid and to the temperate zone. The Contortæ belong to the tropical region, where they form from the fortieth to the fiftieth part of the whole. In the temperate zone they diminish in number, until, towards the polar circle, they almost entirely cease.

401.

It is interesting also to know the limits within which the cultivation of the useful plants is confined. The cultivation of Cocoa, Coffee, Anatto, Cloves, and Ginger, is limited to inter-tropical regions. The Sugar Cane, Indian Figs, Dates, Indigo, and Battatas, pass the tropics as far as the 40th degree of N. Lat. Cotton, Rice, Olives, Figs, Pomegranates, Agrumæ, and Myrtles, grow in the open air, as far as the 45th and 46th degree. The Vine succeeds best with us within the 50th degree of N. Lat.; this, also, is the limit, especially in the West of Europe, of the cultivation of Maize, Chesnuts, and Almonds. Melons also succeed to the same latitude in the open air.

In the West of Europe, the cultivation of Plums, Peaches, Wheat, Flax, Tobacco, and Gourds, ceases at the 60th degree of N. Lat. In the East of Europe, the cultivation of Apples, Pears, Plums, and Cherries, terminates at the 57th degree; but Hops, Tobacco, Flax, Hemp, Buckwheat, and Pease, succeed there even under the 60th degree. Hemp, Oats, Barley, Rye, and Potatoes, are planted by the Norwegians under the polar circle; and the Strawberry flourishes, even at the North Cape, under the 68th degree.

———

CHAP. V.

HISTORY OF THE DISTRIBUTION OF PLANTS.

Linné, de Telluris habitabilis incremento : in Amœn. Acad. vol. ii.
Zinn, Vom Ursprung der Pflanzen : im Hamburgishen Magazine.
Bergman, Jordklot. Phys. beskrifn. ii.
Zimmerman, Geographische Geschichte des Menschen.
Schouw, Diss. de Sedibus Plantarum originariis.

402.

We come now to answer the questions, in what manner
plants have originated, and how they have distributed themselves. Are we to admit, that plants have been distributed
from one point on the surface of the earth, to all its parts?
or must we believe that they belong properly to every country in which they grow? The founder of Scientific Botany has defended the former of these opinions at a great expence of ingenuity, acuteness, and learning ; but we appre-
hend that we must adopt the latter conclusion, with some li-
mitations.

403.

When we examine the remains of the primeval world, we
find the first traces of vegetable impressions in the slate formation. These remains of the former vegetable world belong almost entirely to the lower families : they consist, for the
most part, of Grasses, Reeds, Palms, and Ferns,—the latter,
however, being almost always destitute of fruit. But although
these forms cannot be referred to any one of the species which
are at present known, they have yet so much the appearance
of tropical productions, that we are forced to admit a very
high degree of heat at the surface of the earth during its

former state, which heat must, at that time, have been diffused over all the zones, because we find the same productions
in the slate formation of all parts of the earth *.   In order
to explain this, it has been supposed, that the plane of the
ecliptic, during the former state of the globe, was completely
different in its position, and that, consequently, our planet had
then another situation in respect to the sun.   But Bode, the
worthy veteran of Prussian astronomers, has shewn, that the
plane of the ecliptic has been, for 65,000 years, between the
20th and 27th degree ; and that at present it is about 23
minutes less, and, consequently, the inclination of the axis of
the earth as much greater, than in the time of Hipparchus,
who lived about two thousand years ago.   The former solution must, therefore, be entirely abandoned , (Neue Schriften der Berlin Gesellschaft naturforschender Freunde).

Shall we then consider as satisfactory another explanation,
which has been advanced by one of the most ingenious and
learned investigators of the ancient state of the globe? According to this author, the Earth, during its primeval
state, was completely surrounded by water.   By slow degrees the sea retired ; the highest mountain tracts were laid
bare, and the lowest and densest atmospherical stratum,
supported by the surface of the sea, now rested upon the
highest primitive land, which, like an island, emerged but a
little way above the ocean.   While the heat was chiefly generated in the lower strata of the air, it must also, at that time,
have been equally diffused throughout all parts.   The naked
summits of the mountains were gradually mouldered by the

It seems, indeed, that all the carbonaceous matter of the more ancient
slate formation ought to be considered as the oldest remains of plants which
had been growing, but which had been stopped in their progress; and
that all calcareous matter ought to be considered as the remains of a begun, but
suppressed creation of animal bodies; (Steffen's Beyträge zur innern Naturgeschichte der Erde, s. 27. Dessen, Handbuch der Oryctognosie, b. ii. s. 353.)
In what manner mineral substances are formed from corrupting vegetables,
we perceive from the production of iron-pyrites, in our Peat Mosses, where it
is found in layers, under the thin, broad, reed leaves, after they have become
putiid.

influence of light, of aerial matters, and of moisture. To the primitive and transition rocks succeeded the horizontal flœtz formation. A multitude of bodies was now formed, which light elicited from the organizing water. These bodies decayed, and left behind them the original materials of carbon and lime, from which still more perfect forms were to spring, until, at length, Ferns, Grasses, and Palms, were produced, which, during the high level of the waters, enjoyed upon the declivities of the mountains, a high and equal temperature. With these forms, creative nature remained contented, till new revolutions of the surface of the earth gave the waters an opportunity of retreating still further, when new formations arose.

This hypothesis, which is favoured by the Geognosy, derives particular support from the concurring testimony of the most ancient inhabitants of the world, respecting universal inundations and floods; as also, more especially, from the Persian Cosmology, in which Albordsch, the highest primitive mountain, the hill of light, or navel of the earth, plays a principal part, while it is yet surrounded by water. The primitive light produces, upon this mountain, during an ever equal temperature, all living things; (Kanngiesser Altherthums wissenschaft, s. 8. u. 18.; Algemeine Encyclopædie, th. ii. s. 375.)

404.

What revolutions the surface of the earth underwent, before it assumed its present state, we know not. But with the period of the alluvial mountains begins the present vegetation of the earth's surface, and those forms which we now see in unvarying perpetuity, have been so for several thousand years, or since the earth assumed its present state, (142.)

Whether now, as Linnæus maintained, one example only of every genus of plant existed at the beginning: whether all these single genera were produced by the hand of Nature upon the highest mountain ridge of the earth, along with the single pair from which the human race has proceeded, and with similar individual pairs of all other animals: whether, at least, the highest mountain ridges may, in general, be regarded as the birth-places of the vegetable world, as Willdenow asserted; and whether, therefore, plants have been distributed from single stations, by their own migrations, and by other means, which nature provides,—these are questions which ought to be examined and answered with the greatest care.

405.

We are not disposed altogether to deny the migration of plants in similar climates, because we know from experience, that the *Datura Stramonium, Erigeron Canadense,* and *Æsculus Hippocastanum,* are not natives of Germany; but that the first was imported, as it is said, by the gypsies,—the Horse-Chesnut, by the Austrian embassy which was sent to Constantinople at the end of the sixteenth century,—and the *Erigeron Canadense,* in consequence of some commercial relations which we had with North America;—as this latter country has also probably received from us, in the course of commerce, the *Agrostis Spica Venti, Trichodium caninum, Anthoxanthum odoratum, Alopecurus pratensis, Poa trivialis, Bromus secalinus, mollis, Dactylis glomerata, Hordeum vulgare, Dipsacus sylvestris,* and many other plants, (392.) We know that a great many plants have passed from India into Italy, along with the cultivation of rice: we know that the West Indian negroes have introduced into the western world a great many plants from Africa, which at present grow wild there, as the *Cassia occidentalis,* and *Chrysobalanus Icaco.*

It is certain, that sea plants have been brought by ships, from the southern into the northern sea, as has been the case with *Fucus cartilagineus,* Turn., *Fucus natans,* and *bacciferus.* West Indian fruits are every year driven upon the coasts of Norway, and of the Faroe Islands, during storms from the south-west; as Cocoa Nuts, Gourds, the fruit of *Acacia scandens, Piscidia Erythrina,* and *Anacardium occidentale.* But after all this has been granted, we are still far from being in a condition to maintain the migration of plants and their distribution from one point over the surface of the earth.

406.

It betrays very limited ideas respecting the laws of vegetation to suppose, that all the plants upon the face of the earth, which require such different climates, such a different constitution of the mountain-rocks, so various a composition of soil and of water, could ever have been assembled upon one and the same high mountain ridge. All testimonies, indeed, confirm the supposition, that the human race, and the domestic animals, descended from the high mountain plains of central Asia, between the 27th and 44th degrees of N. Lat. We may also conjecture, that the different kinds of grain grow wild in these latitudes, as it is also probable that the domestic animals are there found in their native state. But the innumerable other wild growing plants, of all quarters of the globe, which are so frequently confined to a single island, or to a single circle of the continent, cannot possibly have their native seats in those regions; otherwise the remains of that vegetation which is now dispersed over the whole face of the earth, would be found in Northern India and Persia, in Thibet, and in the Mogul empire. It is physically impossible, that plants, which in Germany grow only upon calcareous soils, or upon basalt and other peculiar mountain rocks, could, at an early period, have flourished upon the primitive granite and gneiss of the Himalaya Mountains.

407.

As little can we assent to the opinion of those who consider the high mountain tracts as so many birth-places, or foci of vegetation, and of the neighbouring Floras. We admit, that when a particular mountain chain stretches into the level country beneath it, its peculiar plants will also appear in the low land. The flœtz limestone of central Germany confirms this conclusion in the strongest manner. But when this is not the case, the low country never partakes of the Flora of the neighbouring mountains. Otherwise, *Seseli Hippoma-*

rathrum, *Teucrium montanum*, *Poa alpina*, and *Stellera Passerina*, would soon diffuse themselves from our calcareous hills to the flat country, and even over our porphyry mountains.

It is true that mountain tracts commonly form the boundaries of Floras. But this happens, not because these mountains are the birth-places of the vegetable world, but because climate and temperature change with them. The Rhætian Alps separate Germany from Italy; on their southern declivities we observe Laurels, Pines, Beeches, Cypresses, Jasmins, and other similar plants, which do not grow on their northern sides. But the temperature on the opposite sides of these Alps is also completely different.

It must also be added, that the limits of Floras are not defined by mountain tracts alone, but that even in a great extent of level country the Floras have their proper boundaries. *Andropogon Ischæmum*, *Asperula cynanchica*, *glauca*, M. B. *Centunculus minimus*, *Lycopsis pulla*, *Bupleurum rotundifolium*, *Peucedanum officinale*, *Cnidium silaus*, *Silene noctiflora* and *conoidea*, and *Centaurea calcitrapa*, seem not to pass beyond the 52d degree North Lat. into central Germany. At that point, *Angelia Archangelica*, *Lonicera periclymenum*, *Andromeda polifolia*, *Arbutus uva ursi*, and other forms begin to appear. In completely level countries, the *Acer campestre*, *Pseudoplatanus*, *Populus alba*, *nigra*, and *Sambucus nigra*, cease to grow at the 56th degree N. Lat. The Myrtle, Mastick, Oak and Cork tree, the flowering Ash, and the Caper tree, pass not beyond the 44th degree North Lat., whether mountain tracts or level countries occur at this limit. The heights of the Wolga, or Alaunian Mountains of the ancients, are said to be the limit between the eastern and western Floras; but according to Pansner's recent examination, the entire Wolga heights are only alluvial land, covered with sea sand. Besides, the eastern Flora is seen a great way on this side of the Wolga heights, (Neue Geographische Ephemeriden, b. v. s. 141.) The Weichsel on the north, and the Oder on the south, seem better entitled to be considered as the limits of the western and eastern Flora. On the farther side of these streams, we find *Plantago arenaria*, Kit. *Anchusa Barrelieri*, Vilm., *Flœrkia lilifolia*, *Angelica pratensis*, M. B., *Acer platanoides*, *Andromeda calyculata*, *Silene tatarica*, *Dianthus serotinus*, Kit. (east from Cracow), *Anemone patens*, *Ranunculus cassubicus*, *Teucrium Laxmanni*, *Dracocephalum Moldavica* (east from Grodno and Jaroslaw), *Bunias orientalis* (east from Lemberg), *Isatis tinctoria* (beyond Warsaw), *Astragalus Onobrychis*, *Melilotus polonica*, *Pentaphyllum Lupinaster*, *Hieracium collinum*, Bess., *Orchis cucullata* (beyond the Niemen.)

408.

Had plants been distributed from single, and, as it is thought, from elevated points on the earth's surface, the Floras of contiguous regions would necessarily have been confounded, and could not have been so distinctly appropriated, as we see them to be. It must be added, that winds and birds, rivers, and the waves of the sea, are far from being able completely to have effected the universal dispersion of plants. There can be no doubt, that the wind is able to diffuse to a certain extent some particular seeds, which are furnished with crowns, hairs, and other appendages. But it is not able to disperse to any distance the *Carduus cyanoides*, which grows on a single grassy hill near Halle, and on the steep banks of the Elb above Tochheim, although the seeds of this plant are furnished with a crown of bristly hairs. The Syngenesious plants, too, the seeds of which can be so easily transported by the wind, are by no means common in the greater number of countries. If the wind favoured the migration of plants, we might determine their correspondence in most countries from the distance. But we have already noticed (397.), that the most distant countries have common plants, whilst the most dissimilar Floras are found in neighbouring lands, and some plants grow quite insulated, (395.) The dispersion of plants over large tracts of country has also been ascribed to birds, because they devour the fruits, and often allow them to pass from them undigested. But no example of this can be produced except the Misletoe, and therefore this

3

assertion deserves no particular refutation. Streams, indeed, can carry down seeds; and plants from those higher regions through which the streams flow are accordingly often found growing on their banks. But the Flora on the banks of one and the same stream, is very different in the different districts through which it passes, as is seen in the clearest manner upon the shores of the Elbe; for in Bohemia very different plants appear from those which spring in the neighbourhood of Dresden,—others, again, make their appearance near Wittenberg and Barby,—and a yet different set near Lauenburgh and Hamburgh.

These considerations lead us to conclude, that the vegetable world has neither descended from one common birthplace, nor diffused itself from one country into another; but that vegetation is in every case the product of the joint influence of temperature, soil, and the particular composition of the moisture of the earth.

Nor is the conclusion of Brown (on Congo, p. 50.), that the native country of a genus is always where the greatest variety of species is found, by any means to be admitted, since the example of Nicotiana shews the contrary. The greatest number of its species are found in South America; yet the *Nicotiana Chinensis*, Lehm. and *fruticosa* are certainly indigenous to Eastern Asia.

CHAP. VI.

ON MALFORMATIONS AND DISEASES OF PLANTS.

Linné, Philosophia botanica.   S. 119, 131.
Jager, Uber die Missbildungen der Gewachse.
Gallesio, Theorie der vegetabilischen Reproduction.
Keith's System of Physiological Botany.

Hopkirk's Flora Anomoia, a general view of the anomalies in the Vegetable Kingdom.

Ginanni, Delle Malattie del Grano in erba.

Tessier, Des Maladies des Grains.

Fabricius, in Norske Vidensk. Selsk. Skrift.

Plenck, Physiologia et Pathologia Plantarum. Vienna.

Seetzen, Systematum de morbis plantarum dijudicatio.

Burdach, Systematisches Handbuch der Obstbaum-Krankheiten. Berlin.

F. Re Saggio, teorico-pratico nelle Malattie delle Plante.

Cir. Pollini, nella Biblioteca Italiana, tom. vi.

F. W. Göthe, Zur Naturwissenschaft überhaupt, besonders zur Morphologie. Tubingen.

409.

We have considered (176, 192.) the Abortion, Degeneration, and Union of Organs, as effects of a constant law of nature. If we would distinguish from these the Malformations or Anomalies, we must consider these latter as variations or degenerations of forms and colours, which are less permanent, and which are not inconsistent with the health of the entire plant. We say that malformations are variations or degenerations which are less permanent, because very often they disappear by reproduction, although there are instances of their being propagated for a short time. We distinguish malformations from diseases by this circumstance, that in the former all the organs continue to be propagated with their due proportions.

410.

With respect to the causes of malformations, we may remark, that most of them arise from cultivation, and from that too great attention which in cultivation is paid to some particular organs; or they arise from too great luxuriancy of growth, which is injurious to the constancy of the fundamental type of forms. We hence observe, that malformations are again lost, when the sterility of the soil, and a coarser method of treatment confine the growth. Even climate has an undeniable influence on many of these variations of form, as we observe in the full or double Hyacinths, which, after bulbs have been brought from Holland, unfold only once in

Germany and Italy the complete magnificence of their double structure, and then pass again to their simple form. But it is also true, that want of fertility in the soil is another frequent cause of malformations, as we observe distinctly in the Stunting and Speckled structure of the stem and branches.

411.

In the stem and in the branches, we observe sometimes the speckled structure, sometimes the witch knots, and sometimes the downward direction of the branches, as consequences of malformation. What we call the Speckled structure, is an admired form of the wood, in which the knots are more numerous and more mixed with each other than is usual, and a great quantity of buds seem to have been but half formed. Many woods, as the Birch, Poplar, and Yew wood, have a particular disposition to become speckled, especially when they grow on dry, stony or rocky soils. Even art can aid in the production of this structure, by a frequent withdrawing of the branches from the light ; (Marten's Theorie uber die Entstehung des Masernholzes.) Connected with these spots, are the witch-knots of the Scots, which are chiefly found upon the Highland birches. They consist of buds intermingled in a great variety of ways, from which, however, no proper branches proceed, but a crowd of thin twigs, in the form of a bush or shrub, shoot out, as we often find in the Pines of the very dry sandy plains of Germany ; (Keith, ii. 278; Hopkirk, p. 62.)

From a similar cause proceeds the downward direction of the branches of the Birch and Ash, which must be considered as a malformation, because, by propagation, it finally disappears.

The stem of herbaceous plants is often fascicular, or has that structure which is called Fasciation. In Asparagus, *Hieracium cerinthoides, Carduus palustris, Celosia cristata,* and *Ranunculus bulbosus,* this structure is observed to be more or less permanent, in so much, that in the case of *Celosia,* it remains almost unchanged, provided the mode of treatment be entirely the same. This fascicular form seems to arise from the union of a number of branches, as may be dis-

tinctly seen in Asparagus. The union of the flowers on the
summit of such fascicular stems favours this explanation;
since the Hieracium already mentioned, as well as the *Ranunculus bulbosus*, shew distinctly this union of the flowers;
(Gilibert Demonstr. elem. de Bot.)

412.

Among malformations, we place also the discoloration of
the leaves, particularly the fascicular streaks, the silvery or
golden margins, and many other varieties of spots which are
common among garden plants,—as in Myrtle, Sage, Ivy,
Holly, the Agave Americana, Sempervivum arboreum, and
many of the Pelargoniæ.   These spots are not diseases, because the whole plant has all the signs of being in a perfectly healthy state.   But neither are they effects of a law of nature, like the spots of *Orchis maculata*, and the red coloured
leaves of *Caladium bicolor* and *Amaranthus tricolor*, because they are not continued by propagation.   But it is likely that such discoloured spots are incapable of performing
their function, namely, the exhalation of oxygen gas, as indeed experiments shew; (Sennebier, Physiologie Vegetale,
iv. p. 273.

413.

Luxuriancy of growth produces a manifold subdivision of
the leaves and a curling of their margin, as we find strikingly
exemplified in *Polypodium cambricum, Scolopendrium officinale, Acer platanoides*, and *Fraxinus excelsior*. We cannot consider these forms as permanent, or form peculiar
species from them, because they are by no means propagated
by seed, but only by buds and layers.   The manifold indentations of the leaves of the common Alder, and of the *Pimpinella saxifraga*, are of the same nature, and, on account of
their little durability, deserve to be considered rather as varieties.

414.

Retrogradations sometimes occur in vegetation (Goethéns
unregelmassige Metamorphose), when the more perfect organs

are not unfolded, but leaves and other lower organs arise instead of them. We thus observe, that the blossoms of the *Juncus subverticillatus,* when it remains as *Juncus fluitans* constantly under water, degenerate into long stem leaves. In the same manner, it is not uncommon, with the *Rubus fruticosus,* when in dark forests it is deprived of the sun's light, to put forth only leaves instead of blossoms. In the double flowers of *Hesperis matronalis,* we frequently remark the transition into the leaf form. In the *Colchicum autumnale,* not only the blossoms, but even the filaments and pistils, have been seen to assume the colour and form of stem-leaves; (Bernhardi in Romer's Archiv. b. 2.) When the garden tulip is very double, the outermost petals are often marked with green streaks, and even the innermost, which have arisen out of the pistillum, shew the same colouring.

When seeds pass into bulbs, as has been observed in *Bulbine Asiatica* and *Morea Northiana,* the same kind of retrogradation in the process of vegetation takes place, as when we observe that the siliques of Clover degenerate into leaves, or that a pear pushes out leaves (Keith, ii. tab. 9. fig. 12.), and that one flower arises from another; a mode of growth which in some cases is a law of nature, but which in our Centifoliæ, and in other instances, is a consequence of malformation produced by luxuriant growth.

415.

These retrogradations in the process of vegetation, explain the circumstance of blossoms becoming multiplicate, and perfectly double or full. Cultivation stimulates the organs of nourishment, and the instruments of propagation pass into these. Yet there are inferior gradations of this disposition to become double, in which the organs of fructification remain unconfined in their evolution, and in the exercise of their functions. When in the *Hydrangea* of our gardens, the parts of the calyx expand, and become of the nature of a corolla, the evolution of the filaments suffers so little by this, that we frequently observe eight of them instead of five. In the same

manner, the fructification of the full Balsam is not injured. In
the more perfect instances of double flowers, not only the filaments, but the pistils, and even the nectaries, when they are present, pass into the corolla. In the common Columbine of our
gardens, this last change happens. Yet the nectaries are often
multiplied to the greatest degree, while the petals remain un-
affected. It seems to be established by a very remarkable
observation, that the nectaries even sometimes supply the
place of the instruments of fructification; (Muller in Ver-
handlungen der Gesellschaft zur Beförderung der Natur-
kunde und Industrie Schlesiens, th. i. s. 214.)

When monopetalous corollæ become full, they are divided
during this process, as is distinctly seen in *Antirrhinum majus*,
and *Jasminum Sambac*. When compound flowers are filled,
they either return, when they are radiatæ, to the original tube-
form of the disc florets, as we see in the quilled China aster
of our gardens, and in *Tagetes*; or the disc florets degenerate
into ray florets, which is almost constantly the case with the
Calendula officinalis, *Pyrethrum Parthenium*, and *Anthemis
grandiflora*. It is remarkable, that the papilionaceous flowers almost never are full. *Spartium junceum* is the only exception with which I am acquainted. Some Japanese flowers,
Anthemis grandiflora, Ramat., *Clerodendron fragrans*, Vent.
and *Keria Japonica*, De Cand., grow always full.

416.

To malformations of the fruit, we refer partly its formerly
remarked retrogradation to the form of bulbs and leaves, and
partly its want of seeds, which also is a consequence of
luxuriant growth, and of the ceaseless propulsion of the juice
increased by art. The Musa almost never bears seeds.
Our Figs always contain only female flowers, the ovaries of
which are consequently abortive. The Italian Azarole, the
Chinese Cedrat, our Ananas, our Barberries and Plums
without seeds, are malformations of the same kind, which may
be considered as the consequences of cultivation. The growth
of one fruit in another, which is partially observed in the
Agrumæ, belongs to the same class of facts; (Linné, in

Annal. Transalp. i. p. 414. ; Meidinger in Beschaftigungen der Berlin Gesellschaft naturforschender Freunde.)

417.

We now come to what are called the Diseases of Plants, or to those varieties in the form and in the composition of their parts, which are injurious to the fulfilment of their functions, and to the continuance of their life. As plants are organic bodies endowed with vital activity, the same causes which affect animal bodies must produce similar effects upon vegetation. They must, therefore, be affected in the same manner as animal bodies, by heat and by cold, by moisture and drought, by a deficiency and by a superfluity of the ingredients of the atmosphere. We have partly stated above (369, 370, 372, 376.), the effects of the great natural agents upon the vegetable world. Among other effects, we have mentioned the blighting of corn during severe storms of lightning. The aerial smoke or dry-fog, which Pfaff (Ueber den heissen Sommer von 1811, s. 52.) calls a dry electrical vapour, and which is apparently an impregnation of the atmosphere with sulphurous acid gas, is likewise very hurtful to plants, because, by means of it, an over excitement and parching are produced, and the lively green of the leaves is changed into a dirty brownish-yellow

418.

But vegetables are subjected to a still greater number of causes of disease than animals, because an innumerable crowd of small parasitic plants and insects beset them, suck out their juices, disturb their functions, and are injurious to their life. In many cases, a diseased tendency in the plant seems to favour the production, or at least the increase of these enemies. We have already (321.) stated the production of what is called Mildew, or of the sprinkling of plants with aphides, to be a consequence of the unnaturally increased evaporation of saccharine matter. In the same manner, we observe a much more rapid and general production of Lichens on gooseberry bushes which grow upon an unfruit-

T

ful soil, than on those which are reared in good garden land. The blight also in Wheat, which is a degeneration of the grain, by which it passes into Coniomyci, seizes for the most part on grains that are not ripened, but not on those that have attained their perfect state. We can hence partly provide against the occurrence of this disease, by permitting the seed to become perfectly ripe and hard before it is taken in ; and even then it ought not to be stacked, but instantly thrashed out to such an extent, that only about two-thirds of the grain may be beaten out, and the less ripe seeds left behind.

That a diseased tendency must, for the most part, previously occur, when diseases are generated in plants, even by parasitic plants, is evident from the growth of Fungi on sickly stems and branches.   On our Alder we find the *Boletus alneus* ; on our Willow, *Boletus adustus, fumosus,* and *suaveolens* ; on our Beech, *Boletus fomentarius* ; on our Birch, *Boletus betulinus* ; on our Ash, *Boletus fraxineus* ; and on the stem of the Oak, *Dædalea quercina,*—as proofs of the diseased state of the plant, and of the tendency of its juices to corruption.   In like manner, a number of Fungi, as also the *Rhizomorpha subcorticalis,* appear upon the roots of trees when they stand in too moist a soil, shewing the diseased disposition of the roots.

419.

The barks of our trees are subject to cracks, to the flowing of resinous matter, to leprosy, and scab,—all of which diseases either pass into others that are still more dangerous, or invite a crowd of parasitic plants and insects, by which the evil is made worse.

The cracking of the bark in our fruit trees is for the most part the consequence of an over luxuriant growth, during which too many layers of alburnum are deposited, so that the inner bark and the rind cannot yield and make room for them. The inner bark being thus rent, usually makes way for the passage of the nutritive sap through the rent, and this sap, when it reaches the air, assumes the consistence of gum.   By this means the tree must naturally be enfeebled, and finally perish. When the bark crack , becomes hard and scaly, without this flowing of resin, it seems to suffer these injuries either from being exposed to too powerful a heat of the sun, from the influence of too dry seasons, or of too barren a soil. The leprosy or scab which we have mentioned, destroys the Olive-trees in Italy ; (Giovene in Opuscoli scelti,. xiii. p. 106.) In our Cherry trees, the *Sphæria pulchella* lodges between the torn rind and the inner bark. But still worse guests are the earwigs, the wood wasps, and the drilling worms, which often make long excavations under the rind, and thus become injurious to the life of the stem.

420.

A superabundance of raw juice in trees, which have been rendered feeble by management or weak by frost, produces the dropsy or jaundice. The bark becomes spongy, and, when pressed, gives out a great quantity of water ; the young shoots are thin and powerless, the leaves pale and yellow, and fruit is seldom produced. Stimulating and powerful nourishing matters from animal dung, mud, lime, and even soot, when they are applied in time, take away the disease, and shew the correctness of the explanation we have given of it.

We may attribute in some measure to the same cause the debility of the alburnum in forest trees ; because, wherever it occurs, either early frost, or other weakening causes, have prevented the concentration of the sap, and the proper formation of wood ; the unformed sap thus remains in the alburnum, (298.) ; (Mezieres, De la Force de Bois, p. 94 ; Slevogt, in Laurop's Annalen.)

421.

We meet with blotches and canker, as diseases of trees, the former of which are manifested by dark spots in the rind and wood, and commonly have their origin in the sterility of the soil, and in other enfeebling causes. This disease chiefly lays waste the Mulberry trees in Italy, and has given rise to manifold and very anxious investigations ; (Scopoli, Ann.

T 2

Hist. Nat. iv. p. 115.; Palletta, in Atti della soc. patriot. di Milano, Re. p. 303.

The canker proceeds mostly from a hardening of the bark, in consequence of which the juices become sharp and corrosive, make their way through the rents and slits of the rind, consume the parts that are below them, and at last completely destroy the wood.

422.

Our Corn crops are injured chiefly by parasitic plants of the lowest class. The rust upon the leaves and stalks is nothing but a *Puccinia*, which closes up the epidermis of the leaves, and thereby destroys their functions. Whether this plant is generated by the *Æcidium Berberides*, is a matter of much uncertainty; (Sir Jos. Banks on the Blight in Corn, in Ann. of Botany, ii. p. 51.)

The flying blight, by which Oats and Maize are chiefly injured, consists, as we formerly remarked, of an innumerable multitude of spherical black Coniomyci (Ustilago segetum), which presuppose a degeneration of the grain, and by which it is completely consumed. The soiling blight (Uredo sitophila, Ditmar.), on the other hand, contains smaller grains within a spherical covering, and is instantly discovered by its disagreeable smell, resembling herring-pickle. What more remote causes, beside the formerly mentioned predisposition of the grain of Wheat, contribute to the production of this evil, is not quite clear. But the infectious nature of this species of Coniomyci cannot be denied. So strong, indeed, is this tendency, that it clings to the glumes or caps of the wheat grain, and even to the fine hairy bodies which rise upon the points of the grain. The steeping of the grain in lime, and in a solution of common salt, cleans the Wheat indeed in most instances from any adhering rust; but the manure itself, if it be mingled with wheat-straw that had been blighted, communicates to the wheat, which is grown upon lands so manured, the diseased quality. It is not impossible, that even the want of a free circulation of air in wheat fields, or the too

deep ploughing of the furrow in iron-shot soils, may contribute to the diffusion of blight or rust.

423.

The innumerable kinds of Coniomyci which are connected with blight or rust, are undoubtedly individual forms ;
but they seem to owe their origin to a peculiar transforma-
tion of the spherical and vesicular bodies, which the generative sap contains. We hence see them appearing in abundance upon healthy leaves, in which we observe either a superfluity of juice, or a perspiration of the generative sap.
This is obviously the case with the *Uredo candida,* which appears in abundance upon the *Thlaspi bursa,* and is covered
with an inflated epidermis. The same thing is distinctly perceived in the *Uredo tremellosa* and *cincta* of Strauss, the production of which is commonly attended by the perspiration of
a fluid having the appearance of a jelly. And the *Nemaspo-
ra,* of which a great many kinds appear upon the branches of
the Poplar, shews, beyond all doubt, that it owes its origin to
the generative mucilage.

Again, we frequently see the degenerating juices of the leaves
becoming hard and producing shapeless masses of a blackish
colour, which are called *xyloma,* and have the appearance of
new shoots that have died. But if, by the influence of the
original vital powers, a new activity is awakened in the juice
which had been thus entirely changed or dead, the primitive
forms again appear in their simplest state. They appear as
spheres or vesicles, as in *Eurotium, Camptosporium, Spodo-
phleum* (Tab. V. Fig. 5, 7.), *Podisma, Phacidium,* and other
Fungi ; among which, however, there are some that seem
but slightly to injure the health of the leaves, because they
seem to be formed from the perspired sap upon the epidermis
only, as in the instance of *Phyllerium.* It often happens,
however, that the epidermis of the leaf is torn at the same
time, and surrounds, in a definite and peculiar form, a crowd
of Coniomyci, which have been generated in the sap, (*Æcidium, Röstelia.*)

424.

A quite different kind of decomposition is that to which Rye is subject, when it degenerates into *clavus*, which has a great resemblance to what in German is called the *stone blight* in Wheat. The grain swells, bends itself, and comes out from its husk. It has a sharp taste, and contains neither gluten nor saccharine matter, but putrid oil, free phosphorous acid, ammonia, and corrupted starch ; (Buchner, Repertorium fur die Pharmacie, b. iii. s. 100.) It is remarkable, that infusory insects, like vinegar eels, are found in it. But whether these were formerly present, and occasioned the disease of the grain, or were produced by the degeneration, has not been well ascertained. Meanwhile it is certain, that very moist years and wet lands contribute in a very great degree to the production of clavus.

425.

Insects occasion a numberless crowd of diseases, and of the causes of death, to plants. Some of these have not yet been sufficiently investigated, as the round navel-shaped bodies, which we find in such quantities on our fading Oak leaves, and which, by some authors, are called *Xyloma pezizoides*, and in the Flora Danica (1492), *Sclerotium fasciculatum*, but which have been best examined by Hopkirk, (Flor. Anom. p. 10. tab. xi. fig. 1.) It is impossible, and it is not indeed suited to our present purpose, to mention all the kinds of injuries which insects occasion to plants. We seek only to present the most important facts, according to the common division of insects.

Among the Coleoptera we mention, first, the May Bug, (Melolantha vulgaris), the larvæ of which are known as grubs, and live four years under ground, where they occasion the greatest devastation among the roots of trees. In their perfect state, too, they lay waste the leaves and buds of orchard and other trees. Equally injurious, but not so common, is the Spring beetle, (Elator Segetis) : the larva continues five years in that state, and is equally hurtful with the former kind to the roots of grain ; (Spence and Kirby,

Introd. to Entomol. i. p. 181.) The *Carabus gibbus*, also, not only in its larva state, but as a perfect insect, destroys the wheat crops throughout great tracts of country; (Germar's Magazine, i.) There is a Staphylinus, the larva of which insinuates itself into the grain of wheat, while it is springing, and kills it; (Walford, in Linn. Transact. ix. p. 156.) The *Bostrychus typographus* lives entirely upon the inner-bark of Pines, and so hollows it out into winding cavities, that six and thirty years ago, a million and a half of the Pinus picea and Pinus sylvestris, upon the Hartz alone, fell a sacrifice to it; (Trebra, in Schriften der Berlin Gesellschaft natur-forschender Freunde, b. iv t. 4) The *Anobium tessellatum*, which is also called the Death-watch, devours both decayed and living wood. Two Attelabi, also, may be mentioned, of which the one, *A. Bacchus*, is destructive to Vines; the other, *A. pomorum*, to the buds of Apple-trees. To the same order belongs the *Buprestis viridis*, the larva of which gnaws the alburnum of the red Beech, and produces the same kind of winding excavations, as the *Bostrychus typographus*; and, lastly, there is the well known Earth-flea (*Haltica oleracea*), which, during dry seasons, is so destructive to Greens, particularly to plants with cruciform blossoms, as the Rape, and Cabbage species. The *Crambus Brassicæ*, the larva of which lays waste also the fields of Caraway and Coriander, is the greatest enemy to the Rape (Brassica oleracea laciniata.)

426.

The second family of destructive insects is the Hemiptera, among which the leaf-lice, or aphides, from their incredible fecundity, and endless increase, destroy most of the plants upon which they alight. Almost every plant has its own kind of aphis, and of these many bring forth twenty times in a year. Even under the bark of Apple trees, a very destructive kind had long been found, the *Aphis lanigera*, which occasioned great devastations, especially in England, (Sir Jos. Banks, in Transact. Horticultural Soc. ii. 162.) The want of a free circulation of air is particularly favourable to the production of *Hemiptera.* They are accordingly produced in hot-houses to which little air is afforded. Cabbage plants are less subject to their depredations in the open field than in gardens. To this family belongs also the *Chermes,* one species of which, *Ch. cacti,* produces the Cochineal; a second is found upon the Oaks of the South, and produces the French *Chermes*; a third kind, *Ch. polaricus,* nestles in the roots of the *Scleranthus perennis;* and all the three kinds produce colouring matter. To this order belong the *Cocci,* which fix themselves almost immoveably, and quite flat, upon the plants of our hot-houses, and suck out the sap of the plants, with their proboscis, which springs from their breast. We are acquainted with two species, *Coccus hesperidum* and *C. adonidum.* There is also the *Cercopis spumaria,* which sucks the juices of Grasses, and especially of Willows, and gives it out again in the shape of foam: it is called Cuckow's Spittal,—and when, as sometimes happens, it falls down in drops, it has given rise to the expression of Dropping-Willows.

The small *Thrips physapus* is also very common in the flowers of many plants, and perhaps assists in the impregnation, but frequently, also, it gnaws the germen.

The flowers of *Juncus obtusiflorus,* and *acutiflorus,* are disfigured by the puncture of the *Livia juncorum,* and the mischief done in corn fields by the *Acheta gryllotalpa,* is known to every person.

427

The innumerable crowds of butterflies, particularly in the caterpillar state, are exceedingly destructive to plants. The greatest enemies to fruit-trees are the caterpillars of *Bombyx dispar, chrysorrhœa, cœruleocephala, Hispaniola, processionea, Neustria,* and of *Noctua brumata.* The caterpillars of *Papilio Cratœgi, Brassicœ, Rapœ,* and *Napi,* suck principally the garden vegetables. In Fir woods, the larvæ of *Bombyx Pini, Hadena piniperda,* and *Phalæna geometra piniaria*; and in Oak woods, the larvæ of *Bombyx monacha,* and *Noctua brumata,* occasion very great devastation. The soft woods of the Willow and Poplar are attacked by *Bomb. cossus, Sesia crabroniformis,* and *Nitidula grysea.* The caterpillars of *B. graminis,* lay waste the meadows, (De Geer, Mem. sur les Insects, ii. p. 341.); and those of *Hepiolus Humuli,* destroy the Hop-gardens.

428.

Among insects of the order Hymenoptera, the Gall Insects are the most remarkable, for they deposite their eggs in plants, which consequently exhibit remarkable excrescences, and these are often distinguished by the most singular shapes and peculiar colours. The mossy and crisped excrescences upon the Wild Rose, which are known by the name of *Bedeguar,* proceed from *Cynips Rosæ ;* the gall-nut, from *C. quercus,* which produces different kinds, however, according as it appears on the leaves, on the leaf-stalk, or on the flower-stalk. *Hieracium sabaudum, Salvia pomifera,* and *Glechoma hederacea,* exhibit similar excrescences. The Wild Figs, too, are punctured by similar insects, and although the swelling of the fruit is thus assisted, the animals have no effect in producing the impregnation of the plant; (Pontedera Anthol. ii. p. 33; Olivier Voy. dans l'emp. Othom. ii. p. 171.)

The origin of what is called the Willow Rose, from the puncture of *C. Salicis,* is in the highest degree remarkable. In spring, this insect deposites its eggs in the leaf-buds of the *Salix Helix, alba,* and *riparia.* The new stimulus attracts the sap,—the type of the part becomes changed, and, from the prevailing acidity of the animal juice, it happens, that in the rose or stock-shaped leaves, which are pushed out, a red colour, instead of a green, is evolved. Superstition is thus frequently cheated in its hopes, but it is also delivered from its fear ; (Grass, in Eph. Nat. Cur. Dec. i. ann. 5. ; Wincler eben Dass. ann. vi. vii. n. 117. 229. ; Albrecht, in Act. Nat. Cur. vol. ix.; Schroter, in Berl. Samml. b. ii.; Sims, in Ann. of Bot. i. p. 374.)

The *Hylotoma Fabr.,* the larva of which are distinguished by two prominent eyes, and eleven pair of feet, are extremely injurious to Pines, especially one species of the insect (*Hylo-*

toma pini). The yellowish green caterpillar shews itself in incredible numbers, and destroys the cones completely.

429.

Among gnats and flies we mention, first, the *Musca pumilionis*, Bierk., or the *Mosillus arcuatus*, Latr. This fly lays hold of the Wheat and also of the Rye crops, while they are young, but these frequently shoot out more luxuriantly afterwards; (Spence and Kirby, Introd. to Entom. i. p. 170.) A small yellow gnat, *Tipula tritici*, eats into the blossoms of Wheat, and destroys them; (Linn. Trans. iii. p. 242.) Lastly, we may enumerate, among the unwinged insects, the small red acarus of our hot-houses (*Acarus telarius*), which, when enough of air is not given to the plants, or when they are kept too warm, overspreads them with a fine web, and so destroys them.

CHAP. VII.

HISTORY OF BOTANY.

I. *Ancient History till the Revival of Science.*

430.

Scientific Botany is indebted for its origin to the philosophical schools of ancient Greece. But it was the physics of plants, much rather than descriptive botany, which was then cultivated, because, in the *first* place, from the small number of plants which were then known, and which, among the Greeks and Romans, scarcely exceeded a thousand, it was not found necessary to think of classifying them,—of forming a theory for this purpose,—of arranging them according to a scientific system,—and of giving them a regular nomenclature; *secondly,* Because the views of the ancients, with respect to

natural bodies, were entirely confined to the explanation of
phenomena, and to the employment of physical substances in
arts and trades: and, in the *last* place, because the physics of
plants, like physics in general, were then derived from mere
processes of reasoning. It is hence that, in the writings and
fragments of the Greek philosophers, we find chiefly some physical notices respecting the life and nourishment of plants,
which they endeavoured to explain by the analogy of the animal kingdom ; and along with these many happy ideas respecting the rank which plants hold in the scale of natural
bodies, and respecting their relations to external animals.

At the time when the Athenian Republic was in its most
flourishing condition, it is true that several persons, who were
called Rhizotomæ, devoted themselves exclusively to the digging of roots and finding of herbs, for the advancement of
arts, and particularly 'of medicine. Some of these persons,
who were called Pharmacopolæ, seem even to have issued from
the schools of the philosophers, and to have acquired for themselves a comprehensive knowledge of plants, whence they
were called Cultivators of Physics. But the greater number
pursued their occupation as market-cryers, and observed a
multitude of superstitious customs, on which account they
are rather to be regarded as traders, than as men who had
been trained in a scientific manner.

431.

The first founder of the natural science of plants, was undoubtedly Aristotle of Stagira, to whom the nick-name of
Pharmacopolist was even given, because, for a long time, he
employed himself in collecting medicinal plants. But his genuine works respecting plants have been lost, and what we
now possess under this name, is but the insipid forgery of an
ignorant Greek of the middle ages.

Aristotle's follower and favourite scholar, Tyrtamus of
Lesbos, to whom his master gave the name of Theophrastus,
on account of his eloquence, drew his principles, undoubtedly,
from the information of his great teacher. He also cultivated
the knowledge of plants entirely after the fashion of the Peri-

patetic School. But he seems to have undertaken few jour-
neys and travels, since he always appeals to the testimony of
the diggers of roots, the cutters of wood, and the inhabitants
of the mountains. But, as he lived between the years 371
and 286 before Christ, the ever-memorable march of Alexander the Great through Asia and Africa, afforded him an
opportunity of becoming acquainted with many foreign plants.
Although he notices these but occasionally, and without of-
fering any exact descriptions of them, yet his works, under
the title of A History of Plants, and On the Causes of Plants,
are immortal memorials of his ceaseless attention to the vegetable world, and of his excellent observation of the phenomena which it presents. But we must not expect from him
either a scientific arrangement of objects, or a systematical
enumeration of the plants known to him; but we must view
the whole as the production of a philosopher, who, almost
without predecessors, endeavoured, for the first time, to employ the reasoning faculty upon the phenomena of the vegetable world. The best edition of his works is that by
Schneider, and was published, in four octavo volumes, at
Leipsig, in 1818. Theophrastus was also the first who kept
a garden for plants, and in his legacy he named some of his
scholars as keepers of this property.

432.

But he found none of his scholars worthy of being a successor
to himself. Notwithstanding the foundation, during his time,
by the liberality of the Ptolemies in Alexandria, of the most
celebrated school of antiquity; and, although, from rivalry
with the kings of Pergamus, the libraries in Alexandria
were raised to the rank of the best in the world; yet the very
liberality of the Egyptian kings produced, by means of the
superfluity of literary helps, such a learned indolence, and
such a predilection for dialectic and grammatical investigations, that the study, as well as the science of nature, were
entirely neglected. Nay, the Pharmacopolæ were again separated from the learned physicians and teachers of that

school, and employed themselves, as formerly, in the digging
of roots,—a low and superstitious trade.

The kings Mithridates, Eupator of Pontus, and Attalus
Philometer of Pergamus, promoted, to a certain extent, the
knowledge of plants, by maintaining botanical gardens, in
which they reared poisonous plants, and made experiments
with other plants, as antidotes to poison. In the courts of
these kings lived the two most learned rhizotomæ of antiquity,
Cratevas, and Nicander of Colophon. The work of the former exists only in manuscript. But Nicander has left us two
very obscure works respecting poisons and antidotes, both of
which have been excellently edited by Schneider, in 1792 and
1816.

433.

After Greece was subdued by the Romans, the knowledge
of the conquered so far passed over to the victors, that the
latter, who always sought out only what was useful, cultivated the study of plants to as great an extent, as it afforded
advantages to the arts and trades.

In the works of the old Romans, Cato, Varro, and Co-
lumella, respecting Rural Economy, the best editions of which
are those published by Schneider, in 1794; as also, in the Geor-
gics and Eclogues of Virgil, we find a multitude of plants
named, which were useful in horticulture and agriculture.
It is much to be lamented, that we no longer possess the writings of the younger Juba, king of Mauritania, whom Cæsar
had caused to be educated in Rome. These works consisted
of a Treatise on the History of Nature,—a Description of the
Canary Isles, which were discovered by him,—Notices respecting Lybia,—and a History of Arabia. According to the
testimony of the ancients, he described plants, on all occasions,
with the most scrupulous care.

434.

The most celebrated writer among the oldest botanists,
is Pedacius Dioscorides, of Anazarbus, in Silicia. He lived
in the middle of the first century of our æra, was a phy-

sician, and followed the Roman armies in their expeditions through the greatest part of the Roman empire. The work of his which we possess, and the best edition of which was published by Sanacenus, at Frankfort, in 1598, is entitled Materia Medica, and contains, therefore, an enumeration of all the medicinal plants which were known to the ancients. These are arranged in rather a capricious order, and are de. signated not only by the common Greek names, but also by the Roman, Punic, or African, and other barbarous names : they are frequently described at great length, their situation assigned, and proofs of their medicinal efficacy produced. This work, next to that of the elder Pliny, has exercised the most enduring dominion over the schools, since it was held, for more than fifteen hundred years, to be the only fountain of all knowledge relating to natural history, and particularly of botanical information.

435.

Caius Plinius Secundus, commonly called the Elder, a commander and statesman during the middle of the first century of our æra, left behind him a Summary of all Science, Knowledge, and Arts, which, for the most part, he had extracted from the Greek and from some Roman writers. The work bears the title of a History of Nature, or of the World, and the best edition of it, in ten octavo volumes, is that published by Franz, after Harduin, at Leipsig, between 1778 and 1791. The plants are treated in it, in alphabetical or. der, according to the descriptions of Theophrastus and Dioscorides. Here and there also, some notices are added, and plants are described, which were unknown to his predecessors ; and he himself has informed us, that in his youth he acquired his knowledge of plants in the garden of Antonius Castor, a son-in-law of the well known King Dejotanus.

436.

Among the later Romans, the number of persons who cultivated the knowledge of nature diminished, in proportion as the night of barbarism descended, and, for a long time, the

remains, even of Greek and Roman learning, were entirely hid.
The Arabians, indeed, after they had instituted schools
of learning, infirmaries, and laboratories, applied themselves
diligently to the study of medicinal plants. But they drew
their knowledge entirely from Dioscorides, whom, however,
they did not peruse in the original, but in a translation which
had been made from a Syrian copy. But, as it is probable
that neither of the translators was a botanist, they could neither avoid the grossest mistakes, nor be of the least advantage
to the science.

Nevertheless, the flourishing trade which this nation carried on, for some centuries, from Madeira to China, made
them acquainted with some remarkable oriental plants, which
had escaped the notice of the Greeks. There were also, in
the western parts of the Arabian empire, some inquisitive students of nature, who endeavoured to correct and to extend
their knowledge by travel, among whom was Ebn Beitar, a native of Mallaga, who flourished in the thirteenth century,
and whose work we possess only in manuscript.

437.

About the beginning of the eleventh century, the Arabians
became the teachers of the other nations of western christen-
dom, who now formed their schools of learning according to
the Mahometan pattern, and translated their books from the
Arabian. In this manner arose a four times repeated translation of Dioscorides, which served as the foundation of the
knowledge of medicinal plants; and we may easily imagine
how completely changed this work must have seemed to be,
and how little advantage science could gain from it.

The first faint spark of a sure knowledge of plants gleamed
during this darkness of the middle ages, when, after the example of the Minorite Monks, whom the Pope sent, in the
thirteenth century, as missionaries into the Mogul empire,
and to the court of the pretended Prester John, several
merchants undertook the same expedition. Among these,
the most illustrious was Marco Polo of Venice. He examined, during fifty years, most of the regions of Middle

and Southern Asia, as well as the eastern coast of Africa, brought from thence many rare fruits and seeds, and, for the first time, described, from actual inspection, the plants of India, and of the islands of the Indian Ocean. This treatise is found in the original, in the second volume of Ramusio's great collection.

Meanwhile, in the cloysters of the West, some know-ledge of medicinal and garden plants had been preserved,—which plants were endeavoured to be made extensively known by what was called the Hortus Sanitatis. This contained an alphabetical catalogue of useful plants, to which miserable plates were added, and which was translated from one lan-guage into another. The Latin edition of Meidenbach, at Mentz, in 1491; the German of Schonsperger, at Augs-burgh, in 1488; and that in the Lower Saxon dialect, by Cube, at Lubeck, in 1492, are well known.

II. *First Establishment of Scientific Botany.*

438.

During the flourishing condition of the free states of Italy, which had been raised to distinction by trade, and by their constitutions, science and art were first established on a pro-per basis, and those Greeks that had been banished by the Turks, namely, Emanuel Chrysoloras, Bessarion, and Theo-dore Gaza, in particular, first made the Italians acquainted with the great masterpieces of ancient Greece. Hence arose a very active and well known rivalry,—in the search for memo-rials of Grecian art and science,—in the multiplication and il-lustration of the genuine works of the ancients by writing and printing,—and even in the imitation of their celebrated works. It was now that Dioscorides and Pliny were, for the first time, studied in the original,—the belief being universal that their works are the only and the abundant fountain of the knowledge of plants. But, at the same time, attempts were made to ascertain what native plants properly bore the names which the ancients had assigned.

439.

The Italians, Hermolaus Barbarus, Marcellus Virgilius, Nicolaus Leonicenus, John Manardus, and Antony Musa Brassavola, became celebrated and useful, indeed, in their age, by such investigations; but they pursued these studies rather as grammarians, than as natural historians.

The proper fathers of the later botany were Germans, who, independent of the ancients, examined and made known the plants of their native country. Among these, the most ancient was Otto Brunfels, schoolmaster in Strasburgh, afterwards a physician, who died in 1534. His Herbarum vivæ Icones were published at Strasburgh, in folio, with wood cuts, in 1532 and 1536.

To him succeeded Leonhard Fox, professor at Ingolstadt, and afterwards at Tubingen. He died in 1565. His Historia Stirpium appeared at Basil, in folio, in 1542. In this work, we find wood cuts, true to nature, of about four hundred German plants, and here also we find the first catalogue of technical terms in botany.

Hieronymus Tragus, schoolmaster at Zweybrucken, afterwards a physician at Hornbach, who died in 1554, had also collected plants on the Hundsruck, the Eyfel, the Ardennes, the Vogeses, on Jura, and in the countries on the Rhine. His book on herbaceous plants appeared in German, at Strasburgh, in 1551.

Valerius Cordus, also, who was taken from the world by an early death, at Rome, in 1544, had carefully examined the plants of Germany. His literary remains were published by Conrad Gesner, at Strasburgh, in 1561.

This Gesner, one of the most learned and excellent men of his time, was schoolmaster and corrector, afterwards physician and professor, at Zurich, and died in 1564. He acquired the highest merit as a botanist, by not only collecting and describing the plants of Switzerland, but also by leaving behind him a great number of excellent designs, wood cuts, and copperplates, of foreign plants, in which he was the first who attended to the parts of fructification. These remains came two hundred years afterwards into the hands of Schmidel,

U

who published them in 1754 and 1771, under the name of Opera Botanica.

The plants of the Hartz were published by John Thal, a physician at Nordhausen, who died in 1587, in a work entitled Sylva Hercynia, at Frankfort, in 1588, in quarto. This publication, after the death of the author, was taken care of by Joachim Camerarius.

A scholar of Tragus, named Jacob Theodore Tabernamontanus, a native of Bergzabern, in Alsace, published a work similar to that of his German predecessor, the best edition of which is that published by Hieronymus Bauhin, at Basil, in folio, in 1731. Although this work contains many things copied from other authors, we also find in it a multitude of plants which were not known to his predecessors.

440.

The inhabitants of the Netherlands, who had been incorporated with the German empire under Charles the Fifth, being urged by the tyranny of his successor, Philip the Second, freed themselves from the Spanish sovereignty, and obtained their independence, after an opposition of many years. This bloody struggle promoted the trade and prosperity of the nation. Arts and Sciences were cultivated in the Netherlands, with German diligence and zeal, and botany prospered, in proportion to the opportunities that were afforded of obtaining plants from foreign countries.

Rembert Dodonæus, a native of West Friesland, an Austrian physician, afterwards a professor at Leyden, who died in 1586, was one of the oldest and most distinguished founders of botany. His Stirpium Historiæ Pemptades VI. were published at Antwerp, in an enlarged edition, in folio, in 1616.

Matthias Lobelius, of Flanders, who was afterwards superintendant of the garden of Queen Elizabeth of England, and died in 1616, not only discovered a multitude of plants during his travels in France, the Netherlands, England, and Germany: he also made the first attempt to arrange them according to a certain natural affinity. His Stirpium Nova Adversaria, published at London in 1570 and 1605, in folio, was succeeded by his Plantarum Historia, published at Antwerp in 1576; and, at last, all the plates of his works, and those of his predecessors, were republished in the Iconibus Stirpium, at Antwerp, in 1591, in quarto.

In ardent zeal for the discovery of plants,—in submitting to sacrifices of every kind,—and in the very successful issue of his labours, Charles Clusius, of Antwerp, excelled all his predecessors. As companion of the noble Fuggerius, through the whole south of Europe, he enjoyed every opportunity of collecting, describing, and drawing, the plants of Germany, France, Spain, and Portugal. He lived several years in England, and also in Vienna, as superintendant of the imperial gardens, from whence he made the tour of Austria and Hungary. At last he was professor at Leyden, and died in 1609. His chief work is the Rariorum Plantarum Historia, published at Antwerp, in folio, 1601.

441.

Among the Italians of the sixteenth century, some also distinguished themselves by an extensive and careful search for plants, especially Anguillara, who was for a long time professor at Padua, afterwards at Ferrara, and died 1570. No person was better acquainted with the plants of his native country, of the large islands in the neighbourhood of Italy, of Greece, also of Dalmatia, and of the Grecian islands. He described them, with a constant reference to their names in Dioscorides and Pliny, in his Semplici, published at Venice, in octavo, 1661.

Peter Andrew Mattioli, a native of Sienna, and an Austrian physician, who died in 1577, was one of the best informed discoverers in botany. His Commentaries on Dioscorides, are either cited according to the edition of Valgrisius, with small figures, published at Venice, 1560, in folio, or according to that of Bauhin, with large figures, at Basil, 1674, in folio.

One of the most active and eminent discoverers of Italian plants, was Fabius Columna, a Neapolitan of high birth, whose bad health was the occasion of his predilection for botany. We possess a work of his called Phytobasanos, published at Naples, 1592, and another denominated Ecphrasis Stirpium, published at Rome, 1616, in quarto, in which the drawings of plants, after the model of Gesner, are connected with representations of the parts of fructification.

442.

The knowledge of Indian plants was promoted in the sixteenth century, by the victories of the Portuguese; and the two Portuguese physicians at Goa, Garcia ab Orto, and Christopher da Costa, published accounts of many medicinal plants, which Clusius translated in his Exoticis, printed at Antwerp, in folio, 1605.

The discovery of America also unexpectedly enriched the science, and the Spanish governor in the West Indies, Gonzalo Hernandez Oviedo, was the first to give a proof of the advantages thus obtained.

The east was investigated by Peter Belon, who was sent to travel at the expence of the Cardinal Tournon; by Leonhard Rauwolf, and by Prosper Alpinus, professor at Padua, who died in 1617. The observations of the first of these are translated in the Exoticis of Clusius. Rauwolf's Travels were printed in German, at Lauingen, 1582, in quarto; and the works of Alpinus, De Plantis Ægypti, in 1640, in quarto, and De Plantis Exoticis, at Venice, 1627, in quarto, contain excellent plates and descriptions of a number of very rare plants.

443.

It has already been mentioned, that Lobelius made the first attempt to establish a definite arrangement of plants. But Andrew Cesalpinus, professor at Risa, who died in 1603, gave a wrong direction to the search after fixed scientific principles; for, in his work De Plantis, published at Florence, 1583, in quarto, he first constructed a system, the foundations of which were the fruit and its parts, especially the embryon, and its situation in the seed.

As, about the end of the century, an almost infinite multitude of plants was discovered, and different names were given to these by each writer, it became a matter of urgent necessity, to review the synonymes, in order to give some certainty to the knowledge of plants. This difficult labour was undertaken by Caspar Bauhin, professor at Basil, who died in 1624. His Pinax Theatri Botanici, printed at Basil, 1623, in quarto, is still a necessary aid in the complete study of the science. The Theatrum Botanicum, which was intended to contain the natural families of plants, has not been fully published, but we possess only the Prodromus, published at Frankfort, 1620, in quarto, with excellent plates, and the first part of the larger work, which was published at Basil, 1658, in folio. Caspar's brother, John Bauhin, physician to the chief of Mumpelgard, who died 1613, collected a great many plants, and arranged them according to a plan similar to that of his brother. But his Historia Plantarum Universalis, which was published in three volumes, at Ifferten, 1651 and 1653, disappointed expectation, both in regard to the arrangement, and to the plates.

III. *First Establishment of the Doctrine respecting the Structure and Systematical Arrangement of Plants.*

444.

We are principally indebted to the establishment of learned societies in the seventeenth century, and to the invention of the microscope, for the first attempts at a more minute examination of the structure of plants. In the Society of London for the Promotion of Science, which was liberally supported by Charles the Second, several men were found, under the management of the King himself, who occupied themselves exclusively with the dissection and microscopical examination of plants. Of these, the most distinguished was Nehemiah Grew, secretary to the society, who died in 1711. His discoveries are recorded in the immortal work, the Anatomy of Plants, London, 1682, in folio. In this work we find the first notice of the twofold sex of plants, which doc-trine he had learned from Thomas Millington, a professor in Oxford.

The same British Scientific Society also published the ex-cellent and peculiar investigations of Marcellus Malpighi, a professor at Bologna, who died 1694, in the *Anatome Plan-tarum*, 1675 and 1679, in folio.

A citizen of Delft, named Antony Leuwenhoek, who died 1723, also contributed very much to the establishment of this science, by his minute investigations respecting the structure of plants.

The French Academy of Sciences, founded in 1665, also distinguished itself by discoveries respecting the structure and nature of plants ; its first members, Claude Perrault, who died 1688, Denis Dodart, who died 1707, and Edme Ma-riotte, who died 1684, having instituted a multitude of in-teresting investigations, especially respecting the nourishment of plants.

445.

Joachim Jung, a German, born in Lubeck, and Professor at Hamburgh, who died 1657, first improved the technical language, and published in his lectures more accurate notions respecting the relations and nomenclature of plants. Al-though his Opuscula were first published a hundred years af-ter his death, at Coburg in 1747, yet copies of his Institutes had been circulated in Great Britain, and the natives of that country, who appeared as reformers of scientific botany, fol-lowed, according to their own confession, these copies.

Among these Britons, the first was Robert Morison, a Scotchman, who, during the usurpation of Cromwell, lived in France, and afterwards was Professor at Oxford. He died in 1688. We have already (211.) mentioned his Præludia Botanica as the work in which the first critique on the ar-rangement which at that time was in use is found. He also laid the foundation in the same work for a better discrimina-tion of genera. He became chiefly meritorious by the publi-cation of his Historia Plantarum Universalis, which appeared in three volumes, at Oxford 1715, in folio, and contains more than 3600 species of plants, arranged according to the natural method, and illustrated with good plates. In the steps of Morison followed John Ray, an English clergyman, who, after having travelled during many years through the whole of Europe, lived without preferment, and died 1705. His Methodus Plantarum emendata, the third edition of which was published in 1733, contains the true principles according to which the genera and species of plants ought to be distinguished. At the same time, a natural method is pointed out in the same work, in which attention is paid as much as possible to all and each of the parts, and no preference is given to one above the rest Ray also distinguished himself with respect to the British Flora, by his Synopsis Methodica Stirpium Britannicarum, published for the third time by Dillenius in 1724. He likewise published a general view of the vegetable kingdom, according to the natural method, under the title Historia Plantarum, in three volumes, London, 1686 till 1704, folio.

Paul Herman, a Professor at Leyden, who died in 1695, attempted to improve this method, by paying more regard to the fruit ; as did also Herman Boerhaave, who was Professor in the same place, and died 1738. The Flores Floræ Lugduni-Batavæ, Leyden 1690—12, of the former author ; and the Index I. and II. Plantarum, quæ in Horto Lugdunensi aluntur, of the latter, Leyden 1720-4, deserve here to be noticed.

446.

Although botanists were now in a fair way of introducing the natural method, attempts were not wanting to lay the foundation of an artificial system, for the sake of beginners. The corolla was the first part that drew attention, and its division and form were the foundations of the earliest artificial system. Augustus Quirinus Rivinus, Professor in Leipzig, who died 1725, set out in his great work, Introductio generalis in Rem Herbariam, Leipzig 1690 till 1699, in folio, from the principle that the corolla, as being the part which marks the perfection of the plant, is the most important part. Hence he divided plants according to the parts of the corolla, but chiefly according to the regularity or irregularity of its form. But he extended the idea of irregularity so far, that he regarded even the bent form of the pistil as an instance of the irregularity of the corolla, without reflecting, that, in the first place, the pistil in some species of the same genus, as Pyrola and Epilobium, is bent downwards, and in others is erect; and, in the second place, that this bent position is often the consequence of dichogamy. This system no doubt suffered much well founded opposition, particularly from Ray and Dillenius; but in Germany it was so great a favourite, that at a later period it was with difficulty overcome by the Linnæan system.

447.

An excellent French botanist Joseph Pitton de Tournefort, who, after having travelled for many years through the South of France, and among the Pyrenees, had also examined the Levant, and died as Professor at Paris 1708, founded a system similar to that of Rivinus, but with more regard to the form of the corolla than to its regularity. He proposed this system in a work which appeared at Paris, under the name of Institutiones Rei Herbariæ, in 1719, in three volumes, with 489 plates, in which the characters of most of the genera are given in a masterly manner. An excellent account of his travels in the east, Relation d'un Voyage du Levant, was published at Amsterdam 1718, in two parts; and the new plants which were found there, 1356 in number, were inserted in the *Corollarium* of his Institutions.

448.

Meanwhile, the knowledge of foreign plants was promoted in various ways. The Dutch took Brazil from the Spaniards; and Count Moritz of Nassau, as governor of the newly conquered territory, took with him a natural historian, William Piso, and an artist, George Marcgraf, whose observations on the plants and animals of Brazil were published at Amsterdam in 1658, in folio, under the title, De Indiæ utriusque re naturali.

In the East Indies, also, the sovereignty of the Dutch was favourable to science. The Governor of Malabar, Henry Adrian van Rheede, commanded the plants of Malabar to be marked and described in a style of kingly magnificence. Hence originated the Hortus Malabaricus, published between 1676 and 1703, in twelve folios. This work was surpassed, not in the number of species, but in the value of the definitions and descriptions, by the Herbarium Amboinense, which was patronized by George Eberhard Rumph, governor at Amboina, and was published by John Burmann, in seven volumes, at Amsterdam, between 1741 and 1751.

The West India plants were investigated by Hans Sloane, a learned Irishman, who was physician to the Governor of Jamaica, and afterwards to the King of Great Britain, and president of the Royal Society. He died 1753. His principal work is entitled a Voyage to Madeira; in two volumes, London, 1707 to 1727, in folio.

449.

Botanical gardens also became extremely common during the seventeenth century. In Padua a botanical garden had been laid out since the year 1533; in Pisa since 1544; in Pavia since 1556; and in Bologna since 1568. About the end of the 16th century, Peter Richier de Belleval laid out the first botanical garden at Montpelier in France, in which he reared the plants of the south of France, and left behind him a multitude of notices respecting them, which, at the distance of two centuries, were published by Villars and Gillibert in the Demonstrations Botaniques of the latter, at Lyons 1796, in quarto. The Royal Garden at Paris was first laid out in 1635. In England, the Royal Garden at Hampton Court, and the garden of medicinal plants at Chelsea, had been richly stocked since the time of Queen Elizabeth. The superintendant of the former was John Parkinson, whose Theatrum Botanicum, published at London 1640, in folio, contains an arrangement of the plants according to their uses and situations. His disciple was Leonhard Pluknet, who became known by drawings of very rare plants in his Almagestum Botanicum, published in London 1697 and 1705, in quarto. The garden for medicinal plants was superintended by Jacob Petiver, who died in 1718, and whose works, published in London 1764, in three folio volumes, also contain a multitude of plates of plants.

In the Netherlands, the most celebrated garden was that at Amsterdam. Its rare plants were ordered to be engraved in copper and described, by the chief councillor John Commelyn. We have thus obtained the work entitled Horti medici Amstelodamensis rariorum plantarum descriptio et icones, published at Amsterdam in 1697 and 1702, in two volumes folio. The garden at Leyden, laid out in 1577 by Bontius, was now superintended by Paul Hermann. His Catalogus Horti Lugduno-Batavi, published at Leyden 1687, in octavo, and Paradisus Batavus, at Leyden 1705, in quarto, are valuable works. The most remarkable plants of the Dutch gardens were ordered to be engraved with great care by Jacob Breyn, a merchant in Dantzig, and he has described them in his Exoticarum plantarum centuria, published at Dantzig in 1678, in folio. Among the Dutch gardens, that which was supported by the Bishop of Eichstadt, under the inspection of Basilius Besler, an apothecary at Nurnberg, was very celebrated. A description of its rare plants is contained in a magnificent work, entitled Hortus Eystettensis, published in 1613 in folio.

Among the gardens of Italy, that at Bologna was most celebrated,—the superintendant of which, Jacob Zanoni, caused to be engraved and described a multitude of rare plants, in his Istoria Botanica, published at Bologna in 1675 in folio. What was denominated the Catholic Garden, the owner of which was the Pope, and its superintendant Francisco Cupani, was remarkable for a multitude of rare plants, natives of Sicily. The great work entitled Panphyton Siculum, which contains plates of these plants, is now only, in some of its fragments, an ornament of libraries.

450.

Native Floras were also objects of very careful investigation during the seventeenth century.

Jacob Barrelier, a Dominican, a native of Paris, who died 1673, had carefully examined the vegetable kingdom throughout the whole of the south of Europe, and made a multitude of discoveries, which were published long after his death, under the title Plantæ per Galliam, Hispaniam, et Italiam observatæ; Paris 1714, in folio, with 1324 copperplates. His labours were rivalled by those of Silvius Paul Boccone, an Italian Cistercian monk, who travelled over the greatest part of Europe, and died in his native town Palermo in 1704. His most important works are his Icones et Descriptiones rariorum plantarum Siliciæ, Oxford 1674, in quarto ; and his Museo di Piante rare, Venice 1697.

The Flora of Prussia found an editor in John Losel, professor at Königsberg, who died 1656, and whose Flora Prussica was published at Königsberg 1703, in quarto.

IV Events preparatory to the Reformation of Linnæus.

451.

During the time which intervened between Tournefort and the publication of the Linnæan reformation, the appearance of the latter author was introduced by some learned men. In particular John Henry Burkhard, a physician in Wolfenbuttel, published in an epistle to Leibnitz, which was again edited by Lorenzo Heister 1750, the passing thought, that plants might be divided according to the number of their filaments. But, as he almost immediately opposes this idea, he can by no means be considered as properly a predecessor of Linnæus.

But the doctrine of the sex of plants, which had been obscurely hinted at by Grew, was experimentally illustrated by Jacob Bobart, and established by John Ray. Rudolph Jacob Camerarius, professor at Tubingen, endeavoured circumstantially to prove it, by observations and experiments, in a letter to Valentini. That letter was again copied by Gmelin, in his treatise De Novo Vegetabilium exortu.

At this time, during the prevailing love for atomic explanations, the discovery of the seminal animalcules, gave an opportunity for their employment in accounting for the fructification of plants. Samuel Moreland, Stephen Francis Geoffrey, and others, maintained that the matter of the pollen penetrated into the ovarium, (381.) But this account was opposed in the strongest manner by Sebastian Vaillant, professor at Paris, who died 1721, in his Discours sur la Structure des Fleurs, Leyden 1718, in quarto. Vaillant also obtained distinction by his disquisitions respecting many families of plants, as well as by his Parisian Flora, Botanicon Parisiense, Leyden 1727, in folio.

452.

The most accomplished predecessors of Linnæus were Jacob Dillenius, John Scheuchzer, and Peter Antony Micheli. The first was early a professor in Giessen, afterwards superintendant of the Sherardian garden at Eltham in England, and lastly professor at Oxford. He died 1747. How little he was attached to the systems of his time, how completely he understood the manner of investigating the parts of fructification, even of Cryptogamous plants, had been already proved by his Catalogus Plantarum, Giessen 1718, in octavo. In England he published the Hortus Elthamensis, London 1732, a work which was intended to combine unexampled beauty of plates, with the most minute investigations and the most careful descriptions. But every thing which had hitherto been done in this department, was surpassed by his Historia Muscorum, Oxford 1741, in quarto.

The merits of John Scheuchzer, professor at Zurich, who died 1737, are chiefly confined to an examination and arrangement of the Grasses, which we find in his Agrostographia, Zurich 1775, in quarto.

Peter Antony Micheli, superintendent of the gardens of the Grand Duke of Florence, and who died 1737, laboured in the same spirit as Dillenius, searching chiefly for the sexual parts of the lower organic bodies. His Nova Plantarum Genera were published at Florence 1729, in small folio.

453.

The boundaries of our knowledge of plants were also uncommonly extended, and the reception of the Linnæan system prepared by travels into foreign countries, undertaken by acute and well informed natural historians. The most remarkable of these travellers was Charles Plumier, a monk of the order of the Minimi, who at different times spent several years in the West Indies, and died 1704. His Nova Plantarum Genera, Paris 1703, contains descriptions and plates of 120 new genera. He described the West Indian Ferns in his expensive work, Traité des Fougeres de l'Amerique, Paris 1705, and five hundred descriptions of plants which he had left behind him were published by John Burmann, under the title Plantarum Americanum fasciculus 1—10; Amsterdam, 1755 to 1760, folio.

Another monk of the order of the Minimi was Lewis Feuillée, who lived two years in Chili and Lima as royal botanist and mathematician. He died 1732. In his Journal, written in French, Paris 1714 to 1725, we find a multitude of rare plants of these regions described and figured.

454.

Asia was very diligently and thoroughly examined by Engelbrecht Kampfer. He was a native of Lemgo, and went with the Swedish deputies to Persia, where he staid some years, and then sailed with the Dutch fleet to the East Indies, remained a year in Batavia and two years in Japan, and at last returned, at the distance of ten years. He died 1716. In his Amœnitates Exoticæ, Lemgo 1712, in quarto, he published excellent descriptions and plates of Japanese and of some Persian plants.

Asia Minor and Armenia were first examined by John Christ. Buxbaum, a native of Merseberg, who was physician to the Russian Embassy at Constantinople. He died 1730.

His principal work is entitled Plantarum minus cognitarum centuria, 1—5, St Petersburg, 1728 to 1740.

Northern Asia was traversed during ten years, at the command of the Empress Anna, chiefly by John George Gmelin. He died 1755. His Flora Sibirica, in four volumes, St Petersburg 1747 to 1769, contains a multitude of the most remarkable and rare plants.

John Burmann, professor at Amsterdam, who died 1780, made use of the collections of other travellers in his Thesaurus Zeylandicus, Amsterdam 1737, and in his Rariorum Africanum Plantarum, dec. 1—10, Amsterdam 1734 to 1739.

A magnificent work of Marcus Catesby, on the Floras of the southern provinces of North America, the Natural History of Carolina, &c. was published in two volumes, London 1731 and 1743.

V. *The Linnæan Period.*

455.

Charles Linné gave their new form to all the parts of Natural History ; but he deserves to be in a peculiar sense called the Founder of the Historical Part ; for he first regulated the artificial language,—fixed the laws of Classification,—unfolded the generic characters,—was the first to settle the idea of species,—invented trivial names and specific characters. He enriched the science of plants by a more exact investigation of exotic Floras, and by a more sure determination of some thousand new species discovered by others. In the last place, he formed a system, the value of which has been already estimated, (133.) If we were disposed to find fault with him and with his system, we might derive occasion from his neglect of microscopical examinations,—from his superficial study of Cryptogamous plants,—from his giving too little attention to the anatomy and physiology of plants,—and from the following circumstances : That he often exhibited in a defective manner the characters of the southern plants, owing to the want of actual inspection ;—that he set a higher value upon the corolla and petals than upon the fruit and seed;—and, lastly, that he overlooked many species, from incorrectly regarding them as subspecies.

He was born at Roshult, in Sweden, 1707, and performed, in 1732, his memorable journey through Lapland, from which he brought, as a sort of botanical booty, his admirable Flora Lapponica, the second edition of which was published, by Smith, at London, 1792. In Hartecamp, in Holland, where he was superintendant of the Clifford Garden, from 1735 to 1737, he first published his Systema Naturæ, Leyden, 1735, folio; then the Hortus Cliffortianus, Leyden, 1736, folio; and, besides other treatises, the Genera Plantarum, Leyden, 1737, in octavo. In 1741 he was professor at Upsal, and published, 1745, his classical Flora Suecica; in 1751, the Philosophia Botanica; and, in 1753, for the first time, his Species Plantarum, in which 7300 species were enumerated. In 1762, the second edition of this work appeared, in which the number of species had been increased by about 1500. His later discoveries were published in the Mantissa Prima and Altera, and he died 1778.

456.

In his own time, a certain degree of opposition continued to be made in Germany and France, to the innovations which he introduced. In Germany this was occasioned, in the *first* place, by the favourers of the system of Rivinus, to whom belonged, in particular, Chr. Gottl. Ludwig, professor at Leipsig, who died in 1773; and by whom the system of Rivinus was always considered as fundamental, in his Definitiones Plantarum, although he endeavoured to connect it with the Linnæan System. In the *second* place, the authority of Haller, who was too much an enemy to the innovations of Linnæus, was detrimental to their extension. And, in the *last* place, attempts were made to substitute other systems in the place of the Linnæan, among which that proposed by John Gottlieb Gleditsch, professor at Berlin, who died 1786, deserves chiefly to be mentioned. This system appeared in 1764, and founded the arrangement of plants simply

upon the situation of the filaments, (140.) In France, the
same thing happened, partly from the too great favour with
which the system of Tournefort was received, partly because
Michael Adanson, of the Academy at Paris, who died 1806,
had again directed the attention of botanists, in his Familles des Plantes, to the natural affinities (164), and Bern-
hard Jussieu, professor at Paris, who died 1777, founded
upon them a Natural Method, which is usually denominated the System of Trianon, because the plants were arranged according to this system in the royal garden at that
place ; (Mem. de l'Acad. de Paris, 1774, p. 175—197.) The
founder of this system took, as the principle of arrangement,
and the bond of the natural families, partly the pretended
number of cotyledons, partly the number of the petals, and
partly the insertion of the filaments on the receptacle, the
calyx, the corolla, or the pistil.

Meanwhile, the principles of the sexual theory were dis-
cussed during the time of Linnæus, and this doctrine was se-
cured against objections and misapplications. Joseph Gottlieb Kolreuter, professor at Carlsruhe, who died 1799, in his
preliminary notices respecting some experiments relating to
the sex of plants, 1761 to 1766, threw great light on the necessity of the co-operation of the two sexes. William Frederick Von Gleichen, counsellor to the Margrave of Anspach,
who died 1783, raised some doubts respecting the actual passage of the pollen, and proposed many objections to the sexual theory, (Das Neueste aus dem Planzenreich, Nurnberg,
1768, folio) ; and Caspar Frederick Wolf, of the academy at
Petersburgh, who died 1794, gave, in his Theoria Generationis, Halle, 1774, the most complete discussion of the phenomena of fructification, as he also gave the first explanation
of the evolution of the organs of plants from one another ;
(Nov. Comment. Petrop. xii. p. 403 ; xiii. p. 478.)

457

The anatomy of plants was neglected in the time of Linnæus. But George Christian Reichel, professor at Leipsig,
who died 1771 ; John Hill, physician in London, who died
1775 ; and Horace Benedict de Saussure, who died 1799,

were celebrated exceptions. The first gave the earliest correct representation of the primitive form of the spiral vessels; (Diss. de Vasis Plantarum Spiralibus, Leipsig, 1758, 4to.) Hill's Construction of Timber, London, 1770, 8vo. contains good investigations respecting the structure of wood, as also respecting the effects of the absorption of coloured fluids,—and it is adorned with good plates. Saussure's Observations sur l'Ecorce des Feuilles et des Petales, Geneva, 1762, contains the first correct researches respecting the slits. Nor must we forget the excellent researches of Charles Bonnet, who died in 1793, respecting the uses of leaves. These were first published at Gottingen, 1754. Above all, however, Henry Ludwig du Hamel du Monceau, inspector of the French Navy, who died 1782, deserves to be celebrated as the greatest writer of that period respecting the physiology of plants. La Physique des Arbres is the title of his immortal work, which was published in two volumes at Paris, 1758.

458.

Foreign countries were examined, during the time of Linnæus, chiefly by his scholars, among whom Frederick Hasselquist, who died 1752; Peter Forskol, who died 1763; Peter Löfling, who died 1756; and Peter Kalm, who died 1779, deserve particularly to be named. Hasselquist's Travels in Palestine were published by Linnæus himself, 1757; Forskol's Flora Ægyptiaco-Arabica was published by Niebuhr, at Copenhagen, 1775; Löfling's Travels in the Spanish Dominions of America were published by Linnæus, 1758; and Kalm's Travels in North America were published in three volumes, at Stockholm, 1753 to 1761.

The botanical treasures of Philibert Commerson, the fellow traveller of Bougainville, who died 1773, could be of no service to Linnæus, because they were deposited with the French Academy, and have been partly employed by Antony Lorenz Jussieu.

Northern Asia was examined in the most careful manner, during the time of Linnæus, by Peter Simon Pallas, who died in 1811. His Travels through the various Provinces of

X

the Russian Empire, in three volumes, St Petersburgh, 1771 to 1776, produced a rich harvest of botanical discoveries. The Indian Flora also became more known, from the labours of Nicholas Lorenzo Burmann, professor at Amsterdam, who died 1793, because, in his Flora Indica, Leyden, 1768, he has formed a great number of new species from the collections of others.

The West Indies were examined in the most complete manner, during the age of Linnæus. Patrick Browne published his Civil and Natural History of Jamaica, London, 1756, in folio; Nicolaus Joseph de Jacquin, his excellent Historia Stirpium Selectarum Americanarum, Vienna, 1763, in folio; and a French apothecary, Fusee Aublet, his incomparable Histoire des Plantes de la Guiane Française, in four volumes, Paris, 1775, in quarto.

Lastly, the treasures of the South Sea Islands were made known by Cook's two companions during his second voyage, namely, John Reinhold, who died 1798, and George Forster, who died 1794. The Characteres Generum Plantarum appeared at London 1776.

459.

Among the native Floras which became known in the time of Linnæus, we may mention the Flora Carniolica of John Antony Scopoli, professor at Pavia, afterwards chief physician in Idria, who died 1788, published at Vienna 1772; but more especially the Flora Austriaca of Nicolaus Joseph de Jacquin, in five centuriæ, Vienna, 1773 to 1776, folio; also the Historia Plantarum in Palatinatu Electorali crescentium of John Adam Pollich, physician to the imperial miners, who died 1780, three volumes, Manheim, 1776, and the Flora Herbornensis of John Daniel Leer, 1775. Albrecht von Haller, originally professor at Gottingen, afterwards landamman of the canton of Bern, who died 1777, published a masterly work, his Historia Stirpium Helvetiæ indigenarum, Bern, 1768, folio.

Among the French Floras of that period, the preference is due to Lewis Gerard's Flora Gallo-Provincialis, Paris, 1761, in octavo. Antony Gouan's Flora Monspeliaca, Lyons, 1765, and the Illustrationes et Observationes Botanicæ of the same author, published at Zurich, 1773, in folio.

The plants of Italy were examined with great diligence by Francis Seguier. His Plantæ Veronenses appeared in three volumes at Verona, 1745 to 1754.

A Spanish Flora was published by Don Joseph Quer y Martinez, professor at Madrid, who died 1764. It appeared in four quarto volumes at Madrid, 1762 to 1764.

An excellent English Flora was published by William Hudson, apothecary in London, who died 1793, a second edition having appeared in 1778; and the valuable Flora Scotica of John Lightfoot, who died 1788, was published in two volumes, London, 1777.

The Danish government did a permanent service to the science, by causing the plants of Denmark to be engraved at its own expence, and by devolving the care of the work first upon George Christian Oeder, and afterwards upon Otto Frederick Müller. In consequence of their labours appeared the masterly work entitled Flora Danica, the four first volumes of which were published from 1761 to 1777. Nor ought we to forget the Flora Norwegica of John Ernst Gunnerus, bishop of Drontheim, who died 1773, two volumes, 1766 and 1772.

460.

Among the botanical gardens which were most celebrated in the time of Linnæus, we may notice particularly that at Vienna, the rare plants of which were marked, by Jacquin, in the Hortus Botanicus Vindobonensis, Vienna, 1770 to 1776. three volumes. The garden at Upsal had been already described by Linnæus himself, in the year 1748.

VI. *Recent History of Botany.*

461.

Since the death of Linnæus, the chief labours of botanists have been employed in perfecting his system, in applying it

to the lower families of plants, in correcting it, and in following out with more attention those views which he had
neglected. Hence their chief attention has been directed to
the improvement of the generic characters,—more care has
been used in the examination of fruits and seeds, and, by degrees, the Linnæan System has come to be regarded simply
as an assistance to beginners, whilst the forming of a Natural
Method has been viewed as the highest object of botany.
Among the individuals who have examined the sexual system, especially in the lower organic bodies, Casimir Christopher Schmidel, professor at Erlangen, who died 1793, John
Hedwig, professor at Leipsig, who died 1799, and Joseph
Gottlieb Kölreuter, deserve to be first mentioned. Schmidel's
Icones et Analyses Plantarum, Nurnberg, 1782, in folio;
Hedwig's Theoria Generationis, Leipsig, 1798; his Funda-
mentum Historiæ Naturalis Muscorum frondosorum, Leipsig, 1782; and his Stirpes Cryptogamicæ, in four volumes,
Leipsig, 1787 to 1797: and Kölreuter's Entdecktes Geheimniss der Kryptogamie, 1787, are the works in which principally the existence of the sexual parts, in the lower organic
bodies, are treated of. Yet that these excellent natural historians only followed out an idea which had formerly been con-
ceived, has been shewn by Samuel Gottlieb Gmelin (Historia
Fucorum, St Petersburgh, 1768), and Philip Cavolini, (On
the Animal Plants of the Mediterranean, translated by William Sprengel, Nurnberg, 1813). The effect of the nectaries
on fructification, was completely developed by Christian Conrad Sprengel, who died 1816; (Das Entdeckte Geheimniss
der Natur im Bau, und in der Befruchtung der Blumen, Berlin, 1793, 4to.)

462.

The Linnæan Species Plantarum has found several editors of
very unequal merit. John Jacob Reichard, physician at Frankfort on the Maine, who died 1789, produced nothing in his
edition, which was published in 1779 and 1780, in four volumes, but the supplements from the Mantissæ of Linnæus,
and here and there some scattered remarks. The Systema
Vegetabilium of John Frederick Gmelin, professor at Got-

tingen, 1791, soon fell into neglect, because it was conducted with little discrimination. Charles Lewis Willdenow, professor at Berlin, who died 1812, edited an excellent edition of the Species, especially in that part which embraces the ten latter classes. His edition appeared between 1797 and 1810, in ten volumes, and goes as far as the end of the Ferns. An early death took from us a still more excellent follower of Linnæus. This was Martin Vahl, professor at Copenhagen, who died 1804, and whose Enumeratio Plantarum includes only the first and second, and a part of the third class. Christian Henry Persoon, in his Synopsis Plantarum, 1805 and 1807, gave an Abridgment of Willdenow, and added to it the recent discoveries of the French. Of late, John Jacob Romer, professor at Zurich, who died 1819, and Joseph August Schultes, professor at Landshut, have commenced an undertaking, to which we cannot but wish a more successful progress, but to which the well-founded objection of too great diffuseness, and of the want of critical discrimination, may be made; (Systema Vegetabilium, vol. i. 4to., Tubingen, 1817 to 1819.)

463.

The science has made the most important advances, since the attention of natural historians was directed to the most essential product of vegetation, namely, the seed and fruit; and since, by this means, the idea of natural relationship has been again awakened. This zeal has been excited in the liveliest manner by the masterly work of Joseph Gærtner, physician at Calw, in Wirtemberg, de Fructibus et Seminibus Plantarum, which appeared in two volumes, at Stutgard, 1788 and 1791, with 180 copperplates, and to which his son added a Supplement, with 45 copperplates, in 1805.

Independent of this author, Antony Lorenzo Jussieu, professor at Paris, in the spirit of his uncle, formerly mentioned, constructed a natural method, which was published under the title Genera Plantarum, 1789, and is distinguished chiefly by its correct and carefully constructed generic characters. Stephen Peter Ventenat, professor at Paris, who died 1808, also was of much service to the natural method, by his Tableau du Regne Vegetal, Paris, 1799, four volumes; as also Augustus John George Charles Batsch, professor at Jena, who died 1802, by his Tabulæ Affinitatum Regni Vegetabilis, Weimar, 1802; and, it is to be hoped, that Augustus Pyramus de Candolle, professor at Geneva, will gain the highest credit, by the continuation of his Systema Naturale Regni Vegetabilis, Paris, 1818.

464.

The anatomy and physiology of plants have gained new life, especially in Germany, France, and Italy, since the structure of plants has been examined, without reference to preconceived ideas. John Hedwig, along with many important truths, had also given currency to some obvious mistakes, especially in the collection of his scattered works, Leipsig, 1793, in two volumes; and the correct view (Prodromo di Fisica Vegetabile, Padua, 1791) of Andrew Camparetti, professor at Padua, obtained little success, at least in Germany. Antony Krocker, (De Plantarum Epidermide, Halle, 1800), and the author of this history, in his Introduction to the Knowledge of Plants, Halle, 1812, endeavoured, indeed, to lay open these mistakes, and to shew the true structure of plants. But more attention was paid in France to the frequently mistaken ideas of C. F Brisseau Mirbel, of the French Academy, in his Traité d' Anatomie et de Physiologie Vegetales, Paris, 1802. Meanwhile, Henry Frederick Link, and Charles Asmund Rudolphi, professors at Berlin, as also Ludolph Christian Treviranus, professor at Breslau, published more correct views; (Link, Grundlehen der Anatomie und Physiologie der Pflanzen, Berlin, 1807; Rudolphi, Anatomie der Pflanzen, Berlin, 1807; and Treviranus, vom Inwendigen Bau der Gewachse, Göttingen, 1806.)   Since then, Mirbel has come nearer the truth; (Exposition et Defense de ma Theorie de l' Organization Vegetale, Amsterdam, 1808.) John Jacob Paul Moldenhawer, professor at Kiel, by his Contributions to the Anatomy of Plants, Kiel, 1815, quarto; and George Kieser, professor at Jena, by his Mémoire sur l' Organization des Plantes, Haarlem, 1813, quarto; and by his Grundzuge der Anatomie der Pflanzen, Jena, 1815, have obtained great distinction. On the physiology of plants in general, several introductory works have appeared, namely, the Traité Theorique et Pratique sur la Vegetation, by Mustel, a French officer, published at Rouen, 1781, in four volumes ; the Physiologie Vegetale of John Senebier, a Genevese clergyman, who died 1809, Geneva, 1800, in five volumes ; the Vegetable Physiology of Darwin, physician at Derby, who died 1802, translated at Leipsig, 1801 ; and the System of Physiological Botany of P. Keith, clergyman at Bethersden, in England, London, 1816, in two volumes.

465.

Native plants have recently been examined with great care.

To begin with Germany.—In the Manual of Botany, by Christ. Schkuhr, mechanician at Wittenberg, who died 1811, published at Wittenberg between 1791 and 1803, in three volumes, with nearly 500 plates, we have a multitude of very excellent drawings and dissections, chiefly of native, but also of many exotic plants. The Flora of Germany, in plates, by Jacob Sturm, artist at Nurnberg, in three parts, and sixty-five numbers, likewise deserves, in every respect, the most honourable mention. The Flora Germanica of Henry Adolphus Schrader, professor at Gottingen, of which the first volume appeared in 1806, is distinguished by the most perfect accuracy and care. It is only to be lamented, that it has not been continued till the present time. Among the Floras of particular districts of Germany, we may mention, especially, the Flora Badensis, in three volumes, by Charles Christ. Gmelin, physician at Baden, published at Carlsruhe, 1805 and 1810; the Flora Cryptogamica Erlangensis of Charles Frederick Philip Martins, published at Nurnberg, 1817 ; and the Prodromus Floræ Neomarchicæ of John Frederick Rebentisch, published at Berlin, 1804.

A general Flora of France was published by Augustus Pyramus de Candolle, in the Flore Francaise, Paris, 1805 to 1816, in six volumes ; and by Loiseleur Deslongchamps, in his Flora Gallica, Paris, 1806 and 1807. Among the

Floras of particular districts we may mention, especially, the classical Histoire des Plantes de Dauphiné, by Villars, professor at Strasburg, who had carefully examined the Alps which divide Italy from Switzerland, the Vogeses, and the south of France, along with Chaix, a clergyman at Gap, and Clapier, physician at Grenoble. He died in 1813. The work was published in four volumes, 1786 to 1789, at Grenoble, with 55 copper-plates. The Histoire Abregée des Plantes des Pyrenees, Thoulouse, 1813, in octavo, also belongs to the class of approved works.

Among Italian Floras, the following deserve the most respectful notice, namely, the Flora Pedemontana of Charles Allioni, professor at Turin, who died 1804, published in three volumes, at Turin, 1785, with 92 copperplates, folio; the Flora Neapolitana, Naples, 1811, in folio, by Michael Tenore, professor at Naples; and the Flora Ticinensis, Pavia, 1816, by Dominicus Nocca, professor at Pavia, and John Baptiste Balbis, professor at Lyons, Pavia, 1816. The Sicilian Flora has found its votaries in Antonin Bivona-Bernardi, who published Plantarum Sicularum cent. 1—2, Palermo, 1806, 1810, in octavo; and Stirpium Rariorum in Sicilia provenientium Manip. 1—4, Palermo, 1813 to 1816, in quarto; and in Constantin Schmalz Rafinesque, who published Caratteri di alcuni Generi di Piante, Palermo, 1810, octavo.

The Flora of Portugal, which had long been neglected, was carefully edited by Felix Avellar Brotero, professor at Coimbra, in his Flora Lusitanica, Lisbon, 1804, two volumes; and by Henry Frederick Link, and the Count Hoffmansegg, in the Flore Portugaise, Berlin, 1809 to 1814, folio.

Great Britain can boast of an excellent work, with copper-plates, on the native Flora, published under the title English Botany, by James Sowerby and Sir James Edward Smith, in thirty-six volumes, from 1790 to 1814. The latter author also published a Flora Britannica, in three volumes, London, 1800 to 1804, which possesses distinguished scientific merit. Not less meritorious is the Muscologia Britannica, by William Jackson Hooker and Thomas Taylor, London, 1818.

With respect to the other northern countries, the Flora Danica, by Martin Vahl and James Wilkin Hornemann, was continued to the end of the ninth volume. In Sweden, J. W. Palmstruch and C. W. Venus, published a Swedish Botany, beginning in 1802, after the model of the English Botany; and George Wahlenberg published a masterly work, entitled Flora Lapponica, Berlin, 1812.

Among the districts of Poland, those which are most southerly were examined in a very complete manner by W. S. J. G. Besser, professor at Krzeminiec, in Podolia, and his Primitiæ Floræ Galiciæ, Vienna, 1809, in two volumes, belong to the class of the most perfect works of this kind.

The rich treasures of Hungary, and of the neighbouring territories, were examined at the expence of Count Francis von Waldstein, by Paul Kitaibel, professor at Pesth, who died 1817, and were made known in his masterly work, entitled Descriptiones et Icones Plantarum Rariorum Hungariæ, Vienna, 1803 to 1812, in three volumes; The Transylvanian Flora also, found an editor in Joh Christ. Gottl. Baumgarten, physician at Schäsburg, whose Enumeratio Stirpium Transylvaniæ was published in three volumes, at Vienna, 1816.

466.

Among foreign countries, which, in recent times, have been examined by botanists, we begin with the Levant. John Sibthorp, professor at Oxford, twice examined Greece and Asia Minor, and was only prevented by death from publishing his discoveries. In consequence of his will, however, a magnificent work, entitled Flora Græca, has been published since the year 1806,— a useful compend of which work has been published, in two volumes, by Sir James Edward Smith, in the Prodromus Floræ Græcæ, London, 1806 to 1813. The Icones Plantarum Syriæ Rariorum of Jacob Julius la Billardiere, published at Paris, 1791, also deserve to be mentioned with applause. The Arabian plants, brought by Forskol, were re-examined by Martin Vahl, and described, along with many other plants brought from Malabar, by

Konig, and the West Indies, by Pflug and Rohr, in his Symbolis Botanicis, vol. 1—3, Copenhagen, 1790 to 1794, folio.

Since the Russian dominion extended itself over Caucasus, this very important country has been examined by several excellent natural historians, and, in regard to its botany, it has been most carefully investigated by Baron Frederick Marshall von Bieberstein. His excellent Flora Taurico-Caucasica appeared in two volumes, at Cracow, 1808.

The dominion of England in the East Indies gave new life to the zeal for the search of plants in these regions. The most important work on the subject was that published by William Roxburgh in his Plants of the Coast of Coromandel, London 1795, three volumes. The Flora Cochin-Chinensis of John de Loureiro, a Portuguese missionary, which was published at Lisbon 1790, is also very valuable. With respect to Japan, we have the admirable Flora Japonica of Charles Peter Thunberg, published at Leipzig 1784, with forty copperplates.

In regard to Africa, the learned companions of the expedition of Bonaparte to Egypt, in the year 1798, have enriched the science with many important contributions, in the magnificent work entitled Description de l'Egypt, Paris 1813. A compend of this was published by A. R. Delile, under the title of Memoires Botaniques, Paris 1813. On the Flora of northern Africa, appeared a valuable work of Renatus Desfontaines, professor at Paris, Flora Atlantica, Paris 1800, in two volumes, with 261 copperplates. One part of the coast of Guinea was examined by A. M. F. J. Palisot-Beauvois, Flore d'Oware et de Benin, Paris 1804 to 1810, in folio. Charles Peter Thunberg described the rich Flora of southern Africa in his Prodromus Plantarum Capensium, Upsal 1794 and 1800 ; and in the Flora Capensis, Upsal 1813. Mascaren's Land and Madagascar were examined by Aubert du Petit-Thouars, Histoire des Vegetaux recueillis dans les Isles Australes d'Afrique, Paris 1806, in quarto.

We possess excellent Floras of North America by Andrew Michaux, Flora Boreali-Americana, Paris 1803 ; by Frederic Pursh, Flora Americæ Septentrionalis, London 1814.

and by Thomas Nuttall, Genera of North American Plants, Philadelphia 1818, in two volumes. Stephen Elliot examined the southern states ; Botany of the Southern States, Carolina and Georgia, Charlestown 1817, 1818. Some contributions were furnished by Henry Mühlenberg, clergyman at Lancaster, who died 1816, Catalogus Plantarum Americæ Septentrionalis, Lancaster 1813, octavo ; by Const. Schmalz Rafinesque, Florula Ludoviciana, translated from the French of C. C. Robin, New York 1817, octavo ; and by C. W. Eddy and J. Torrey ; (A Catalogue of Plants growing spontaneously within thirty miles of the city of New York, Albany 1819, octavo.)

Olaus Swarz, professor at Stockholm, who died 1817, described in a very complete manner the West Indian Flora in his Flora Indiæ Occidentalis, Erlangen 1797 to 1806, in three volumes.

The Spaniards Hippolytus Ruiz and Joseph Pavon, have made us acquainted with a multitude of new genera and species belonging to Peru and Chili, in their Flora Peruviana et Chilensis, Madrid 1798, in folio. But Alexander von Humboldt has gained immortal honour by his numerous botanical discoveries, the fruits of his travels through Spanish America. He published them along with Amatus Bonpland, in his Plantes Equinoxiales, and with Charles Kunth in his Nova Genera et Species Plantarum, Paris 1815, in three volumes.

Lewis Née, the companion of Malaspina, examined with much care the South Sea Islands. Antony Joseph Cavanilles, professor at Madrid, who died 1804, availed himself of his treasures, and described them in his Icones et Descriptiones Plantarum, Madrid 1791 to 1799, in six volumes. Jacob Julius la Billardière published an excellent Flora of New Holland, under the title Novæ Hollandiæ Plantarum Specimen, Paris 1804, in folio, with 265 copperplates. We owe the greatest obligations to the ingenious Robert Brown, whose Prodromus Floræ Novæ Hollandiæ, London 1810, very rare, and his General Remarks on the Botany of Terra Australis, London 1814, are uncommonly valuable.

467

Some particular families and genera have been examined in recent times with the most perfect care, and the science has thus been extended.

The Fungi have been represented in good plates by Augustus John George Charles Batsch, professor at Jena, who died 1802, (Elenchus Fungorum, Halle 1783, 1784); by Bulliard, (Herbier de la France, div. ii. 1791, in folio); by Jacob Bolton, (History of Funguses growing about Halifax, vol. i. iii. suppl. tab. 1.—182, Huddersfield 1788,—1791, quarto); and by James Sowerby, (Coloured Figures of English Mushrooms, n. 1—29, London 1799 to 1814, folio.) The scientific knowledge of Fungi and their genera was first established by Henry Julius Tode, clergyman in Mecklenburgh, who died 1799, (Fungi Mecklenburgensis Selecti, fasc. 1. 2. Luneburg 1790, 1791, quarto.) He was followed by Christ. Henr. Persoon, whose system was long the only one; (Synopsis Methodica Fungorum, Gottingen 1801, octavo; Observationes Mycologicæ, vol. i. ii. Leipzig 1796, 1798.) J. B. von Albertini and L. D. von Schweinitz extended the knowledge of species; (Conspectus Fungorum in Agro Nieskiensi crescentium, Leipzig 1805, octavo.) But Henry Frederick Link (Berlin Magazin, iii. s. 1—42, vii. s. 25—45), and C. G. Nees von Esenbeck (Das System der Pilze und Schwamme, Wurzburg 1817, quarto), were the founders of entirely new views and divisions of these families.

Our knowledge of the Algæ was chiefly extended by Albert Wilkelm Roth, physician at Vegesack; (Catalecta Botanica, fasc. 1—3, Leipzig 1797, 1806.) Afterwards John Peter Vaucher, clergyman at Geneva, examined the fresh water Confervæ, and divided them into natural genera; (Histoire des Conferves d'Eau douce, Geneva 1803, quarto.) Lewis Weston Dillwyn published excellent plates of the same, (Synopsis of the British Confervæ, fasc. 1—20, London 1802, and following years); and J. A. P. Ducluzeau described them as they grow in the south of France, (Essai sur l'Histoire Naturelle des Conferves des Environs de Montpel-

lier, 1805, octavo.)   The Fuci have lately been more correctly classed by J. B. F. Lamouroux, professor at Caen,
(Ann. de Mus. xx. p. 21, 116, 267.) ; by Joh. Stackhouse,
(Nereis Brittanica, ed. ii. Oxford 1816, folio ; by Charles
Agardh, (Synopsis Algarum Scandinaviæ, London 1817, octavo) ; and by H. Christ. Lyngbye, (Tentamen Hydrophy-
tologiæ · Danicæ, Copenhagen 1819, quarto, with 70 copperplates.)   Dawson Turner published valuable plates of
the known species, (Fuci, or coloured figures and descriptions of the plants referred by botanists to the genus Fucus,
vol. i.—iv. London 1807—1811.)

George Francis Hofman, professor at Göttingen, now at
Moscow, published good plates of the Lichens, (Plantæ Li-
chenosæ, vol. i.—iii, Leipzig 1789 to 1801, folio).   But for
the system of this family, we are indebted to Erick Acharius, physician at Badstena, (Lichenographia Universalis,
Gottingen 1810, quarto ; Synopsis Methodica Lichenum,
Land. 1814, octavo.)

The Jungermanniæ were chiefly studied by William Jackson Hooker, (Jungermanniarum Icones, fasc. 1—20, London 1813, quarto.)

The Musci Frondosi were investigated in a very excellent
manner by John Hedwig, and a system proposed by him,
which deserved and obtained almost universal approbation,
because the parts of fructification in particular were examined
and represented with great skill, (Fundamenta Historiæ Naturalis Muscorum frondosorum, vol. i. ii. Leipzig 1782—4 ;
Descriptiones et Adumbrationes Muscorum frondosorum, vol.
i. ii., Leipzig 1789 to 1797, in folio ; Species Muscorum
frondosorum, Leipzig 1801, quarto.)   His worthy follower,
Fr. Schwagrichen, enriched botany with a great number of
new species, (Supplementum ad Species Muscorum, vol. i. ii.
Leipzig 1811, 1816, quarto).   We are also much indebted
to W. J. Hooker for our knowledge of British (Muscologia Britannica, London 1818, octavo,) and exotic Mosses,
(Musci Exotici, vol. i. ii. London 1818, 1819, octavo).   A
new system of this family has lately been constructed by Sam.

1

El. von Bridel, (Methodus Nova Muscorum, Gotha 1819, quarto.)

Sir James Edward Smith had already given much attention to Ferns, and has established the principles according to which they ought to be divided, (Mem. de l'Acad. de Turin, vol. v. p. 401.); But Olaus Swartz completed this system, (Synopsis Filicum, Keil 1806, octavo.)

Christ. Schkuhr published a good monograph on the Reeds, (Beschreibung und Abbildung der Ried-grasser, Wittenberg 1801 to 1806, octavo, with 93 copperplates. The Austrian Grasses were excellently delineated by Nicolaus Thomas Host, an Austrian physician, (Icones et Descriptiones Graminum Austriacorum, vol. i.—iv. Vienna 1801 to 1814) ; and John Gaudin, clergyman in Waadtland, described the Grasses of Switzerland, (Agrostologiæ Helveticæ, vol. i. ii. Paris 1811, octavo.) John Christian Daniel Schreber, professor at Erlangen, who died 1810, instituted excellent investigations respecting particular grasses, and published very good figures of them, (Beschreibung der Grasser, th. 1—3, Leipzig 1769 to 1810, in folio.) An entirely new and peculiar system was formed by A. M. F. J. Palisot de Beauvois, (Essai d'une Nouvelle Agrostographie, ou Nouveaux Genres des Graminées, Paris 1812, octavo.)

Among the Pine tribe the genus Pinus was described by Aylmer Bourke Lambert, (A Description of the Genus Pinus, London 1803, in folio).

The Coronariæ and Liliaceæ were studied by Augustus Pyramus de Candolle and P. J. Redouté, (Histoire de Plantes Grasses, i.—xxii. Paris 1799 to 1811, in folio ; and Les Liliacées, vol. i.—viii. Paris 1802 to 1816, in folio.) The text of the latter work was furnished for the first four volumes by De Candolle, and for the fifth, sixth, and seventh, by F. de la Roche, and for the eighth by A. R. Delile. The genus Aloë was thoroughly and carefully examined by Charles Lewis Willdenow (Berlin Mag. v. s. 163. f.), and by the Prince of Salm-Dyck, (Veizeichniss der verschiedenen Arten des Geschlechtes Aloë, 1817, octavo.) These, and other succulent plants were described by Adrian Hardy Haworth in several works, (Synopsis Plantarum succulentarum, adjungitur Narcissorum revisio, London 1819, octavo ; Miscellanea Naturalia, London 1803, octavo ; Transactions of Linn. Soc. vol. vii. p. 1. s.) The Spathaceæ were divided by John Bellenden Gawler, now Ker, into several Genera, (Ann. of Bot. i. p. 216.)

The Scitamineæ were subjected by William Roscoe (Trans. Linn. Soc. vol. viii. p. 330.), and by William Roxburgh (Asiat. Research. vol. xi. p. 200.) to a new revision. The same was done by Olaus Swartz with the Orchideæ, (Act. Soc. Scient. Upsal, vi. p. 59. s.; Stockh. Acad. Handl, 1800, p. 202. s.; Schrader's Journal, 1799, st. ii. s. 201; and Neues Journal, st. i. s. 1. f.)

Of the Chenopodeæ, Peter Simon Pallas studied the salt plants.of the East ; (Illustrationes Plantarum imperfecte aut nondum cognitarum, Leipzig 1803, folio.)

Among the Thymelææ, the genus Daphne was examined by John Em. Wikstrom ; (Diss. de Daphne, Upsal 1817, quarto.)

A complete revision of the Proteaceæ was published by Robert Brown ; (Trans. Linn. Soc. vol. x. p. 15.—226.)

Of the Amentaceæ, the Willows were examined by George Francis Hoffman, (Historia salicum, iconibus illustrata, vol. i. ii. Leipzig 1785 to 1791, folio) ; and by W. C. Seringe, teacher at Bern, (Essai d'une Monographie des Saules de la Suisse, 1815, octavo).

The North American Oaks were studied by Andr. Michaux ; (Histoire des Chênes de l'Amerique, Paris 1801, in folio.)

Of the Tricoccæ, the genus Croton was made the subject of an inquiry by Edward Ferdinand Geiserl ; (Crotonis Monographia, Halle 1807, octavo.)

The genera Primula and Nicotiana were studied by J. G. C. Lehmann, professor at Hamburgh ; (Monographia generis Primularum, Leipzig 1817, quarto ; Historia generis Nicotianarum, Hamburg 1818, quarto.) The same author published a classical revision of the Asperifoliæ ; (Plantæ e familia Asperifoliarum nuciferæ, vol. i. ii. Berlin 1818, quarto.)

Michael Felix Dunal, professor at Montpellier, examined the genus Solanum; (Histoire de Solanum, Montpellier 1813, quarto; Solanorum Generumque affinium synopsis, Montpellier 1816, octavo.)

A complete revision of the Contortæ was published by Robert Brown, (Transact. Wern. Soc. vol i. p. 40. s.); and excellent monographs of the genus Stapelia have been drawn up by Francis Masson (Stapeliæ Novæ, London 1796, folio), and by Nicolaus Joseph von Jacquin (Stapeliarum Descriptiones, iconibus illustratæ, Vienna 1806, folio.)

The species of the genus Erica have been described and figured by H. C. Andrews, (Coloured Engravings of Heaths, vol. i.—iii. London 1803 to 1809, folio); and by John Christopher Wendland, (Ericarum Icones et Descriptiones, fasc. 1—17, Hanover 1798 to 1806, quarto.)

A new tribe of the Compositæ was constructed by Marian Lagasca (Amœnidades Naturales de las Espanas, Orihuela 1811, quarto), and by Augustus Pyramus de Candolle, (Ann. du Mus. xix. p. 59.; Recueil de Mém. sur la Botanique, Paris 1813, quarto). But Henry Cassini (Journ. de Bot. Nouv. iv. p. 231. s.; Dictionnaire des Scien. Nat. x. p. 131. s.), and Robert Brown (Trans. Linn. Soc. vol. xii. p. 75.) have published interesting observations on the whole family.

The Valerianæ were examined by P. Dufresne, (Histoire Naturelle de la Famille des Valerianées, Montpellier 1811, quarto.)

Of the Rubiaceæ, the genus Cinchona was described by Aylmer Bourke, (A Description of the Genus Cinchona, London 1797, quarto.)

The umbelliferous plants were examined after Peter Cusson, professor at Montpellier, who died 1783, (Hist. et Mem. de la Soc. Roy. de Médec. 1781, 1782, p. 275.), by George Frederick Hoffman (Genera Umbelliferarum, ed. nov. Moscow 1816, octavo,) and the author of this history, (Plantarum Umbelliferarum denuo disponendarum Prodromus, Halle 1813, octavo; Species Umbelliferarum minus cognitæ, Halle 1818, quarto.) The genus Eryngium was described by Francis de la Roche, (Eryngiorum Historia, Paris 1808, folio.)

Caspar Count von Sternberg gave excellent figures of the Saxifragæ, (Revisio Saxifragarum, iconibus illustrata, Regensburg 1810, folio.)

The Anoneæ were arranged by Michael Felix Dunal, (Monographie de la Famille des Anonacées, Paris 1817, quarto.)

The cruciform plants were revised by Robert Brown (Aiton. Hort. Kew. ed. 2. vol. iv. p. 71—130.), and by Desvaux (Journ. de Bot. Nouv. iii. p. 145. s.); the genus Biscutella by De Candolle, (Recueil de Mem. sur la Botan. 1813, quarto); the Papavereæ by L. G. A. Viguier, (Histoire Naturelle des Pavots, Montpellier 1814, quarto) ; the Ranunculeæ by J. A. J. Biria (Histoire Naturelle des Renoncules, Montpellier 1811, quarto.)

Of the leguminous plants, the Clover species were more accurately determined by Joh. Christ. Dan. Schreber, (Sturm's Deutsche Flor. No. 17.); and by Gaetan Savi, (Observationes in varias Trifoliorum Species, Florence 1810, octavo). Augustus Pyramus de Candolle (Astragalogiæ, Paris 1802, folio), and Peter Simon Pallas (Species Astragalorum descriptæ et iconibus illustratæ, Leipzig 1800, folio,) revised the genus Astragalus. Humboldt and Kunth are now publishing the American Mimosæ.

Charles Lewis l'Heritier (Geraniologia, Paris 1787, 1788, folio,) and H. C. Andrews (Geraniums, or a Monography of the genus Geranium, London 1805,) examined accurately the Geraniæ.

Nicolaus Joseph von Jacquin published an excellent monograph of the Oxalidæ, (Oxalis Monographia, Vienna 1794, quarto.)

The Malvaceæ, Melieæ, Passifloræ, and Malpighieæ, were revised by Antony Joseph Cavanilles, (Monadelphiæ Classis diss. x. Madrid 1790, with 296 copperplates, quarto) : the Ochneæ and Simarubeæ were revised by Augustus Pyramus de Candolle, (Recueil de Mem. sur la Botan. 1813.)

Of the Aizoidæ, the Mesembryanthema were examined by Adrian Hardy Haworth, (Observations on the genus Mesem-

Y

bryanthemum, London 1794, octavo; Miscellanea Naturalia, London 1803.)

The Melastomeæ, collected during the travels of Humboldt, were revised by Amatus Bonpland in a magnificent work, (Monographie des Melastomes, fasc. i.—vi. Paris 1809, quarto.)

Of the Roseaceæ, the Potentillæ were subjected to a new revision by C. G. Nestler, professor at Strasburg (Monographia de Potentilla, Strasburg 1816, quarto), and by J. C. G. Lehman. The splendid works on Roses by Miss Lawrence (Collection of Roses from Nature, London 1799, folio,) and by P. J. Redouté (Les Roses, liv. i.—xi. Paris 1816, folio,) have not been of any particular advantage to the science.

468.

Among the botanical gardens of recent times, that at Schonbrun near Vienna deserves to be first mentioned. Its treasures were described by the elder Jacquin, in the Hortus Schonbrunnensis, Vienna 1797, in four volumes. To this department belong also his Icones Plantarum rariorum, in three volumes, Vienna 1781 to 1795, and the Eclogæ Plantarum rariorum of his son Joseph Francis von Jacquin, Vienna 1811 to 1816. Next to this, the garden at Berlin has become celebrated by the zeal of Willdenow, and the rare talents of its present superintendant Otto. The Hortus Berolinensis of Willdenow, Berlin 1809, folio, rivals the work of Jacquin; and his Enumeratio Plantarum Horti Berolinensis, Berlin 1809, is a very rich, scientific, and useful catalogue.

Among the botanical gardens of France, that established by Josephine Bonaparte at Malmaison has been celebrated from the splendid work which the superintendant Ventenat began to publish in 1803, in folio, under the title Jardin de la Malmaison, and which Amatus Bonpland continued, (Description des Plantes rares cultivées à Malmaison et à Navarre, livr. i.—x. Paris 1803 to 1816, in folio.) Ventenat published figures of the rarer plants in a nursery-garden, and described them under the titles, Description des Plantes dans le Jardin de M. Cels, Paris 1800, folio, and Choix des Plantes dans le Jardin de Cels, Paris 1803, folio.

The gardens of Great Britain are the richest and most celebrated of modern times. In the first rank stands the Royal Garden at Kew, of which a description was published by its first superintendant William Aiton, who died 1793, under the title Hortus Kewensis, three volumes, 1789. The new edition of this work was published by the younger Aiton, with the assistance of Robert Brown, in five volumes, from 1810 to 1813. The treasures of the English gardens had been previously figured and described by Charles Lewis L'Heritier, who died 1800, in his Sertum Anglicum, Paris 1788, and in his Stirpes novæ aut minus cognitæ, fasc. i.—vii. Paris 1784, folio; also by Sir James Edward Smith in his Exotic Botany, London 1804 to 1808, and by Richard Antony Salisbury in his Paradisus Londinensis. But the finest works in this department, are the expensive copperplate works of Henry Andrews, The Botanist's Repository, London 1797 to 1808; The Botanical Magazine of Curtis and Sims, and the Botanical Register, edited by Ker. An intelligently conducted Catalogue of the Plants in the English Gardens was published by Robert Sweet, under the title Hortus Suburbanus Londonensis, 1818.

Among the other garden catalogues, the following deserve to be particularly noticed: Agustus Pyramus de Candolle's Catalogus Plantarum Horti Botanici Monspeliensis, 1813, octavo, and Marian Lagasca's Elenchus Plantarum quæ in Horto Regio botan. Matritensi colebantur, Madrid 1816, quarto; Genera et Species Plantarum, quæ aut novæ aut sunt nondum recte cognoscuntur, Madrid 1816, quarto.

PRACTICAL PART.

PRACTICAL PART.

CLASS I.

1.

Hippuris vulgaris.

Kazenzahl, Weisse Seetanne, Tannwedel.—Fren. *Pesse commune.*—Eng. *Mare's-tale, Paddow-pipe.*—Ital. *La Corregiola femmina.*—Swed. *Hastswants.*

There are few of the higher plants which have so simple a structure as this, or which have so much simplicity in the relations of the essential parts.

In deep ditches, in standing and running waters, there arises in spring a round stem, from one to two feet above the water, straight, perpendicular, almost stiff (41.), and simple (45.) Its colour is reddish; and its circumference is in size about that of the quill of a pigeon's or hen's feather. It is also jointed (40.), and shoots out under the water, first, horizontal fibrous roots, and, next, pellucid, linear-lanceolate leaves. On dissection, we discover in the circumference a compound cellular texture, with distinct regular interstices (299.); and in the centre there is a firm cord, composed of spiral vessels and sap-tubes (280.)

Above the water, there spring from each joint of the stem commonly from eight to ten leaves, arranged in a circle (36.); sometimes also, especially under the water, fashioned into a spiral shape (37.) horizontal, linear-lanceolate, somewhat ob-

tuse (55.), quite entire, without nerves, opake, of the length
of a finger-nail, or of the thumb, (16.) These leaves, especially on their lower surface, have pretty large slits, (309.)

At the base of the leaves, or in the axis of the leaves, there
appear in spring, in the first place, some larger bodies, which
we may consider with Haller (*Hist. Stirp. Helv.* 1752), as
buds; and, secondly, small ovaria, surmounted by a simple
pistil, on both sides of which, the rudiments of the two-lobed
anthers stand directly over the germen. The anther afterwards is elevated upon a simple filament, about the length of
the pistil, by the side of which it stands, and continues till the
seed is almost fully ripe. The external cover of the germen
may be considered as the calyx, which in that case will be a superior one, (34.) The germen shews as its fruit, whilst the external cover swells, a smooth, oval nut, with a similar shaped
cavity, in which the seed lies. This contains, amidst the nearly consumed albumen, a filamentous embryo in the centre,
having its radicle directed upwards.

Geographical Distribution.

With respect to the geographical distribution of the plant,
it seems to thrive only in the waters of the Northern Hemisphere. It grows as far as the Polar circle, (*Wahlenb. Flor.
Lapp.* p. 1.), in Europe, Asia, and America, (*Pursh. Amer.
Sept.* p. 3. ; *Nuttal,* p. 3.) But the American form is considered as a distinct variety, since it has almost always only
six leaves in the circle. But it does not grow farther south
than the north of Italy, the south of France, and the north
of Spain. Hence its southern limit in Europe seems to be the
44° N. Lat. In Asia, it scarcely passes beyond the 50°. But
in North America it is found even under the 35°.

Synonymes and Figures.

Polygonum femina, *Lobel. ic.* 792.—*Matth. Valgris,* 485.—
    Doden. 113.—*Dalech.* 1072.—*J. Bauh. Hist.* 3. 732.
Cauda equina femina, *Gerard. Emac.* 1114.

Equisetum palustre, brevibus foliis polyspermum, *C. Bauhin,*
15. *Parkin's Theatr.* 1200.

Pinastella, *Dill. Giess. App.* 168. *Buxb. Hal.* 261.

Limnopeuce, *V Cord. Hist.* p. 150. *Vaill. Mem. de Paris,*
1719, p. 15. t. 1. f. 3. *Hall. Stirp.* n. 1572.

Hippuris vulgaris, *Linn. Fl. Dan.* 87. *Engl. Bot.* 763.
Gärtn. Fruct. t. 84. *Richard in Ann. du Mus.* 3. t. 30.
f. 3.

Affinity.

The first authors who arranged this plant in a natural order, placed it beside *Ceratophyllum, Myriophyllum, Zannichellia,* and immediately after it they placed *Pilularia,* (*Ray,
syn.* p. 136.) Linnæus also placed it along with similar
plants in his 15th tribe, which he called *Inundatæ,* (*Giseké
Ord. Nat.* p. 327.) Batsch followed the same plan (*Tab.
Affin. Regn. Veget.* p. 161.), denying at the same time, very
improperly, to the Inundatæ, the albumen in the seed. And
Jussieu opened with this plant his family of the Naiadæ,
placing *Chara* immediately after it, (*Jussieu, Gen. Plant.*
p. 18.) But Adanson was the first who thought of a higher
place, for he placed his *Limnopeuce* among the Elæagnæ between *Thesium* and *Cynomorium,* (*Adanson, Famill.* 2.
p. 80.) More lately Jussieu has approved of this arrangement, (*Ann. du Mus.* 3. p. 323.); and De Candolle has
placed the plant among the Onagræ, (*Fl. Franc.* 4. 415.)
Lastly, Nuttal thinks it has no affinities.

But when we direct our attention to the existence of a central bundle of spiral vessels, and to the slits on the surface,
and recollect the law, that essential differences of internal
structure almost always correspond with differences in the
formation of the seed, and in external relations (170.), Adanson's idea gains considerable strength. If we compare *Elæagnus* and *Hippophäe* with this plant, the formation of the
seed, and the position of the embryon, on which most depends, agree completely, (171.) Instead of the nut in our
plant, these others have a drupe. The calyx, which in
the *Hippuris* is not unfolded, consists in *Elæagnus* of four

parts, and in *Hippophae* of two; but, in both cases, it stands above the fruit.   It is true, that in the two latter plants, the filaments are four; but we know that a uniform abortion accounts for the want of organs of the same kind, (180.)

Uses.

Only goats eat this plant.   It contributes to the purification of air in standing water.

2.

Agardhia cryptantha.

Char. Gen. Cal. 3-sepalus, inferus.   Cor. 5-petala, convoluta.   Stam. 1, anthera magna.   Drupa 3-locularis, 3-valvis.

Ramum florentem habeo e Brasilia, cortice fusco. *Folia* opposita, ovata, inæqualia, coriacea, integerrima, acuta, nervoso-venosa, duos pollices cum dimidio longa, duos fere pollices lata, utrinque glaberrima; *petioli* fusci; canaliculati, semipollicares, patentes. *Racemus* terminalis, aphyllus, nutans. *Flores* flavidi.   *Calyx* trisepalus, subimbricatus. Æstivatio convolutiva.   *Corolla* 5-petala, extus sericea. *Receptaculum* villosum.   *Stamen* unicum laterale.   *Anthera* ratione reliquarum partium maxima, curvata, bilocularis. *Pistillum* triquetrum.   *Drupa* ovalis pollicaris, atra, trivalvis, trilocularis.

Planta affinis Cryptostomo Schreb. (Moutabea Aubl. Guian. t. 274.) ipsique mangiferæ.   Sed hæc tum calyce 5-sepalo, tum drupa monosperma, illa antheris quinque stamini unico insidentibus, et calyptra corollæ differt.   Videtur tamen Terebinthaces adnumeranda esse.

Dixi in honorem Agardhii, prof. Lundinensis, qui algarum historiam egregie illustravit.

CLASS II.

3.

Circæa Lutetiana, *L.*

Hexenkraut, Stephanskraut, Waldkletten.—Fren. *Herbe de St Etienne.*—Engl. *Enchanter's Night-shade.*—Swed. *Gulsirse.*

About St John's day, an interesting plant appears in our dark and moist woods. It has a woody, articulated, creeping root, which is commonly thickly entwined with the roots of the tree, and is hence difficult to be torn up. From this root a small stem arises, about the thickness of a corn-stalk, straight, round, simple, with fine and undistinguishable hairs, green in its colour, and from a foot and a half to two feet in length.

From the stem, leaf-stalks shoot out, opposite to one another, at distances which are about an inch from one another, having undistinguishable hairs, angular, almost marginated (28.), an inch long, and standing open ; and upon these are leaves completely oval-shaped, almost in the form of a heart, with undistinguishable hairs, obtusely pointed, with notches at intervals on the margin, nerved and veined, an inch and a half long, and an inch broad.

On the top of the stem stands the many-blossomed bunch (84.), the principal and secondary stalks of which are more strongly ciliated than the lower part of the stem. The secondary stalks, from three to four lines in length, stand almost horizontal, afterwards they are reflex ; they bear at their lower part the ovaria, of a round shape, afterwards pear-shaped, studded with hooked bristles (80.) ; over the ovaria are two calyx-leaves, of an oval or oblong form, reflex, of a reddish colour, and two petals of a pale red, of an

3

inverted heart-shape, shorter than the two filaments. The
filaments stand before the calyx leaves, and alternate like
them with the petals, (196.) The stigmata are two. The
æstivation is valvular.

The capsule consists of two loculi, opens from below, and
contains in each loculus a seed, which, without any albuminous matter, contains a completely unfolded embryon, with
two erect, thickish cotyledons, and a radicle scarcely distinguishable.

A subspecies of this plant is produced in mountain forests,
and especially in North America, with a smoother stem, with
the leaf-stalks completely linear and smooth, and with leaves
more softly ciliated. This is the *Circæa intermedia*, Ehrh.
beytr. 4. s. 42. *Sturm Deutsch. Flor. Heft.* 23. *C. Lutetiana Canadensis Michaux*, bo. ram. 1. p. 17. *Pursh, Amer.
Sept.* 21. *Nuttall*, 18.

The *Circæa alpina*, again, is essentially distinguished
by a smooth, branchy stem, which is never longer than a
small span, and by small cordate leaves. There is also a
small bractea under every flower, which is wanting in our
species.

Geographical Distribution.

The *Circæa lutetiana* appears to be diffused in the northern
hemisphere from 37° to the 64° N. Lat. For Sibthorp found
it in the Bithynian Olympus, and Marshall of Biberstein in
Tauris. On the other hand, it is wanting in Lapland, where,
instead of it, the *Circæa alpina* grows. Its limits in North
America have not yet been exactly determined. But it seems
to grow from 40° to the 50° N. Lat.

Synonymes and Figures.

Circæa lutetiana, *Lob. Hist.* 137, ic. 206. *Ger. Emac.* 351.
  Ocymastrum verrucarium, *J. Bauh. Hist.* 3. 977. Solani-
  folia Circæa dicta, *C. Bauh. Pin.* 68. *Park. Theatr.* 351.
Moris. s. 5. t. 34. (C. lutetiana, *Dalech.* 1338. is *C. alpi-*

na.) *Fl. Dan.* 210. *Engl. Bot.* 1056. *Sturm Deutsch. Flor. Heft.* 23.

Affinity.

Ray was completely mistaken when he placed *Circæa* along with *Callitriche, Stratiotes,* and *Hydrocharis,* (*Ray Syn.* p. 289.) Linnæus also shewed no great insight into its affinities, when he placed *Circæa* along with *Boerhaavia* and *Valeriana,* among the Aggregatæ, (*Ord. Nat.* 48.) Adanson first perceived its true relations, when he placed the plant among the Oenagræ, (*Famill. des Plantes,* p. 85.) Later botanists have followed Adanson more readily, since we have become acquainted with a Mexican plant which is very like the *Circæa,* namely *Lopezia,* Cav. in which we find only a different numerical proportion. *Ditmaria,* too, (*Erisma Rudg.*) lays claim to a still greater resemblance. The numbers 2, 4, 8, prevail in this family ; and although *Œnothera* and *Epilobium* are removed from *Circæa* by different proportions, yet *Gaura* and *Haloragis* furnish intermediate members of the series ; and *Escallonia,* Sm., although the number 5 prevails in it, opens its capsule exactly like the *Circæa.*

Uses.

The name *Circæa* was given to the plant by *Lobelius,* because superstition regarded it as a charm ; hence, too, the English name *Enchanter's Night-shade.* The well known witch *Circe* is understood to have made use of this charm, and Gerard affirms, that the *Mandragora* had been confounded with this plant. At present no other use of it is known, except that in America a yellow dye is procured from the root.

4.
Salvia Brasiliensis.

S. calyce ampliato colorato tridentato corollam excedente, foliis ovatis serratis acuminatis glabris basi cuneatis.

Habitat in Brasilia.

Caulis herbaceus, quadrangularis, glaber, nodosus. *Folia* opposita, longe petiolata, petiolis sesquipollicaribus, angulatis, glabris, ovata, acuminata, basi cuneata, inæqualiter obtuse serrata, nervoso-venosa sesquipollicem longa, supra pollicem lata. *Racemi* terminales, pubescentes. *Flores* subverticillati, ebracteati. *Calyx* puniceo-roseus, unguicularis, ampliatus, nervosus, pubescens, apice tridentatus. *Corolla* inclusa, sordide rubra, bilabiata, labio superiori fornicato, inferiori trilobo. *Stamina* duo, basi appendiculata. *Pistillum* apice fissum. *Achenia* quatuor.

Proximæ *S. Regla*, Cav., *et S. galeata R. et P.*, sed differunt colore calycis, et corolla calycem excedente.

———

CLASS III.

5.

Poa trivialis, *L.*

Gemeines Wiesen-, Vieh- oder Rispengras.—French, *Pâturin rude.*—Ital. *La fienarola comune.*—Engl. *Roughish Meadow-grass.*—Swed. *Angs-gräs.*

This grass, although a very common one, may easily be confounded with others which are nearly related to it. We have, therefore, subjoined an exact description and comparison of it with these others.

From a fibrous root there rises a round stalk, sharp to the touch, about the length of an ell or arm. Where the leaves rise from the sheath, a long ligula remains, (77.) The leaves are small, and, at the same time, sharp to the touch. The flowers are produced on a uniform, spreading panicle, the subordinate stalks of which are horizontal, or even reflex, and also sharp to the touch. The individual ears consist of three glumes, the valves of which are observed to have five fine nerves; on the margin they are of a reddish colour, and at the base they are united by tufts of hair, (94.)

The generic character of the *Poa* is constituted by the oval shape of the ear, and by the valves being destitute of awns.

The species most nearly related to this are the *P. pratensis, serotina* Ehrh. and *nemoralis.* The former is distinguished by its creeping root, its smooth stem, its short, truncated, and unprojecting ligula, and by its earlier season of flowering. For it is in full flower in May, whilst the Poa trivialis begins to blossom in June.

The *Poa serotina* Ehrh., is distinguished from our grass by a root slightly creeping; by a ligula, which projects but a little, and which is truncated; by a panicle of a more pyramidal shape, having its subordinate stalks open, but not reflex; by its ear being smaller and more of a spear-shape, its glumes being commonly five, and coloured yellow at the point. This species also blossoms later than the *P. trivialis.*

The *Poa nemoralis* is distinguished by a stalk as smooth as that of the *P. trivialis*, being, at the same time, a little compressed. The ligula is also truncated. The panicle is not uniformly spreading, but tapering, and inclining towards one side. The ears are lanceolate, contain commonly three glumes, which are open, and have long projecting points. At their base, too, they are almost completely free, although some small hairs appear here also.

Geographical Distribution.

The *Poa trivialis* is one of the plants which have the most extensive distribution in Europe. Towards the north it extends beyond the polar circle, and constitutes the principal produce of the meadows in Lapland. It is also very common in Northern Asia, and in North America. Towards the south it constitutes the chief riches of the meadows in Peloponnesus. All the countries of Europe, Asia, and America, which lie between the 36° and 68° N. Lat., furnish this grass. But towards the south it does not appear to grow beyond the 36° N. Lat. It is wanting entirely in Africa, and every where between the tropics. In like manner it is not found in New Holland, nor in the islands of the South Sea.

Synonymes and Figures.

Gramen pratense, *Lob. ic.* 1. (Gramen minus D°. est P. serotina, and Gramen miliaceum 3. Poa pratensis), Gramen pratense, 1. *Dodon.* 560. *Gerard Emac.*, 2. *Parkins. Theatr.*, 1156. Gramen pratense paniculatum medium, *C. Bauh. Pin.* 2. *Scheuchz. Agrost.* 180. Gramen pratense vulgatius, minus, *Moris.* sect. 8. t. 5.
Poa scabra, *Ehrh.*
P. dubia, *Leer's Herborn.* t. 6. f. 5.
P trivialis, *Linn. Engl. Bot.*, 1072. *Host. Gram. Austr.*, 2. t. 62. *Fl. Dan.*, 1444.

Uses.

It is one of the most productive of the meadow and feeding grasses. In England, a great deal has been said, since the time of Ray, respecting what has been called the Orcheston Grass. Ray cites it under the name of *Gramen caninum supinum longissimum* in the *Indiculus plant. dubiarum* of his Synopsis: he mentions, as its habitat, a meadow near Maddington, two miles from Salisbury, and says, that the plant is twenty-four feet long. Maddington and Orcheston St Mary, from which latter place the grass has its name, lie close by one another. It has lately been established by Swayne and Maton, that this remarkable grass is nothing but a mixture between the *Poa trivialis* and the *Alopecurus pratensis*, and that its extraordinary length arises from the richness of the soil, and from the annual irrigation of the meadow with the water of a spring, which, having the high temperature of 48° or 49° Fahr., necessarily gives an uncommon stimulus to vegetation; (*Withering's Arrangement of British Plants*, 2. p. 190.)

Sir Humphry Davy and Sinclair state the produce of this grass to be equal to that of the *Anthoxanthum odoratum*, of the *Bromus tectorum*, of the *Lolium perenne*, of the *Melica coerulea*; (*Sir Humphry Davy's Elements of Agricult. Chemistr.* p. 100.; Landwirthschaftl. Zeitung, 1815, vid. 303.

6.

Tontelea trinervia.

T. foliis oblongis acuminatis trinerviis integerrimis utrinque glabris, panicula terminali dichotoma pubescente.

Hab. in Brasilia.

Rami teretes, glabri, fusci, cicatricula tuberculosa albida sub quovis ramulo et petiolo. *Petioli* sparsi, erecto-patentes, pubescentes, subpollicares. *Folia* subdigitalia oblonga, basi rotundata, acuminata, integerrima, trinervia, venosa. *Panicula* terminalis, erecta, pubescens, dichotoma, aphylla. *Cal.* quinquefidus, minimus. *Cor.* quinquepetala, calyci affixa, extus pubescens. *Stamina* tria e membrana urceolata progredientia. *Pistillum* unicum. *Bacca* uni-locularis tetrasperma.

Genus hoc, Tonsella a Vahlio dictum, idem cum Anthodonte Ruiz et Pav., satis adfine Chætocrateri R. et P. Salaciæ Lour. et Hippocrateæ L., in systemate artificiali locum obtinet prope ab Hippocratea. Hace vero differt potissimum capsulis tribus bivalvibus, medio dehiscentibus.

In ordine naturali Caseariæ, Samydæ, Athenaeæ Schreb. et Cedrelæ Aubl. cognatum, cum his constituit tribum singularem Samydearum, quæ cum Melieis et Malpighieis multa habet communia. Stamina urceolo aut membranulæ peculiari insidentia, et embryo cotyledonibus plicatis, constituunt characterem tribus.

Nostra species Tonteleæ scandenti Aubl. t . 10. vicina. Hæc tamen scandit, ramos habet divaricatos, folia haud trinervia. Reliquæ species gaudent vel foliis serratis aut denticulatis, vel pedunculis lateralibus congestis.

Z

CLASS IV

7.

Asterocephalus canescens.

Graue Scabiose.   Apostemkraut.—French, *Scabieuse grise-
âtre.*—Engl. *Greyish Scabious.*—Swed. *Grå vädd.*

This remarkable species was confounded with *Ast. Columbaria*, till Kitaibel first taught us to distinguish them; (*Plant.
Hungar.* 1. p. 53. t. 53.)
The brown perennial root creeps almost horizontal, and
pushes out first oblong leaves, tapering at both ends, quite
entire, rarely having two teeth, set with indistinguishable
hairs on both sides, ciliated on the margin: the leaves are
nearly an inch long, four lines broad, with a strong nerve in
the middle, and some lateral veins.

By the side of this first shoot, and often at a later period quite separated from it, there springs up a simple stem,
about a foot long, round, thickly beset with greyish reflex
hairs: below, this stem is set with half pinnated, ciliated
leaves, having indistinguishable hairs, the tufts of which are
lanceolate, linear-shaped, and stiff at the point. The upper
leaves become always finer, stand at wider distances, lose the
pinnated shape, and, at last, become entirely simple. The
joints of the stem have the same greyish colour.

On the tip of the stem stands the compound flower, of a lavender blue colour, and having a fine smell, almost like that
of *Orchis nigra*. The common calyx consists of about
twelve very small uniform obtuse leaves, which are much
shorter than the ray. The flowers are five-lobed, dissimilar,
radiated on the margin, externally set with fine white hairs.
The receptacle contains chaffy leaves, which upwards become broader, and are thickly set with white hairs. Between

the germen and the flower there stands a membranaceous crown, and within it five white bristles, which are much shorter than the germen, and not much larger than the membranaceous crown. Four white filaments are connected with the tube of the corolla, and carry reddish anthers. The pistillum is linear, and has a pretty thick stigma. The nectary is the upper surface of the germen. This swells to an eight-cornered, strongly haired achenium, which continues to exhibit the five white short bristles, and the pappus, and contains the evolved embryon in consumed albumen, with the radicle turned upwards.

Diagnosis.

If we compare the related species, particularly *Ast. Columbaria*, with this plant, we find this other species, in the first place, much taller, somewhat ciliated also, but by no means of a greyish colour. The leaves of the root are mostly lyreshaped, very seldom ovate, and deeply serrated. The joints of the stem are reddish, the leaves of the calyx pointed, and a little shorter than the ray, the chaffy leaves of the receptacle are finely pointed, and, above all, the bristles of the pappus are brown, and almost as long as the germen. The colour of the flower is violet or sky blue.

Ast. agrestis, also, (Scabiosa agrestis Kit. Pl. Hungar. 3. t. 204.) may be confounded with our plant. But that species has always lyre-shaped root-leaves, and bi- or tri-pinnate stem-leaves,—the stem is branchy, and sprinkled with grey hoar ; the leaves of the common calyx are lanceolate and ciliated ; the flowers are of a lilac colour, and the bristles of the pappus of a brown colour, and nearly as long as the germen. Sc. pyrenaica All., which some confound with our plant, is completely distinguished from it by white soft *tomentum,* and by broader *laciniæ. Bertol. Amœn. Ital.* p. 12.

Geographical Distribution.

This species grows on the calcareous soil of central Germany, France, Austria, and Hungary. It seems not to pass beyond the 54° N. Lat. We cannot determine with certainty, on account of the inaccuracy of synonymes, how far it extends towards the south, and whether it passes beyond the 43° N. Lat.

Synonymes and Figures.

Scabiosa minor, 1. 3. ix. xi. *Tabern.* 443. Scab. media, *Gerard emac.* 720.

Sc. capitulo globoso minor, *Moris.* sect. 6. t. 14. *Buxbaum Hal.* 295. *Tournef. Inst.* 465. *Hist. des Plantes aux Environs de Paris,* p. 141, (probe jam distinxit speciem). *Dalibard Flor. Pariss.* 45.

Asterocephalus foliis ad terram ovalibus, *Haller Gött.* 352.

Scabiosa descripta in *Wiegel Flor. Pomer.* p. 25. n. 91. *C. F. ejus Obs. Bot.* p. 23.

Sc. asterocephala, *Thuill. Flor. Pariss.* p. 72. (var flore albo).

Sc. suaveolens, *Desfont. Cat. Hort. Pariss.* ed. 2. p. 131. *De Cand. Fl. Franç.* 4. p. 229.

Sc. canescens, *Kit. Plant. Hung.* 1. p. 53. t. 53. *Pohl. Bohem.* 1. p. 133. *Pers. Syn.* 1. p. 22. *Willd. Enum.* 1. p. 146. *Baumgart. Trans. Sylv.* 1. 77. *Wallroth. Ann. Bot.* p. 143. *Röm et Schult.* 1. p. 66. *Dierbach, Flor. Heidelb.* 1. 39.

Sc. Columbaria, α. γ. *Poir enc.* 6. 711. var. β. *Gmel. Bad.* 1. p. 323. var. ε. *Marsch. Bieb. taur. cauc.* 1. 96.

? Sc. tomentosa Sibth. *Smith, Prod. Fl. Græc.* 1. 85.

On the Family of the Scabiosæ.

The Scabiosæ, otherwise called Aggregatæ, evidently constitute a peculiar family, which are nearly related to the Syngenesistæ, by their compound flowers, by their common calyx, by the chaffy leaves on their receptacle, by their monopetalous flowers, to the tube of which the filaments are fixed,—by their fruit being *inferior*, forming acheniæ, or caryopses, and carrying pappi. But they are distinguished by the free situation of the anthers, by the number of the filaments, which are always but four, by the simplicity of the stigma, and by the direction of the radicle of the embryon upwards, whilst, in the Syngenesistæ, the radicle is directed downwards.

Sebast. Vaillant, long ago, divided the Scabiosæ into several genera, which I have adopted with the names given by Vaillant; (*Vaillant in Mem. de Paris*, 1722. Anleit. zur Kentn. der Gewash. 2. te Aufl. th. 2. s. 584.) Other writers have unnecessarily given new names to them.

Asterocephalus Vaill. is one of those genera the character of which consists in its many-leaved, almost simple calyx, in having its flower divided into five parts, and in its double pappus, which is partly formed by a membranaceous circle, and partly by five bristles. The *Scabiosa*, again, has its flower divided into four parts, and the pappus consists simply of chaffy leaves. *Sc. arvensis*, succisa Vaill. has a scaly calyx, a flower divided into four parts, and chaffy leaves, passing into bristles, as its pappus. *Sc. Succisa*, Pterocephalus Vaill. has a bristly receptacle, and a pinnated pappus. (*Sc. papposa.*)

———

CLASS V

8.

Phyteuma spicatum, *L*.

Wald-Rapunzel, Taubenkropf.—French, *Raiponce en epi.* —Engl. *Rampion, spike-flowered.*—Dan. *Traevlekrone.*

This plant blossoms in our woods in June. From a root about the thickness of a finger, whitish, tuberculous above, and fusiform below, which continues for several years, there springs a herbaceous, smooth, simple stem, reddish at the under part, green above, round, or slightly angular, commonly about a foot and a half high, frequently, also, of the length of an arm. The leaf-stalks are a small span long, alternate, open, far from one another, smooth, marked by furrows, and embracing the stem at their base: higher up they become shorter, and, at last, entirely disappear. The lowermost leaves are cordate, oblong, unequally crenate and dentate, smooth on both sides, three inches long, and one inch broad. The upper leaves are smaller, and, at last, are entirely lanceolate, and without stalks. On the upper part of the stem stands the spike, frequently of a finger's length, the yellowish white, sometimes blue flowers of which open from below upwards, (84.) Immediately under the spike a pair of small stem-leaves are found, which are quite entire.

The calyx surrounds the germen, and terminates at the upper part in five pointed teeth. The corolla consists of five long pointed petals, which, with their upper greenish ends, at first are united around the pistil, and form a short tube, but at their under part they are open. Afterwards they spring from one another, and stand quite open. The filaments are broad and hairy below, form a kind of arch over the upper part of the germen, by degrees become pointed above, and pass into long yellow anthers. The pistil is simple, upwardly hairy, terminates in two convoluted stigmata, and is much larger than the corolla and the anthers. On this account, and because androgynous dichogamy takes place here (103. 331.), the stigma cannot be impregnated by the anthers of the same flower, but from the superior flowers, which blossom later. The nectary is the surface of the germen. The fruit is a capsule of two loculi, surrounded by a calyx, containing in each loculus, on a separate free-standing pillar, a number of fine seeds, which inclose, in the middle of the albuminous substance, the erect embryon, with its two cotyledons.

Diagnosis.

The nearest to this species is the *Ph. betonicæfolium, Vaill. (Delph.* t. 12.), which is distinguished only by longer and smaller leaves, and by an oblong obtuse spike. The *Ph. cordatum Vill.* (t. 11. *nigrum Schmidt.*), also is nearly related, but its blossoms are constantly of a dark violet colour, and the bracteæ are subulate. The other species are more distantly related.

Geographical Distribution.

This species is found throughout the whole of Germany, Poland, Hungary, Transylvania, Upper Italy, and France. It does not grow either in Great Britain or Denmark (the German States excepted), or in Sweden and Russia. As little does it appear to exist further south than the 44° N. Lat., since it is neither found in Sicily nor Spain, neither in Greece nor in Tauris.

Synonymes and Figures.

Rapunculum sylvestre, *Trag*. p. 277. a. (ed. 1551.)
Rapunculum alopecuron. *Dodon.* 165.   *Barrel,* ic. 892.
  alopecuroides, *Clus. Hist.* 2. 171.
Rapuntium majus, *Lobel. Hist.* 178. ic. 329.   *Gerard, Emac.*
  453.
Rapunculus major, Dodon.   *Dalech.* 641.
Rapunculus V. nemorosus 1.   *Tabern.* 794.
Rapunculus spicatus, *C. Bauh. Pin.* 92.   (R. spicatus cœru-
  leus, *C. Bauh. Prodr.* 32, is evidently Ph. betonicæ-
  folium.)   *J. Bauh. Hist.* 2. 809.   *Parkin's Theatr.* 648.
  Tourn. Inst. 113.
Rapunculus corniculatus, folio urticæ *Moris.* sect. 5. t. 5.
  Rap. corn. spica longiore, *Riv. Irr. Monop.   Hall, Hist*
  Stirp. n. 684.
Phyteuma spicata, *Linn. Fl. Dan.* 362.   *Sehk.* t. 39.

Affinities of the Genus.

When we compare the most important organs, in particular the structure of the filaments, the form of the capsule and of the seed, we cannot but be struck with the resemblance of this genus to the *Campanula.* But the distinction between them lies in this, that the laciniæ of the corolla are at first united at the top, that two stigmata are found, and that the capsule consists of two loculi; whilst, in the *Campanula,* there are three stigmata, and three loculi in the capsule. But these, and some related genera, constitute a peculiar

family, the Campanuleæ, which stands between the Ericeæ
and Lobeliæ ; (Anleit. 2. s. 522.)

Uses.

The root affords a milky substance, and is edible ; hence
the name *Rapunculus*, instead of which they use the term
Wild Rape, in Germany. In some countries the plant is
reared in gardens, and the root and young leaves are used in
spring as greens.

9.
Gentiana pneumonanthe, *L.*

Lungenblume, schmalblattriger Herbst-Enzian.—French,
  Gentiana Pneumonanthe.—Engl. *Marsh Gentian, Cala-
  thian violet.*—Swed. *Kläck genzian, Höst-kläckor.*

This plant grows on peat-mosses and moist meadows in
August, and is distinguished, at first glance, by its beautiful
Berlin blue flowers, which are of considerable size.

From a small yellowish-brown fibrous root, there rises a
stem of the thickness of a straw, of a large span long, simple,
somewhat angular, sharp to the touch, but otherwise smooth.
The leaves, which are almost linear, somewhat obtuse, quite
entire, of an inch long, of a shining green on their upper
surface, sharp to the touch on their lower, embrace the stem
with their tapering base, stand opposite to one another, but
so that the nearest pairs form a right angle with one another,
and are thus cruciform, (36.)

On the uppermost axillæ spring the flowers upon short
stalks. The flowers consist of a tubular five-toothed, or
five-lobed calyx, the teeth of which, at an after period, are
reflex,—and of a monopetalous corolla, bell-shaped, of a
Berlin blue colour, internally adorned with yellow points.
The margin of the corolla has five larger, and as many smaller teeth. The æstivation is twisted, (99.) Five filaments
are united at their lower parts, with the base of the corolla.
The yellow, almost arrow-shaped anthers, also stand at first
in a cluster around the simple pistillum ; afterwards they se-

parate from one another, but always stand lower than the
margin of the corolla. The stigma is two-lobed. The fruit
is a superior, simple, two lobed capsule, the valves of which
being bent inwards, form, with their inner margins, the apart-
ment for the numerous seeds. These contain the erect
embryon, with its evolved cotyledons, in the middle of the
albuminous substance.

Diagnosis.

The most nearly related to this species is the *Gentiana tri-
flora* Pall. ; but, in this last species, the linear leaves are about
two inches long, and pointed. The flowers, though of the
same size, are properly without stalks, and are placed by
threes on the uppermost axillæ. This species grows in eastern Siberia. *G. algida* Pall. has much broader, longer, lanceolate, three-nerved leaves. *G. asclepiadea* is much larger,
more branchy, and has ovate-lanceolate leaves. The other
species are still more unlike.

Geographical Distribution.

In Europe this species seems to grow from 45° to 64°
N. Lat., for it is not found more southerly than Greece, nor
more northerly than Lapland. In North America it seems
to be confined to the space between the 40° and 50° North
Lat.

Synonymes and Figures.

Campanula autumnalis, *Dodon.* 168.

Pneumonanthe Cordi, *Lobel. Adv.* 130. *Hist.* 166. *Ic.* 309.
   Tabern. 1176. *Gerard. Emac.* 438. *Parkins. Theatr.*
   406. *Barrel. Ic.* 52. 122.

Gentianæ iv Species, *Clus. Hist.* 313

Gentiana minima, *Matth.* ed. C. Bauh. p. 481. *Barrel.
   Ic.* 51.

Calathiana viola et Campanula pratensis, *Dalech.* 484. *Barrel. Ic.* 54.

Gentianæ species, calathiana quibusdam, *J. Bauh. Hist.* 3.
   527.

Gentiana angustifolia autumnalis major, *C. Bauh. Pin.* 188.
Moris. sect. 12. t. 5.   *Tournef. Inst.* 31.   G. palustris
angustifolia, *C. Bauh.* l. c.
Κυανη *Renealm. spec.* 68.
Gentiana alis floriferis, *Hall. Hist. Stirp.* n. 641.
Gentiana Pneumonanthe, *Linn. Flor. Suec.* n. 288.   *Flor.
Dan.* 269.   *Engl. Bot.* 20.   *Bot. Mag.* 1101.   *Lam. Ill.*
t. 109.

Affinity.

The Gentians constitute, along with *Chironia, Erythræa,
Swertia, Chlora, Exacum,* and *Menyanthes,* a peculiar family, which is distinguished by the numerical proportion of
the essential parts, by the situation of the fruit, by the insertion of the seed,—and which stands between the Jasmineæ and
Contortæ.   (Anleit. 2. 471.)

Uses.

As the composition of the juices corresponds with the family character, we may suspect that those ingredients, which
are found in one genus, or species of gentians, will also be
found in the others, (170.)   Bitter extractive matter is that
by which *Gentiana lutea, Erythræa Centaurium,* and *Menyanthes trifoliata,* are distinguished.   The same bitter extractive matter exists in our species.   Formerly the root was
used as a tonic for the stomach.   It has also a powerful effect upon the urine, and hence the Mecklenburghers call the
plant, *Sta up unn gah weg*; (*Wredow's ökon. Flor. von
Meklenb.* 1. 456.)   It used to be employed in France as a
cure for sprains; (*Commerce. Lit. Nor.* 1743, hebd. 7.)

10.

Viburnum opulus, *L.*

Schneeball, Wasserholder, Hirschholder, Schwelken-baum,
Kalinen, Drosselbeeren, Wasserahorn, Schlingen-baum, *in
Pomerania,* Goosfleder.—French, *Viorne obier, Rose de*

Gueldre, Pain blanc, Pomme de niege.—Ital. *Sambuco aquatico, Maggio.*—Engl. *Common Guelder-Rose, Water-Elder.*—Swed. *Olvon, Häls-bär.*

This is a shrub, which grows from the size of a man in height to twice that size: its stems, which are about the thickness of an arm, having a grey, rifted, but, in other respects, smooth bark, and white spongy pith. The branches stand opposite to one another. The sulcated smooth leaf-stalks, an inch in length, are also placed opposite to one another, and have from four to six kidney-shaped knobs. At the base of each leaf-stalk there are two deciduous pointed stipulæ. The leaves are about four inches long and broad, having three short lobes, somewhat round at the base, cuneiform, with sharply dentated margins, smooth, and having deep nerves and veins on the upper surface, and slightly furnished with hairs on the under surface. The flowers grow on a stalked false umbel, at the top of the branch.

The calyx is very small, and has five teeth. The corollæ are not uniform. At the circumference of the flowers we observe large, white, wheel-shaped, five-lobed corollæ, without ovaria, commonly without anthers, although sometimes these last parts are seen. The central flowers are yellowish, and also divided into five parts. Five filaments, longer than the corolla, are united with its base, and carry yellow anthers, consisting of two parts. The germen stands below the calyx, and is surmounted by three reddish stigmata. Its uppermost part is also the nectarium. It passes into a red, edible, one-seeded berry. The embryon stands in a small hole of the albuminous matter, with the radicle turned upwards.

In the wild state, the marginal flowers alone are unfruitful, because the sexual parts are abortive, (181.) and the corolla is unfolded instead of them. Luxuriant growth in gardens makes the abortion general, all the flowers become unfruitful, and the false umbel is contracted into a ball.

Diagnosis.

The essential character of the species consists in the g.andular leaf-stalks, and in the three-lobed leaves, which are somewhat rounded at the base, but which also taper, and have their lobes short and unequally dentated. The most nearly related to this species are two North American species, *V. Oxycoccos* Pursh, and *edule* Pursh ; for both of them have the same glandular leaf stalks, the same general form of the leaves, the same unfruitful marginal flowers in the false umbel, and the same colour of the berries. But in *V. Oxycoccos* the leaves taper at· the base into a long wedge shape, have three distinct nerves, in the axillæ of which are some hairs : the lobes of the leaves are drawn out to a great length, and have few teeth. In *V. edule* Pursh, the lobes of the leaves are also shorter than in *V. Opulus*, but the teeth are produced into a fine point. *Viburnum acerifolium* also agrees in the three-lobed form of the leaves, but the glands are wanting on the leaf-stalks, these last named parts being furnished, instead of glands, with long white hairs. The leaves are more rounded at the base, have stronger hairs on their lower surface, and the flowers are all alike. Other species have still more numerous differences. The case is the same with *V. orientale* Pall., which has quite the appearance of *V. acerifolium*, and is distinguished from it simply by its larger and more obtuse teeth, and by the oval form of the seed, which, in *V. acerifolium*, is heart-shaped.

Geographical Distribution.

Viburnum Opulus grows throughout all Europe, from 40° to 60° N. Lat. Its southern limit seems to be Constantinople, its most northerly Upsal. In America, the three species already mentioned, *V. Oxycoccos*, *edule*, and *acerifolium*, and in Asia, *V. orientale* supply its place. It grows in great abundance on Caucasus, along with this latter species.

Synonymes and Figures.

Sambucus aquatica, *Trage*, f. 378. b. *Matthiol. ed. C.*
Bauh. 874. *Dalech.* 270. *Tabern.* 1440. *C. Bauh. Pin.*
456. *S. palustris Dodon.* 864. *S. rosea, Lob.* ic. 2.
201. *Ger. Emac.* 1424. *Tabern.* 1440. *J. Bauh. Hist.*
1. 552. *Park. Theatr.* 209.
Opulus Ruellii, *Tourn. Inst.* 607. O. glandulosus, *Monch.*
Meth. 505. *Baumg. Transylv.* 1. 261.
Viburnum Opulus, *Linn. Fl. Dan.* 661. *Engl. Bot.* 332.
Schk. t. 81. *Kerner Baumz.* t. 35.

Affinity of the Genus.

The old writers perceived distinctly the affinity of this genus with Sambucus, and hence they chose for it the names
Sambucus aquatica and *palustris*. Indeed *Viburnum* is distinguished from *Sambucus*, simply by its berries having but
one seed, while the latter genus contains three seeds in the
berry. If we pass by the unessential parts, and attend only
to the relations of the most important, both the genera belong, with *Lonicera*, to one family, which we called the *Ca-*
prifoliæ; (Anleit. 2. vid. 617.)

Uses.

The wood is used in Norway for making weavers combs.
The shoemakers use it for pegs. The stronger branches are
used for tobacco pipes. The leaf is eaten only by cows.
The berries are favourites with the thrush. In Courland
they make vinegar with these berries, and the inhabitants of
Northern Asia form an intoxicating drink from them.

CLASS VI.

11.

Leucoium vernum, *L.*

Grosses Schneeglockchen, Sommerthierchen, Frucklings-Kno-
tenblume, Marzglockchen.—French, *Perce-neige, Nivéole
printannière.*—Engl. *Great Snow-drop.*—Ital. *Primestro,
fior marzajuolo.*—Swed. *Varhviter.*

This lovely flower appears in April among bushes, and on
the pastures. From a white bulb of the size of a walnut,
there first spring several lanceolate leaves, quite entire, obtuse, smooth on both sides, having the upper surface shining, about the length of a finger, and nearly an inch broad.
In the middle of these arises an even divided stalk, about a
small span in length, at the top of which there stands a two-
lobed calyx, of a whitish green colour, and streaked. From
this arises a round flower-stalk, nodding or bent downwards,
smooth, about an inch long, which supports, over the obtuse
three-cornered, smooth, streaked, green germen, a six-leaved,
bell-shaped, downward-hanging corolla, the divisions of which
are callous at the points, and marked by a small green spot.
The exterior integument of the corolla is of the nature of
a calyx, and has slits, (90. 312.) On the receptacle stand
six short white filaments, which carry on their summits yellow, long, bilocular antheræ, containing an oval pollen. In
the middle of the filaments stands the club-shaped green pistil, with a stigma somewhat tapering. The thick part of it
secretes the nectar, (Sprengel's entd. Geheimniss, vid. 182.)
The fruit is a capsule of three loculi, the round seeds of
which contain the unevolved embryon opposite to the umbilicus in the albuminous substance. The strophiolus continues as a withered, folded membrane.

Diagnosis.

This species is distinguished from the other species of the same genus by its single blossomed flower, the *L. œstivum* and *autumnale* having several blossoms,—and by its club-shaped pistil, the *L. autumnale* and *roseum*, Mart. having a linear pistil. *Galanthus*, or the small snow-drop, is indeed very like this plant, but is distinguished by its double three-leaved corolla, the interior petals of which are emarginated, and by its anthers which pass at the summit into a bristle.

Geographical Distribution.

This plant grows throughout the greatest part of Europe, from Upper Italy to Upland in Sweden, and from Spain to Transylvania. It is remarkable, that it is not found it Great Britain, nor does it appear in Tauris, or in Greece. In these countries its place is supplied by *L. œstivum.* It is therefore confined (with the exception of Great Britain) be-45° and 60° N. Lat.

Synonymes and Figures.

Narcissus *Brunfels,* 1. 129, n. vii.   *Matth. ed. Bauh.* 860.
Viola alba, *Fuchs.* 486. 487.   *Dalech.* 1527.
Leucoium bulbosum hexaphyllon, *Dodon.* 230.   *Clus. Hist.*
     1. 168.   L. bulbosum, i.   *Tabern.* 1005.   *J. Bauh. Hist.*
     2. 590.   *Ger. Emac.* 148.   (L. bulbosum serotinum, qua-
     si esset L. æstivum, sed est vernum.)
Leuconarcissolirion, *Lobel. Ic.* 123.
Leucoium bulbosum vulgare, *C. Bauh. Pin.* 55.
Narcisso-leucoium vulgare, *Tourn. Inst.* 387.
Galanthus uniflorus, *Hall. Helv.* n. 1253.
Leucoium vernum, *Linn. Hort. Upsal.* 74. *Jacqu. Fl. Austr.*
     t. 312.   *Batsch. Annal. Flor.* t. 2.   *Schk.* t. 89.   *Sturm,*
     Heft. 11.

Affinity.

Its affinity with *Galanthus, Narcissus, Pancratium,* is obvious, and therefore this genus belongs to the Coronariæ, and indeed to that subdivision which has the fruit below the flower, (Anleit. ii. vid. 247.)

Uses.

Formerly the bulbs of this plant were used instead of the squill, but they have now fallen into disuse.

CLASS VII.

12.

Trientalis Europæa.

Schirmkraut, Sternkraut, Meyerblume.—French, *Trientale d'Europe.*—Engl. *Chickenweed Winter-green.*—Swed. *Duf-kulla.*

On the northern declivities of our forest hills, there appears in June among the bilberries and heath, a gentle, handsome plant, which is particularly interesting, from the numerical relations of its parts.

From a very small woody tuber, which is surrounded by fibrous roots, there rises, about the length of a small span, a simple, even, round stem, about the thickness of a strong linen thread. At the lower part of the stem stand, in a sparse state, very short, roundish or oblong, obtuse leaves. On the upper portion, there grow from five to seven leaves, an inch and a half long, smooth, oblong, tapering at the base, with short stalks, quite entire, or indistinctly crenated, somewhat obtuse, with numerous nerves and veins. From their axillæ arise two or more linear, smooth flower-stalks, two inches in length, having single flowers on their summits.

The calyx consists commonly of seven small sharply pointed leaves; the corolla, which is entirely white, of seven oblong open parts. Opposite to the latter there stand, upon a peculiar membranaceous, glandular circle, commonly seven filaments, almost of the size of hairs, somewhat shorter than the corolla, and bearing curved anthers. The germen is superior, and has a simple pistil. The fruit is a spherical, dry berry, the seeds of which are fastened to a spherical cavity. The embryons stand upright in the albumen.

Affinity.

Although the external appearance of this plant corresponds with that of the Stellariæ, the situation of the embryon, however, is completely different. In the Stellariæ, the curved embryon surrounds the central albumen; in the *Trientalis*, it stands upright, almost in the axis of that substance. By this circumstance, our plant becomes associated with the Lysimachiæ, *Anagallis*, and the family of the Primuleæ. The peculiar numerical proportion probably arises from an imperfect union of two flowers (185.); in consequence of which, instead of ten, there are only seven divisions of the calyx, seven petals, and seven anthers. However, there are examples of five, six, and ten filaments.

Synonymes and Figures.

Herba trientalis, *Valer. Cord. Sylv. Obs.* p. 222.   *J. Bauh. Hist.* 3. 537.

Alsinanthemon, *Thal. Herc.* 15.

Pyrola alsines flore, *C. Bauh. Prodr.* 100.   *Moris. Sect.* 12. t. 10.

Pyrola Brasiliana, alsines flore, *Park. Theatr.* 509.

Pyrola longifolia, flore albo, *Barrel. Ic.* 1156.

Trientalis Europæa, *Linn. Fl. Dan.* 84.   *Engl. Bot.* 15. *Schk.* t. 103.   *Sturm, Fl.* 17.

Geographical Distribution.

Few plants are found so far north as this. For it is found at the North Cape, and in Siberia on the Lena, under the 70° N. Lat. It grows also in North America as far as Hudson's Bay, but with smaller, and more sharply pointed leaves.

Carinthia and Transylvania seem to be its farthest southern limits.

CHAP. VIII.

13.

Vaccinium oxycoccos, *L.*

Moosbeere, Sumpfbeere. In Pomerania, Kramsbeeren.—Fren. *Coussinet, Canneberge* —Engl. *Cranberry.*—Swed. *Tranbar.*

In our marshes and moors, where much *Sphagnum* grows, this plant is found on the rising plots which have been formed by shrubs. The feeble stems, frequently a foot in length, are completely level with the ground, send out here and there fine fibrous roots into the moss, and push forth sparse, bent twigs, which also lie low, but at the points where the flowers spring, are somewhat raised. The leaves stand alternate on short stalks, are evergreen, ovate, quite entire, having the margin somewhat reflex; on the upper surface they are splendent, and of a dark green; on the lower surface greyish, from nearly three to four lines long, and two lines broad. On the points of the shoots appear the flower-stalks, an inch long, reddish, even, set with small red bracteæ: on the end of the stalk rises the calyx above the germen; it consists of four obtuse, ciliated lobes. The corolla consists of four lanceolate, red, curved petals. The eight filaments are flatly compressed and hairy. The anthers are deeply cleft, of a brown colour, and pour out their pollen from the pores of the two white points. The pistil is simple, and projects above the anthers. The nectary is the surface of the germen, and is completely protected by the filaments which stand together, and are hairy: as also by the hanging position of the flower.

The pollen is spherical, surrounded by three rings. The fruit is a beautiful red berry, of many loculi, in each of which there are numerous seeds, and these contain the embryon in an erect position in the middle of the albumen. It is ripe in October.

Diagnosis and Affinity.

There is no species in Europe which is very nearly related to this; but in North America there grows the *Vaccinium macrocarpon*, which resembles our species in its creeping stem, resembling a root,—in its evergreen leaves, which are of a light-grey colour on the lower surface,—and in the general form of the flowers. But the leaves are oblong-ovate, and obtusely rounded at the extremity; the petals also are smaller and longer, and the berries larger. Michaux, however, thinks he has detected transition forms, and hence he considers *V. macrocarpon* as a subspecies of *V. oxycoccos*. A North American species, *V. hispidulum* L., has also a creeping stem, but this, like the back of the leaves, is furnished with reddsh brown stiff hairs. The small, almost stalkless flowers, are placed in the axillæ of the leaves; the anthers are included within the bell-shaped corolla; and the berries are of a snow-white colour.

V. oxycoccos and *macrocarpon* are distinguished from the other species by the shape of the corolla, which in these is deeply divided, quite open, and seems to have the petals reflex, whilst the other species have a pitcher or bell shaped corolla; hence Tournefort long ago formed them into a peculiar genus, under the name *Oxycoccos*, in which arrangement he has been followed by Persoon, Pursh, and others. Roth named them *Schollera*. Sometimes, however, this distinction is not so marked; and in *V. stamineum* L., and *meridionale* Sw., we observe transitions from the one form to the other; whilst in these last mentioned instances the bell-shaped corolla is deeply divided.

The affinity of this genus with *Bæobotrys* and *Empetrum* is striking; through *Styphelia* and *Epacris* it is connected with *Erica*, to which family it belongs; (Anleit. ii. 521.)

Synonymes and Figures.

Oxycoccos, *Val. Cord. Hist. ed. Gesner.* f. 140. b.

Vaccinia palustria, *Dodon.* 790. *Lobel. Ic.* 2. 109. *J. Bauh. Hist.* 1. 525. *Gerard Emac.* 1419. *Park. Theatr.* 1229.

Erica vi. baccifera, *Dalech.* 187.

Vaccinium oxycoccos, *Linn. Fl. Dan.* 80. *Engl. Bot.* 319. *Schk.* t. 107. a. *Lam. Ill.* t. 286. f. 3. *Batsch Anal. Flor.* t. 7.

Oxycoccos palustris, *Pers. Syn.* 1. 419. *O. vulgaris, Pursh Amer. Sept.* 1. 263.

Schollera oxycoccos, *Roth. Fl. Germ.* 2. p. 442.

Schollera paludosa, *Baumgart. Transylv.* 1. 381.

Geographical Distribution.

It is chiefly in the highest northern latitudes that this plant grows, namely, Greenland, Iceland, Lapland, Siberia, Kamtschatka, Unalaska, and North America as far as Baffin's Bay. Its most southern limit seems to be 46° N. Lat.; for it is found in the Floras of Switzerland, of Carinthia, and of Transylvania, but it does not grow in Italy and Greece.

Uses.

Cranberries are an article of food; but in their natural state they are too sour, and they must have been subjected to frost, in order that they may become palatable. If they are kept during the winter in snow, they become a pleasant article of food in spring. In Scotland they constitute so considerable an article of commerce, that at Longtown, on the borders of Cumberland, from L. 20 to L. 30 Sterling worth are sold every day during five or six weeks. The English prepare them with sugar, and use them in tarts; (Lightfoot, Flor. Scot. 1. p. 203.) This luxury, however, cannot be enjoyed by every body, for the Cranberries have a peculiar taste. In Petersburg they are sold in spring, water is poured on them, which thus assumes the colour of madder, and a cooling drink is procured, (Gorter Flor. Ingr. p. 59.) The Swedish apothecaries make syrups and jellies from these ber-

3

ries. The goldsmiths also use the powerful acid of this berry to give silver a white colour, because the copper, with which silver is alloyed, is soluble in it.

———— ◆ ————

CLASS IX.

14.

Butomus umbellatus.

Blumenbinse, Deutsches Blumenrohr, Wasserviole.—French, *Jonc fleuri.*—Ital. *Giunco florido,*—Engl. *Flowering Rush.* —Swed. *Blomwass.*

This beautiful plant grows in our streams, pools, and other deep waters. The root is a horizontal lying tuber, about the thickness of a thumb, frequently still thicker, covered with a blackish rind, from which the fibres of the root, about the thickness of pack-thread, pass downwards, and the flower-shoots rise upwards. The leaves spring from the tuber; they are triangular below, but towards the point they become flat; throughout nearly their whole length they are of the same size; and though their length is about four or five feet, they are scarcely an inch broad.

Between these leaves, and surrounded by them, rises the round, even flower-stalk, four, five, and six feet high, and about the thickness of a finger. This contains a white, spongy pith, which consists of a compound cellular texture, and shews some scattered bundles of spiral vessels. At the top of the stalk stand the flowers in an umbel, which contains about fifteen flowers on stalks; and at its base there is a membranaceous sheath of four, five, and six membranaceous, dry, pointed leaves, which contained the flower before its evolution, but after it has flowered are reflex. The flower consists of six ovate, rose-coloured petals, which externally are of the nature of a calyx, and have slits; but internally are of the nature of a corolla, and have the requisite integu-

ment.   On the receptacle there stand nine reddish, uniform
filaments, with yellow bilocular anthers, which are more early
developed than the stigmata, and are thus an example of the
androgynous dichogamy.   Six ovaria stand together in the
centre, of a star-shape, and in their corners we perceive
ssx honey-drops oozing out.   These ovaria are surmounted by flat stigmata, somewhat emarginated, reflex, and
warty.   The withered flower continues after the decay.   The
germen consists of six simple capsules, which open laterally
and contain the oval, furrowed seed, fixed in two rows at the
sides.   The seeds contain a great deal of mealy albumen,
and at one end is the unevolved embryon, in the shape of a
point.

Affinity.

The nearest related to this plant are the genera *Hydrocleis*, Commers. (*Richard in Mem. du Mus.* i. t. 18.), and
Hydrogeton, Pers., which are distinguished from it almost
solely by the numerical proportions.   In the latter genus, there
is but half the number of petals and ovaria, and but six antheræ.   The seed contains no albuminous substance, which is a
remarkable circumstance, (Enc. Suppl. iv. p. 237.)   *Alisma*,
Sagittaria, and *Limnocharis*, are related to this plant, although the embryon in them lies folded together in a pud-
ding shape, without albumen.   Hence, if we attend to the
character of the embryon only, we cannot exactly place our
plant with the Alismeæ of De Candolle.   But if, as is proper,
we take all the other marks into consideration, we must place
it among the Hydrocharidæ, from which, however, the genus
Nectris Schreb. must be banished, because, according to later observations, its embryon is unfolded, and has distinctly
two cotyledons, (*Rafinesque in Siliman's Journ. of New
York*, vol. i. p. 374.)   Respecting the Hydrocharidæ, vid.
Anleit. ii. s. 262.

Synonymes and Figures.

Gladiolus palustris, *Val. Cord. Hist.* 100.   *Ger. Emac.* 29.
Gl. Aquatilis Dodon. 601.

Juncus floridus, *Matth. ed. Bauh.* 731. J. cyperoides floridus, *Lobel. Adv.* 44. ic. 86. *Dalech.* 989. *J. Bauh. Hist.*
2. 524. J. cyperinus floridus, *Tabern.* 567. *Parkins.*
1197.

Calamagrostis, I. *Dalech.* 1006.

Sedo affinis juncoides palustris, *Moris.* sect. 12. t. 5.

Butomus flore roseo, *Tourn. Inst.* p. 271.

B. umbellatus, *Linn. Fl. Dan.* 604. *Schk.* t. 111. *Engl.
Bot.* 651. *Sturm, Fl.* 18.

Geographical Distribution.

Rudbeck and Linnæus found this plant (*Fl. Lap.* n. 159.)
in the Kemi-elf, therefore under the polar circle. Wahlenberg,
however, who searched this stream in the year 1802, has not
mentioned this plant, as if it did not grow in Lapland. But
it is certain, that in Europe and Asia it passes beyond the 63°
N. Lat., for J. G. Gmelin found it on the banks of the Obi
and Jrtisch, (*Fl. Sib.* 1. p. 77.) How far towards the Equator it stretches, has not been determined ; but Sibthorp found
it in Asia Minor, as far as the 38° N. Lat. (*Smith, Prodr.
Fl. Græc.* 1. p. 269.) From east to west, it stretches from
Ochozk to Lisbon. It is not found in America. Between
the tropics, *Hydrogeton* and *Hydrocleis* seem to supply its
place.

Uses.

Baskets are made of its leaves, and in Holland matts are
formed from it. The tuber of the root, rich in starch, is prepared and used as an ingredient in bread, (Wredow's Oeko-
nom. Flor. Mecklenb. 2. 209.)

CLASS X.

15.

Pyrola secunda, *L.*

Birnbaumchenkraut, Einseitiges Wintergrun, Wald-mangold
—French, *Pyrole unilaterale.*—Engl. *Serrated Winter-green.*—Swed. *Hult-vintergrön.*

This plant grows in moist shady spots in our pine forests,
during June and July, not singly, but in numbers. The root
is woody, yellowish, creeping; it strikes fibres here and there
into the soil below, and pushes out an ascending stem, about
a small span's length, simple, even, roundish, about the thickness of a pack-thread. Over the whole of the stem, small,
green, lanceolate, and stalkless bracteas are found scattered.
The leaf-stalks also stand sparse and open, are even, and about
half an inch long. The leaves are ovate, oblong, unequal at
the base, serrated on the margin, having a herbaceous spine
at the point, even on both surfaces, full of nerves and veins,
and of a beautiful light-brown colour. The end of the shoot is
void of leaves, but the stipulæ appear as bracteæ. The flowers
stand in a one-sided cluster, are of a greenish-white colour,
and consist of a small quinque-partite calyx, membranaceous on
the margin, somewhat indented, and of five oblong, somewhat concave petals. The red filaments, ten in number, surround the germen, and are at first bent doubly, whence the
bilocular antheræ have their pores turned downwards. Afterwards the filaments become erect, and the antheræ then
stand with their pores turned upwards. The germen has
five furrows; the pistillum is simple, and stands perpendicular; the stigma is shield-shaped and five-lobed. The capsule is superior and consists of five loculi, which burst at the

corners. The seeds are surrounded with a reticulated, spongy membrane. The embryon stands erect in the albumen.

Diagnosis and Affinity.

This species could only be confounded with *Pyrola minor*. But *P. minor* has its leaves more rounded, obtuse, slightly serrated; it has pale red flowers, which are not turned to one side only, but to several. The other species are still more dissimilar. The genus *Pyrola* is most nearly related to the *Chimaphila* Pursh, which is distinguished merely by a thick, circular stigma, which, without a pistillum, is placed immediately upon the germen. It borders on *Gaultheria, Clethra, Diapensia, Andromeda, Monotropa,* and *Erica,* along with which it forms the family of the Ericeæ.

Synonymes and Figures.

Pyrola II. *Clus. Pannon.* 506. *Hist.* 2. 117. *Gerard Emac.* 408.

Ambrosia montana, *Dalech.* 1148.

Pyrola folio serrato, *J. Bauh. Hist.* 3. 536.

Pyrola tenerior, *Park. Theatr.* 509.

Pyrola folio mucronato serrato, *C. Bauh. Pin.* 191. *Moris.* sect. 12. t. 10. *Riv. Pentap. Irreg.*

P. racemo unilaterali, *Hall. Stirp. Helv.* n. 1008.

P. secunda, *Linn. Fl. Dan.* 402. *Eng. Bot.* 517. *Sturm, Fl.* 13.

CLASS XI.

16.

Asarum Europæum, _L._

Haselwurz, Weihrauchkraut.—French, _Cabaret, Rondelle, Oreille d'homme._—Engl. _Asarabacca._—Ital. _La bacchera, spigo salvatico._—Swed. _Hasselört._

This small, unostentatious, but interesting plant, grows in our forests, particularly under hazel bushes. Its stem is of a brown colour, about the size of a pigeon's quill, lies low, and throws out fibrous roots. These last are externally brown, internally white, and have a strong smell and a sharp taste, which may be compared to that of pepper or ginger. The ascending shoots, like the leaf-stalks, are set with hairs, and at the base of the leaf-stalks are two membranaceous, brownish _stipulæ_ or sheaths. The leaf-stalks are roundish, set with hairs, and always grow in pairs. The leaves are kidney-shaped, very obtuse at the point, hairy, but of a shining dark-green colour, ciliated, quite entire, intersected by veins an inch long, and two inches broad. Between two of the leaf-stalks springs the flower-stalk, also set with hairs, nodding, half an inch in length, and carrying over the germen a brownish coloured calyx, externally hairy, of the nature of a corolla, and terminating in three pointed, upright standing lobes. Twelve pointed, reddish filaments surround the pistillum, are longer than it, but shorter than the calyx. A little below their summits, the bilocular yellow antheræ are as it were stuck to them. The pistillum is a thick column, which carries at its top a six-lobed, radiated, reddish stigma. The capsule is six-celled, and in each of the cells it contains two seeds concave on one side, which consist chiefly of albumen, and contain the unevolved embryon, like a point, lying at one end. During germination two cotyledons are unfolded, but continue under the earth.

Diagnosis and Affinity.

The other species grow in North America, and are distinguished by the following circumstances.—*As. Canadense* has more cordate, pointed, and hairy leaves, and the lobes of the calyx are open and reflex. *As. Virginicum* has cordate, rounded, even, white spotted, coriaceous leaves, and an almost stalkless even flower. *As. arifolium* Mich., has leaves that are almost spontoon-shaped, coriaceous, white spotted, and a tubular calyx, with a very short margin.

The affinity of this genus with *Aristolochia* is striking in its numerical proportions, in the general shape of the leaves, and in the structure of the seed. The colour of the flowers, and even the composition of the juices, in the case of most of the Aristolochiæ, agree with the corresponding qualities of this plant. But we must not push this idea of affinity so far, as, with Pursh, to place *Asarum* in the class Gynandria; because in *Asarum* the filaments stand on the germen, and consequently the antheræ are not united with the female parts, as must be the case in Gynandria. But between *Aristolochia* and *Asarum*, although more nearly related to the former, stands the *Rhopium* Schreb. or *Meborea* Aubl., because in this plant each of the three pistilla carries two double antheræ below the stigma. *Tacca* is more distantly related to them. These genera constitute together the family of the Aristolochiæ, which stands between the higher and lower forms, and are conterminous with the Polygoneæ, (Anleit. ii. 301.)

Synonymes and Figures.

Asarum, *Braunschw.* f. 68. a.   *Brunf.* 1. 71.   *Fuchs*, 10. *Dodon.* 358.   *Matth.* 36.   *Tabern.* 1129.   *Dalech.* 914. *Ger. Emac.* 836.   *J. Bauh. Hist.* 3. 548.   *Park. Theatr.* 266.
Nardus agrestis, *Frag.* f. 24. b.
Asarum baccharis, *Lob. Hist.* 328. ic. 601.
Asarum vulgare, *Moris.* sect. 13. t. 7.

Asarum Europæum, *Linn. Fl. Dan.* 633.   *Engl. Bot.* 1083
Schk. t. 127.   *Sturm, Fl.* 2.

Geographical Distribution.

The geographical limits of this plant cannot be determined
for want of precise information.   How far *Asarum Europæum* extends towards the north, may perhaps be conjectured
from the fact, that Linnæus, Liljeblad, and the Flora Danica,
place it in Smaland and Jutland, and Gorter in Ingria, but
that it is wanting in the Catalogues of the Northern, Iceland-
ish, and Lapland plants.   It must grow, therefore, as far as
the 60° N. Lat., but not beyond this.   Towards the south,
it grows in Peloponesus, where Sibthorp found it.   According
to him, it extends as far as the 37° N. Lat.

Uses.

The root, when distilled, gives out a volatile oil, which
smells like camphor, and has the same relations, as camphor,
to re-agents.   On being dryed, however, it passes into the
atmosphere, and becomes invisible.   We also obtain from it
a resinous extractive matter, by means of which the root,
when pulverized, or dissolved in wine, serves as an emetic.
But its efficacy is not certain, because, when the root is become old, little good can be expeeted from it.   The inhabitants of Britain use the leaves, pulverized, and mixed with
marjoram and lavender, as a sternutatory, in doses of five
and six grains, in the case of violent inflammation of the eyes,
or headachs proceeding from catarrh.

CLASS XII.

17.

Rubus. (Brombeere, Himbeere.)

This genus stands, in the Linnæan System, between Rosa and Fragaria—a situation which is natural enough, when
we recollect, that all the three genera possess the family character of the Rosaceæ ; for the genus Rubus is shrubby, like
the Rosa, yet there are several herbaceous species, by means
of which they are connected with the Fragaria. In all of them
the leaves are compound ; only some species of Rubus, *Rosa
berberifolia* Pall., and *Fragaria monophylla* Willd., are an exception. In all of them the calyx is quinquepartite, internally
of the nature of a corolla, and carries five petals, and an indeterminate number of filaments. The ovaria are also indeterminate in number, and each of them has its pistillum on
its summit. But there is this remarkable difference, that in
the genus Rosa the lower part of the calyx swells into the
form of a berry, and contains the seed within it, whilst, in
the Rubus and Fragaria, the seeds lie upon the receptacle,
and are often surrounded by the lower calyx. The two latter species are essentially distinguished by the following
circumstances. In the *first* place, the calyx of Fragaria
has five subordinate leaves between its principal divisions,
and may therefore be called decempartite, whilst in Rubus,
on the other hand, the calyx is only quinquepartite. In the
second place, the receptacle of Fragaria swells, and contains the naked caryopsis imbedded in its surface, whilst
Rubus carries compound, one-seeded, juicy berries. All the
Rosaceæ, however, agree in this, that the seed contains no
albumen, but only the evolved embryon, with its cotyledons
turned downwards, and its truncated radicle directed up-

wards. All these relations assign to the Rosaceæ a place in the highest families of plants. (Anleit. 2. 859.)

The diagnostic character of *Rubus* also consists in the simple quinque-partite calyx, and the compound berries, which stand above the calyx. *Dalibarda* Mich., which formerly was reckoned of this genus, is distinguished by its having from three to five, and ten caryopses, which stand on the dry receptacle. We arrange the species of Rubus, according as they are shrubby or herbaceous, and as their leaves are compound or simple. The following is a view of the species that are present known.

* *Fruticosi.*

† *Aculeati.*

a. *Foliis compositis.*

1. R. *idæus* L., foliis quinato-pinnatis ternatis supra glabris subtus albido-tomentosis, aculeis rectis, petiolis canaliculatis, floribus racemosis, laciniis calycinis reflexis.

R. idæus, *Trag.* f. 367. a. *Dodon.* 743. *Matth.* 715. *Clus. Hist.* 1. 117. *Lobel.* ic. 2. 212. *Dalech.* 123. *Ger. Emac.* 1272. *Tabern.* 1298. *J. Bauh.* 2. 59. *Park. Parad.* 559. *Fl. Dan.* 788. *Engl. Bot.* 2442. *Duham. Arbr.* 2. t. 56.

In sylvaticis montosis per omne hemisphærium boreale a circulo inde arctico ad 37° lat. bor. et Kamtskatka inde et Japonica ad fretum Nutka, Sinum Hudsonis, per Lapponiam ac Siberiam, per omnem Europam, usque ad Olympum et Parnassum provenit.

Himbeere.—*Gall.* Framboise.—*Angl.* Raspberry.—*Ital.* Lampione.—*Suec.* Hallon.

Fructus vulgo rubri, interdum et albi: occurrit etiam varietas inermis.

2. R. *cæsius* L., caule repenti tereti cæsio-pruinoso, aculeis subrecurvis, foliis ternatis subpubescentibus, calyce erectoglanduloso pubescente, petalis obovatis emarginatis.

R. minor, *Dodon.* 742.

R. cæsius, *Fl. Dan.* 1213. *Engl. Bot.* 826. *Schk.* t. 136
Hayne Arzneygew, 3. t. 9.

Ad margines agrorum ab Upsala inde per totam Europam ad Thessaliam usque, denique in ipsa Japonia, (60°—
48°).

Brombeere.—*Gall.* Ronce bleuâtre.—*Angl.* Dewberry.—
Ital. Rovo turchino.—*Suec.* Blä Hallon.

3. **R.** *fruticosus* L., caule erecto quinquangulari subtomentoso, aculeis recurvis, foliis quinatis ternatisque petiolatis
supra glabris subtus tomentosis, calyce subtomentoso reflexo,
petalis obovatis integris.

Rubus *Fuchs.* 152. *Dodon.* 742. *Matth.* 714. *Lobel.*
Hist. 619. ic. 2. 211. *Dalech.* 119. *Ger. Emac.* 1272.

R. fruticosus, Linn. *Flor. Dan.* 1163. *Engl. Bot.* 715.
Hayne Arzneygew, 3. t. 12.

Brombeere.—*Gall.* Ronce.—*Angl.* Bramble.—*Ital.* Rovæ.
Suec. Brombär.

Ab Uplandia Sueciæ inde ad Algeriam usque per omnem
Europam et Asiam borealem in sylvis et nemoribus. Occurrit variis formis, caule glabriusculo, minus aculeato, vel plane
inerme, foliis magis minusve incisis, petalis rubicundis, floribus plenis, fructibus albis.

4. **R.** *corylifolius* Sm., caule erecto teretiusculo, aculei,
confertis rectiusculis, foliis quinatis subtus pubescentibus, lateralibus sessilibus, calyce fructus subreflexo.

R. major fructu nigro, *Schmidel*, ic. t. 2,

R. corylifolius, *Smith*, *Fl. Brit.* 542. *Engl. Bot.* 827.
Svensk. Bot. 187.

R. nemorosus, *Hayne Arzneygew*. 3. t. 10. *Willd. Berl.*
Baumz. ed. 2. p. 411.

R. suberectus, *Engl. Bot.* 2572.

In sepibus et nemoribus passim. Præcocius florens quam
R. fruticosus, fert albos et fructus bruneo-nigriusculos.

Per omnem Europam.

z

5. R. *tomentosus* Willd., caule angulato, aculeis recurvis, foliis ternatis obovatis acutis inæqualiter serratis utrinque tomentosis, lateralibus subincisis, calyce tomentoso reflexo.

R. creticus triphyllus flore parvo, *Tourn. Cor.* 43.

R. sanctus, *Schreb. Dic.* t. 8. p. 15. *Willd. Sp. Pl.* 2. 1083.

R. tomentosus, *Willd. Sp. Pl.* 2. 1083. *Enum.* 548. *Berl. Baumz.* ed. 2. p. 409. *Nocc.* et *Balb. Fl. Ticin.* p. 235. t. 9.

? R. tomentosus, *Thuill. Paris.* 253. Dubitatur, eamdem esse speciem, *Poir. Enc. Suppl.* 4. 694.

R. triphyllus, *Bellard. Art. Taurin.* 3. 231.

R. argenteus, *Gmel. Bad.* 2. 434.

R. canescens, *Cand. Monop.* 139. R. collinus, *ib.* et *Nocc. Fl. Ticin.* p. 238. t. 10.

Passim per omnem Europam et Asiam Minorem. R. *obtusifolius, Willd. Berl. Baumz.* ed. 2. p. 409. et R. *agrestis, Kit. Hung.* 3. p. 297. t. 268, huc pertinere videntur, licet aculei rectiusculi sint.

6. R. *glandulosus* Bellard., caule angulato, aculeis rectiusculis subreflexis, foliis ternatis subrotundo-ovatis acuminatis mucronato-serratis glabris ciliatis, venis subtus pubescentibus, caule, petiolis, pedunculis calycibusque glanduloso-hispidis.

R. glandulosus, *Bellard. App. Ad. Fl. Pedem.* p. 24. *Willd. Enum.* 548. *Berl. Baumz.* ed. 2. p. 410. *Baumg. Transylv.* 2. 56.

In Germania, Italia superiore, et Transylvania.

7. R. *hirtus* Kit., caule subangulato, aculeis subrecurvis setisque confertis rubicundis, foliis ternatis cordato-ovatis acutis inæqualiter serratis hirtis, nervo medio subtus aculeato, pedunculis inermibus calycibusque glanduloso-pilosis.

R. hirtus Kit., *Hung.* 2. p. 250. t. 241. *Willd. Enum.* 549. *Berl. Baumz.* ed. 2. p. 413. *Baumg. Transylv.* 2. 55.

In sylvis Banatus et Transylvaniæ.

8. R. *laciniatus* Willd., caule subangulato, aculeis recurvis, foliis quinato-digitatis ternatisque subtus pilosis, foliolis
pinnato-incisis, calycibus reflexis aculeolatis, petalis trilobis.

R. laciniatus, *Willd. Enum.* 550. *Hort. Berol.* t. 82.
Berl. Baumz. ed. 2. p. 416.

Patria ignota.   E seminibus educatus idem manet.

9. R. *occidentalis* L., caule tereti glabro pruinoso, aculeis
recurvis alternis, petiolis teretibus aculeatis, foliis ternatis
ovato-acuminatis incisis argute duplicato-serratis subtus tomentosis.

R. idæus, fructu nigro, Virginianus, *Dill. Elth.* 327.
t. 287. f. 319.

R. occidentalis, *Willd. Sp. Pl.* 2. 1082.   *Berl. Baumz.* ed.
2. p. 407.   *Pursh, Amer. Sept.* 347.

In rupestribus et montibus a Canada ad Pennsylvaniam.
Virginian Raspberry.   Fructus nigri aut rubri.   Petiolis
teretibus et foliis acuminatis duplicato-serratis præcipue differt a R. idæo et cæsio.

10. R. *hispidus* L., caule sarmentoso procumbente tereti,
aculeis recurvis setisque sparsis, foliis ternatis inciso-inæqua-
liter dentatis basi subcuneatis glabriusculis, racemi pedicellis
elongatis setosis, petalis obovatis.

R. hispidus, *Willd. Sp. Pl.* 2. 1083.

R. trivialis, *Mich. Bor. Am.* 1. 296.   *Willd. Enum.* 549.
Berl. Baumz. ed. 2. 414.   *Ait. Hort. Kew.* ed. 2. vol. 3.
p. 269.   *Pursh, Amer. Sept.* 347.

R. flagellaris, *Willd. Enum.* 549.   *Berl. Baumz.* ed. 2.
412.   *Pursh, Amer. Sept.* 347.

R. procumbens, *Mühlenb. Catal.* p. 50.

In arvis a Canada ad Pennsylvaniam.   Fructus nigri.

11. R. *heterophyllus* Willd., caule procumbente subangulato glabriusculo, aculeis raris recurvis, petiolis pedunculisque racemosis inermibus villosis, foliis ternatis glabriusculis
profunde serratis, calyce tomentoso reflexo, petalis integris.

R heterophyllus, *Willd Berl. Baumz.* ed. 2. p. 413.

B b

R. villosus, *Torrey, in Catal. Nov Eboræc.* p. 47.
Ad Novum Ebcracum. R. *triphyllus, Thunb. Fl. Japon.*
215 est sola varietas, foliis subtus pubescentibus.

12. R. *villosus* Ait., caule hispido, aculeis reflexis, foliis
digitatis ovato-oblongis acuminatis serratis utrinque villosis,
petiolis aculeatis, pedunculis racemosis laxis.
R. villosus, *Ait. Hort. Kew.* ed. 1. vol. 2. p. 210. *Willd.*
Sp. Pl. 2. 1085. *Mich. Bor. Amer.* 1. 297. *Pursh, Amer.*
Sept. 346. *Poir. Enc.* 6. 243.
In arvis a nova Anglia ad Carolinam. *Blackberries.*

13. R. *cuncifolius* Pursh., ramis petiolis pedunculisque tomentosis, aculeis sparsis recurvis, foliis digitatis obovatis apice
inæqualiter dentatis plicatis basi revolutis subtus tomentosis,
pedicellis paniculæ divaricatis nudiusculis.
R. parvifolius, *Walt. Carol.* 149.
R. cuneifolius, *Pursh, Amer. Sept.* 347. *Nuttall, Gen.* 1.
308.
In Nova Cæsarea.

14. R. *rosæfolius* Smith, caule teretiusculo piloso, aculeis
recurvis, foliis pinnatis pilosis, foliolis ovato-lanceolatis dupli-
cato-serratis, pedunculis sub-unifloris terminalibus.
R. rosæfolius, *Sm.* ic. ined. 3. p. 60. t. 60. *Willd. Sp. Pl.*
2. 1080.
R. borbonicus, *Pers. Syn.* 2. 51.
R. Commersonii, *Poir. Enc.* 6. 240.
In insulis Mascarenis et Java.

15. R. *pinnatus*, Willd., ramis villosis, aculeis recurvis,
foliis quinatis ternatisque rugoso-venosis duplicato-serratis
utrinque glabris, nervo medio aculeato, pedunculis racemosis
calycibusque villosis.
R. pinnatus, *Willd. Sp. Pl.* 2. 1081. *Ait. Hort. Kew.*
ed. 2. vol. 3. p. 270.
Ad C. B. S. et in insula S. Helena.

16. **R.** *australis* Forst., caule glabro teretiusculo, aculeis ramorum secundis recurvis, foliis ternatis utrinque glabris ovalibus argute dentatis subcoriaceis, floribus racemosis diœciis, sepalis obtusis patentibus.

R. australis, *Forst. Prodr. Fl. Austr.* p. 40. *Willd. Sp. Pl.* £. 1081.

In Nova Zellandia.

17. **R.** *roseus* Poir., ramis flexuosis glabris, foliis ternatis ovato-lanceolatis crenulatis utrinque glabris, nervo medio acueato, stipulis ovalibus, pedunculis solitariis, petalis calyce glabro minoribus.

R. roseus, *Poir.* in *Enc.* 6. 245.

R. coriaceus, *ib.* 237.

In Peruvia.

18. **R.** *parvifolius* L., caule tereti tomentoso, aculeis recurvis confertis, foliis ternatis ovatis subtus albo-tomentosis, floribus racemosis.

R. Moluccanus, *Rumph. Amboin.* 5. p. 88. t. 47. f. 1.

R. parvifolius, *Linn. Sp. Pl.* 707. *Willd. Sp. Pl.* 2. 1083.

In insulis Moluccis.

19. **R.** *Jamaicensis* L., caule glabriusculo, aculeis recurvis, foliis quinato-ternatis inciso-serratis subtus villoso-tomentosis, paniculis terminalibus diffusis.

R. foliis longioribus, *Sloane, Jam.* t. 212. f. 1.

R. aculeatus, *P. Brown, Jam.* 242.

R. Jamaicensis, *Linn. Mant.* 75. *Sw. Obs.* 205. *Willd. Suppl.* 2. 1084.

In Jamaica et Antillis.

20. **R.** *urticæfolius* Poir., ramis angulatis setosis, aculeis raris rectiusculis, foliis ternatis simplicibus utrinque sericeotomentosis, paniculæ ramis hirsutissimis, calyce albido.

R. urticæfolius, *Poir.* in *Enc.* 6. 246.

In Peruvia.

21. R. *fraxinifolius* Poir., ramis glabris teretibus, aculeis raris, foliis septenato-pinnatis utrinque glabris, foliolis ovato-acuminatis inciso-serratis, paniculæ ramis filiformibus glabris.
R. fraxinifolius, *Poir* in *Enc.* 6. 242.
In Java.

22. R. *apetalus* Poir., ramis teretiusculis pubescentibus, aculeis sparsis, foliis septenato-pinnatis subtus albido-tomentosis, foliolis ovalibus serrulatis, racemi axillaris ramis pubescentibus, floribus apetalis.
R. apetalus, *Poir.* in *Enc.* 6. 242.
In Insula Franciæ, *Commerson.*

β. *Foliis simplicibus.*

23. R *Moluccanus* L., ramis hirsutis, aculeis recurvis, foliis cordato-lobatis serratis subtus tomentosis, pedunculis subracemosis axillaribus.
R. Moluccanus latifolius, *Rumph. Amb.* 5. p. 88. t. 47. f. 2. *Linn. Sp. Pl.* 707. *Thunb. Fl. Jap.* 219. *Willd. Sp. Pl.* 2. 1086.
R. alceæfolius, *Poir.* in *Enc.* 6 247. Calycibus inflatis differre dicitur.
In insulis Moluccis, Java et Japonia

24. R. *microphyllus* L. fil., ramis teretibus flexuosis glabris, aculeis sparsis recurvis, foliis cordatis trifidis inæqualiter dentatis glabris, venis subtus pubescentibus, pedunculis solitariis, calyce villoso.
R. microphyllus, *Linn. Suppl.* 263. *Willd. Sp. Pl.* 2. 1086.
R. palmatus, *Thunb. Fl. Jap.* 217.
In Japonia. Fructus lutei.

25. R. *incisus* Thunb., ramis glabris, aculeis sparsis subrecurvis, foliis cordato-subrotundis inciso serratis utrinque glabris, pedunculis solitariis glabris.

R. incisus, *Thunb. Fl. Jap.* 217.
In Japonia.

26. R. *corchorifolius* L. fil., ramis tomentosis, aculeis recurvis, foliis cordato-ovatis acutis sublobatis serratis villosis, nervo medio supra aculeato, pedunculis axillaribus solitariis tomen tosis.

R. corchorifolius, *Linn. Suppl.* 263. *Willd. Sp. Pl.* 2. 1087.
R. villosus, *Thunb. Fl. Jap.* 218.
In Japonia.

27. R. *elongatus* Smith, ramis viscoso-pubescentibus, aculeis sparsis, foliis cordato-acuminatis duplicato-crenatis subtus tomentosis, paniculæ ramis glomeratis, calyce obtuso.

R. elongatus, *Smith Ic. Ined.* 3. t. 62. *Willd. Sp. Pl.* 2. 1087. *Poir.* in *Enc.* 6. 248.
In Java.

28. R. *pyrifolius* Smith, ramis flexuosis, aculeis sparsis, foliis oblongis acuminatis serratis utrinque glabris, paniculæ ramis corymbosis, bracteis incisis, petalis calyce minoribus.

R. pyrifolius, *Smith, Ic. Ined.* 3. t. 61. *Willd. Sp. Pl.* 2. 1088. *Poir.* in *Enc.* 6. 248.
In Java.

†† *Inermes.*

29. R. *strigosus* Mich., caule tereti hispidissimo, foliis quinato-ternatis ovatis acuminatis inæqualiter serratis subtus lineatis candido-tomentosis, pedunculis subtrifloris calycibusque hispidis.

R. strigosus, *Mich. Bor. Amer.* 1. 297. *Willd. Berl. Baumz.* 408. *Pursh, Amer. Sept.* 346.
R. Pennsylvanicus, *Poir.* in *Enc.* 6. 246.
In montosis a Canada ad Virginiam.

30. R. *Canadensis* L., caule purpureo glabriusculo, foliis digitatis denis quinis ternatisque, foliolis lanceolatis argute serratis utrinque nudis, stipulis linearibus subaculeatis.

R. Canadensis, *Linn. Sp. Pl.* 707. *Mill. Ic.* t. 223. *Willd. Sp. Pl.* 2. 1085. *Pursh, Amer. Sept.* 347.

In rupestribus sylvaticis a Canada ad Virginiam.

31. R. *inermis* Willd., caule procumbente glauco-tomentoso, foliis ternatis subtus albido-tomentosis, foliolis ovatis acutis subincisis inæqualiter serratis, stipulis subulatis.

R. inermis, *Willd. Enum.* 548. *Berl. Baumz.* 410. *Pursh, Amer. Sept.* 348.

In Pennsylvania.

32. R. *obovalis* Mich., caule hispido, foliis ternatis obovato-subrotundis serratis nudis, stipulis setaceis, racemis subcorymbosis paucifloris, bracteis ovatis, pedicellis elongatis

R. obovalis, *Mich. Bor. Am.* 1. 298. *Pursh, Amer Sept.* 349.

In paludibus sphagno abundantibus a Novo Cæsarea ad Carolinam.

33. R. *spectabilis* Pursh, caule ramisque flexuosis glaberrimis, petiolis pubescentibus subaculeatis, foliis ternatis ovatis acutis angulatis inæqualiter duplicato-serratis subtus pubescentibus, pedunculis terminalibus solitariis unifloris, petalis ovatis calyce longioribus.

R. spectabilis, *Pursh, Amer. Sept.* 348.   t. 16.

In ora occidentali Americæ borealis.   Flores speciosi punicei.

34. R. *odoratus* L., caule erecto, petiolis pedunculisque glanduloso-pilosis, foliis simplicibus quinquelobis inæqualiter dentatis, venis subtus pubescentibus, calycibus appendiculatis.

R. odoratus, *Corn. Canad.* t. 150. *Mill. Ic.* 223. *Linn. Hort. Cliff.* 192.   * *Willd. Sp. Pl.* 2. 1085. *Berl. Baumz.* 416. *Bot. Mag.* 323.

In sylvis Americæ borealis.   Folia suaveolentia.   Flores rubri.   Fructus flavi.

35. R. *parviflorus* Nuttall, foliis palmato-lobatis, pedunculis subtrifloris, calycibus villosis acuminatis, petalis calyce brevioribus ovato-oblongis.

R. parviflorus, *Nuttall, Gen.* 1. 308.
In insula Michillimakinak lacus Huronum.

** *Herbacei.*

36 R. *saxatilis* L., flagellis reptantibus, caule obtusangulo, foliis ternatis rhombeis acutis inciso-dentatis nudis, pedunculis subternis elongatis, petalis linearibus.

R. saxatilis alpinus, *Clus. Pannon.* 116.   *Hist.* 1. 118.
R. alpinus humilis, *J. Bauh. Hist.* 2. 61.
R. saxatilis, *Ger. Em.* 1273.   *Park. Theatr.* 1014.   *Fl. Dan.* 134.   *Eng. Bot.* 2233.

In rupestribus per Europam, Asiam, et Americam. In veteri orbe a Caucaso (45°), inde ad Islandiam et Lapponiam (66°); in America a Virginia (38°) ad Canadam (50°).

37. R. *arcticus* L., caule simplici glabro, foliis ternatis ovatis obtuse-dentatis glabris, pedunculis solitariis, petalis obovatis emarginatis.

R. humilis flore purpureo, *Buxb. Cent.* 5. t. 26.
R. arcticus, *Linn. Fl. Lap.* t. 5. f. 2.   *Fl. Dan.* 488.   *Eng. Bot.* 1585.   *Bot. Mag.* 132.

Arcticus jure dicitur: namque in Suecia et Norvegia haud citra 60° provenit; vera patria est regio ab occidente Sinui Bothnico contermina, (Helsingeland, Medelpad, Angermanland.)   In Scotia tamen et Sibiria a Kamtschatka ad 56° descendit; in America ad 52°.   In Suecia Boreali nomine *Åkerbär* fructus flavidus sapidissimus fragrans delicias summas constituit.

38. R. *pistillatus* Sm., caule unifloro, foliis ternatis argute serratis glabris, petalis oblongis integris, stylis approximatis.

R. pistillatus, *Smith, Exot. Bot.* 2. t. 86. *Pursh, Amer. Sept.* 349.

R. acaulis, *Mich. Bor. Amer.* 1. 298.

In paludibus Canadæ et in ora Americæ Borealis occidentali,

39. R. *radicans* Cav., caule prostrato sarmentoso aculeato, foliis longe petiolatis ternatis orbiculatis sublobatis crenatis utrinque villosis, pedunculis solitariis elongatis, calycis laciniis dentatis, petalis ovalibus.

R. radicans, *Cav. Ic.* 5. t. 413.

In sylvis Chili. Nisi flores essent rubri et foliorum forma aliena, *Duchesneam fragiformem* Sm. subesse crediderim.

40. R. *chamæmorus*, radice repente, caule simplice unifloro, foliis simplicibus subreniformibus rotundo-lobatis plicatis, calycis laciniis oblongis incisis, floribus diœciis.

Chamæmorus, *Clus. Pannon.* 118. *Hist.* 1. 118. *Ger. Emac.* 1273. 1420. *Park.* 1014. *Pontoppid. Norg. Naturl. Hist.* 1. 215.

R. Chamæmorus, *Linn. Fl. Lap.* t. 5. f. 1. *Lightf. Scot.* 266. t. 13. *Fl. Dan.* 1. *Eng. Bot.* 716.

In Suecia Boreali, præsertim in paludibus Lapponiæ, per omnem Norvegiam, Islandiam, insulis Faeröer, Scotia boreali et Cambria, in Pomerania, ad Fontes Albis, in Meisnero monte Haffiæ, in Curonia, Livonia, Ingria, Sibiria, Kamtschatka, Canada et Nova Anglia. In Europa, si citra 60° occurrit, amat uligines alpestres. In summo jugo Sudetum (50° 50') fere 4000 pedes altam habet sedem. In peninsula Dars Pomeraniæ (54° 30') paludem mari æqualem habitat. In America Boreali sub 44° in paludibus montosis invenitur. Sueci dicunt *Hjortron*, Dani et Pomerani *Multebär*, Scoti *Cloudberry*. Flos albus, fructus aurantiacus.

41. R. *stellatus* Sm., caule erecto unifloro foliis simplicibus cordatis trilobis rugoso-venosis, petalis lanceolatis.

R. stellatus, *Smith, Ic. Ined.* t. 64. *Willd. Sp. Pl.* 2. 1089. *Pursh. Amer. Sept.* 349.

In ora occidentali Americæ Borealis.   Flores purpurei.

42. R. *trifidus* Thunb., caule glabro simplici, foliis simplicibus cordatis trifidis serratis utrinque glabris, pedunculis solitariis.

R. trifidus, *Thunb. Jap.* 217. *Willd. Sp. Pl.* 2. 1089.
Rubus pedatus Sm., geoides Sm. amandandi sunt ad genus *Dalibardam.* Rubus Japonicus, Linn. Willd. et Chorchorus Japonicus Thunb., eamdem constituunt plantam, *Keriam* a De Candollio dictam.

CLASS XIII.

18.

Papaver dubium, *L.*

Ackermohn, Klatschrosen, Feldmohn.—French, *Pavot douteux, Coquelicot.*—Ital. *Rosolaccio.*—*Papavere silvatico.* —Eng. *Red poppy, Smooth-headed poppy.*—Swed. *Systervallmo.*

This plant blossoms in our fields in June and July. From a fibrous, whitish root, there rises a round, herbaceous stem, of the thickness of an oat-straw, two feet high, and, like the leaves, furnished on the lower part with distant, pretty stiff hairs. The leaves are doubly pinnated, and embrace half the stalk; the lobes are lanceolate-linear, and distant from one another. The flower-stalks are nearly a foot long, and furnished with thick accumbent stiff hairs or bristles. The flowers stand single, and have a two-leaved deciduous calyx; four broad, pale scarlet-coloured petals, crenated on the margin, and smooth; an indefinite number of linear filaments, which stand on the receptacle; yellow antheræ of two loculi; a superior, longish, smooth, angular germen, sur-

mounted by the six or eight rayed, shield-shaped stigma,
without a pistillum.   The capsule is unilocular, and the
seeds are placed in several flat cavities, which proceed in a radiated form from without half-way inwards.   It opens under
the persistent stigma with as many holes as there are rays in
the capsule.   The seeds have a wrinkled covering, and contain the small, evolved embryon on the side of the albuminous
matter.

Diagnosis and Affinity.

This species, which is very like the common Poppy (*Papaver rhœas*), is distinguished from it by the thickly accumbent bristles on the flower-stalk,—these bristles, in the other
species, standing horizontally from each other.   The flowers,
too, are by no means of so fiery a red, but are somewhat paler.
P. argemone, which also, although more rarely, grows among
the corn, has a club-shaped, bristly capsule, small and still
paler petals, and bluish antheræ, and filaments which are
thick above.   *P. hybridum*, which grows still more rarely
in Germany, has an almost spherical, sulcated, bristly capsule,
dirty dark red flowers, and bright blue antheræ.

The affinity between the genus *Papaver* and *Chelidonium*
and *Glaucium* is striking, although in the two latter the
form of the fruit is different.   Among exotics, it is still more
nearly related to Argemone, and this affinity is expressed by
the peculiar juices, which in the Poppies are white ; in *Chelidonium*, *Glaucium*, and *Argemone*, yellow ; and in *Abatia*,
black.   The Papavereæ are related, through *Hypecoum* and
Fumaria, to the Cruciform plants, and through *Actæa* to the
Ranunculeæ, (Anleit. ii. 727.)

Synonymes and Figures.

Argemone capitulo longiori glabro, *Moris.* sect. 3. t. 14.
Pavaver erraticum, *Tourn. Inst.* 238.   *Rupp. Ien. ed. Haller*, p. 79.   *Haller, Stirp Helv.* n. 1065.
P. dubium, *Linn. Engl. Bot.* 644.   *Schk.* t. 140.   *Fl. Dan.*
902.

Geographical Distribution.

This species seems to extend, as well as P. *rhœas,* from 60° N. Lat. towards the Tropics. In Lapland it is as seldom met with as in countries between the Tropics. It is also wanting, together with its related genera, in America.

CLASS XIV.

ORDER I.

This is called *Gymnospermia,* because four naked caryopses stand around the pistillum, at the bottom of the calyx, (Tab. III. Fig. 17.) But there are transitions to capsules. In *Verbena,* for example, the four seeds, as long as they are not quite ripe, are included in a membranaceous bladder, which disappears when the seeds are matured. In this instance, the pistillum stands directly on the germen, whilst in the proper Gymnospermæ, it stands in the middle of the four caryopses. These latter bodies lie, in this family, on the thickened and swollen germen (*gynobasis*). The Asperifoliæ, again, which also have four caryopses, have no gynobasis, but the caryopses are surrounded by a ring of nectaries. Of the four filaments, two are commonly shorter than the others, (Tab. IV. Fig. 16.) The pistillum is cleft. The corolla has a disposition to become irregular, and exhibits frequently two lips ; hence this division forms the natural family of the Labiatæ ; (Anleit. ii. 427.)

19.

Galeobdolon luteum, *Huds.*

Gelbe Hanfnessel, Goldnessel, gelber Hahnenkopf.—French, *Agripaume jaune, Lamier des bois.*—Ital. *Ortica morta gialla.*—Engl. *Yellow archangel, Weasel snout.*—Swed. *Sug-plister.*

This plant appears in our forests at the end of May, with a knotty root, from which rises a four-cornered stem, its lower part being of the form of a root, furnished with reflex hairs, and from one foot to a foot and a half high.   The leaf-stalks stand opposite to one another, are furnished with long, soft, white hairs, and are half an inch, and even somewhat longer. The leaves are ovate, unequally crenated, and the teeth are furnished with a small point.   At the base they are almost cordate, in other parts they are hairy, of a dark green, sometimes spotted with white, an inch long, and not quite so broad.   The uppermost are much smaller, and without stalks.   They surround the flowers, which stand by fours and sixes in verticilli, and are surrounded, beside the stem leaves, by smaller bristle-shaped leaves.   The calyx has a short stalk, is smooth, and properly has two lobes.   The upper lip forms an almost erect, long pointed tooth.   The lower lip consists of four teeth, which are also pointed, and almost bristle-shaped.   The yellow corolla is two-lipped: the upper lip is arched, without a stalk, and furnished with jointed hairs: the under lip consists of three flat, small, and spotted laciniæ.   These brownish red spots are the *nectaro-stig-mata* : the nectary is the surface of the gynobasis: the lower part of the tube of the corolla, furnished with hairs, forms the *nectarilyma.*   Four soft haired filaments, broader below, stand on the tube of the corolla: two are shorter than the other two.   They carry four antheræ with double loculi, and are nearly as long as the upper lip of the corolla. The pistillum is cleft at the top: the stigmata are pointed. Four three-cornered, longish caryopses are persistent in the bottom of the calyx, and contain, like all the Labiatæ, the erect embryon, without any albuminous matter.

Affinity.

This plant is so nearly related to the genera *Leonurus, Galeopsis,* and *Lamium,* that it is sometimes classed with the

one, and sometimes with the other of them. But *Leonurus*
is distinguished by the five-awned teeth of its calyx, two of
which form the upper, and three the under lip; and by the
three rounded lobes of the lower lip of the corolla, of which
the central one is the largest. *Galeopsis* has a calyx similar to
that of *Leonurus;* but the lateral lobes of the lower lip of
the corolla form, at the entrance of the tube, a pair of knots,
the upper lip is shortly cleft, and the antheræ open with
fringed valves. *Lamium* has a similar calyx to the former,
only with long-pointed teeth : the tube of the corolla is inflated above: the lateral lobes of the lower lip pass into a
pair of reflex teeth, the central lobe is emarginated : the lobes
of the antheræ are furnished with long hairs. We must,
therefore, regard *Galeobdolon* as a peculiar genus, the character of which is to be sought in the structure of the calyx
and corolla. The single long-pointed tooth, which forms
the upper lip of the calyx, and the three small laciniæ of
the under lip of the corolla, form this character.

Synonymes and Figures.

Urtica iners, III. *Dodon.* 153.
Lamium luteum, *Lob. Adv.* 223. *Ic.* 521.   *Tabern.* 923.
 Gerard. Emac. 702. *Park. Theatr.* 606. folio oblongo.
 Moris. sect. 11. t. 11. *Riv. Monop. Irr.*
Galeopsis s. Urtica iners, *J. Bauh. Hist.* 3. 323. *Tourn*
 Inst. 185.
Galeobdolon, *Dill. Giess.* 49.
Leonurus foliis ovatis acutis serratis, *Linn. Hort. Cliff.* 313.
Galeop is Galeobdolon, *Linn. Fl. Suec.* ed. 2. p. 205. *Fl.*
 Dan. 1272.
Leonurus Galeobdolon, *Scop. Carn.* n. 705 *Willd. Sp.*
 Pl. 3. 115. *Schk.* t. 157
Lamium Galeobdolon, *Crantz, Austr.* 262.
Pollichia Galeobdolon, *Roth. Germ.* 2. 26.
Galeobdolon luteum, *Huds Fl. Angl.* 258. *Smith, Fl. Brit.*
 631. *Engl. Bot.* 787.

Geographical Distribution.

The temperate part of Europe is the native region of this plant. Its most northern limits, as far as is yet known, are Wasa in Finland, Jamteland in Sweden, and Drontheim in Norway, (63°.) Its most southern limit is Hæmus in Rumilia, (41°.) Only thus far, too, the plant grows towards the east; but in Lithuania it grows as far as the Wolga Heights, (33° E. Lat.) Westward it extends as far as the Pyrenees.

ORDER II.

This order is called *Angiospermia,* because the fruits are capsules, or drupes. In the natural arrangement, the plants of this order belong to the Personatæ, Acantheæ, Bignonieæ, and Viticeæ; (Anleit. 2. 390—426.)

20.

Alectorolophus Crista Galli, *M. B.*

Hahnenkamm, Klapperkraut, Wiesenrodel.—Fren. *Crête de coq. Cocriste.*—Ital. *Crista di gallo.*—Engl. *Yellow-rattle.*—Swed. *Hö-skaller, Penninge-gräs.*

This is one of the most common weeds in our meadows and fields, blossoming during summer, and withering entirely during harvest. From a soft fibrous root there rises a four-cornered stem, commonly simple, sometimes considerably branched, smooth, or somewhat sharp to the touch, and sometimes marked with dark red spots. It is about a foot in height. The leaves are set opposite to one another, without stalks, lanceolate, rough, sharply serrated, and cordate at the base, an inch, or an inch and a half long, and from three to four lines broad. In the neighbourhood of the flowers they supply the place of bracteæ, are ovate, and somewhat membranaceous, but otherwise are as much serrated and rough to the touch as those farthest down. The flowers stand oppo-

4

site to one another, and compose together a richly furnished
ear. The calyx is almost stalkless, inflated, membranaceous,
having reticular veins, with a contracted four-toothed opening. The corolla is yellow, about one-half longer than the
calyx, with two lips, and almost personate. The upper lip
is arched, compressed, externally set with short hairs, with
an obtuse, often an emarginated, sometimes a violet-coloured
beak. The lower lip has three short, yellow lobes, which
press upon the upper lip. Four filaments of unequal length
are fixed in the tube of the corolla, and carry four antheræ of two loculi, pointed laterally, and ciliated, which
never rise above the upper lip. The germen has on its
lower margin the nectary, as an insulated gland, and carries a simple pistillum, with a somewhat thickened stigma.
The fruit is a double, compressed capsule, the partition of
which goes right across the loculi. On this partition are
placed the flat compressed, marginated seeds, containing the
embryon opposite to the umbilicus in the albuminous substances.

Diagnosis and Affinity.

We find, according to difference of soil, many subspecies
of this plant, of which *Alect. hirsutus* Allion. (*Rhinanthus
Alectorolophus*, Pollich.), is distinguished by its size and
hairs. It is about two feet high; the stem is furnished with
red spots and with soft hairs: so also is the calyx. We are
prevented from constituting it a peculiar species, by its marked transitions, and by its want of uniformity. Still less could
we form a new division in regard to *Rhinanthus minor*
Ehrh., and *alpinus* Baumg. The former is distinguished
simply by its less size, smaller leaves, and inclosed pistillum.
But in this species, as in several plants, the pistillum has a
different length according to the different age of the blossom.
At first it is inclosed, afterwards it projects a little. Lastly,
if the *Rhinanthus alpinus* Baumg. is to be distinguished by
its variegated, violet, and yellow flowers; this colouring is
also found in the common yellow rattle, and we can only,
therefore, regard these different forms as subspecies.

Very similar, although essentially different, is *Rhinanthus versicolor* Lam., which grows on the shores of the Mediterranean, in Italy and Africa. It has small, linear leaves, the serrated teeth of which are somewhat obtuse, and go very deep, so that the leaves seem to be half-pinnated : the uppermost leaves are dentated only at the base. The corolla is of a reddish purple, and is much longer than the calyx. This latter part is not inflated, but it has also four teeth. The seeds are not flatly compressed, but angular ; hence this species properly belongs to the genus Bartsia, which, very like the *Alectorolophus*, is distinguished from it by its tubular, coloured calyx, and by its angular seeds. The same is the case with *Rhinanthus Trixago* Linn. This plant of the South of France, with a stem of an ell long, branched, and lanceolate, and with large and deeply serrated leaves, has its calyx and seeds similar to those of the former species, and belongs, therefore, to *Bartsia*, with which it is classed by De Candolle.

Allioni and Marschall of Bieberstein, have very properly separated the genus *Alectorolophus* from *Rhinanthus*. The latter genus is distinguished by its two-lipped calyx, and by a tubular, beak-shaped upper-lip of the corolla, having a broad appendage at the point. To it belong *Rhinanthus orientalis* Mill., and *Rh. elephas* L. These genera are arranged in the natural method with the Personatæ, and, indeed, with the subdivision of the Rhinantheæ; (Anleit.2.397.)

Synonymes and Figures.

Crista galli, *Dodon.* 556. *Lobel Adv.* 227. *Hist.* 285. *Ic.* 529. *Dalech.* 1073. *Mas et Fœmina. J. Bauh. Hist.* 3. 436. *Ger. Emac.* 1071. *Park. Theatr.* 713. *Riv. Monop. Irr.*

Pedicularia lutea, *Tabern.* 1180. Pratensis, *Moris.* sect. 11. t. 23. *Tourn. Inst.* 172.

Alectorolophus calycibus glabris et hirsutis, *Hall. Stirp. Helv.* n. 313. 314. *Allion. Pedem.* n. 205. 206. Al. Crista galli, *Marsch. Bieb. Taur. Cauc.* 2. 68.

4

Rhinanthus Crista galli Linn., *Willd. Sp. Pl.* 3. 188. *Fl. Dan.* 981. *Engl. Bot.* 637. *Schk.* t. 169. Mimulus Crista galli et Alectorolophus, *Scop. Carn.* n. 751. 752.

Geographical Distribution.

This is properly a northern plant, which grows in Europe as far as the North Cape (70°), in Iceland, Siberia, Kamtskatka, and North America; as far as Hudson's Bay, and scarcely passes, towards the south, beyond the 44° or 43°; for Tauris, Transylvania, and the South of France, appear to constitute the most southern limits of its distribution.

Uses.

I am not acquainted with any useful property of it. On the contrary, the rattle is a hurtful weed in our meadows, and is not eaten by any animal. Its seed, mixed with meal, gives it a dark appearance, and makes the bread indigestible.

CLASS XV.

ORDER I.

This Order is called that of the *Siliculosæ,* or silicle-bearing plants, because their fruit is almost as long as broad, (96.) Yet here, as every where else, there are transitions. *Farsetia* R. Br., on the one side, and *Braya* Sternb., and *Nasturtium* R. Br., on the other, are so nearly related to this order, that we may sometimes call their fruit a Silicula, sometimes a Siliqua. Several of these silicles are fruits of another species, nuts, in particular, which do not burst, as in *Bunias, Crambe, Cakile, Succovia, Mönch.* But, as the plants correspond with this order in their other relations, we cannot separate them from it.

21.

Teesdalia nudicaulis, *R. Br.*

Sand-Bauernsenf, Taschelkraut, Felsenkresse.—Fren. *Tabouret à tige nue.*—Eng. *Naked-stalked Candytuft.*—Swed. *Sand-iber.*

This plant blossoms with us in spring, on high sandy, open places. From a soft fibrous root, a number of lyre-shaped, smooth leaves, expand themselves into a circle, having their margins sometimes ciliated, but undivided, and about half an inch long. In the midst of them, perpendicular, smooth, round flower-stems arise, about three inches, or, at most, half a foot in length, and of the thickness of a thread. These stalks are furnished with a few lanceolate, or oblong, coloured, scaly leaflets : in other respects they are entirely void of leaves. On the upper part of these appears the small white flower-bunch, the upper part of which resembles an umbel. The single flower-stalks are scarcely two lines in length.

The calyx consists of four pieces ; the corolla of four petals, which, for the most part, are dissimilar, the two outer being larger than the two inner. The filaments are six, and stand on the receptacle. Two of them are longer than the others. On each of the filaments, at its lower part, and turned inwards, there is a whitish leaflet, which is larger in the longer filaments, and smaller in the shorter. The filaments carry bilocular, yellow antheræ. The germen is superior, emarginated above, and carries a short thick pistillum, with a warty stigma. The fruit is a bilocular silicle, with boat-shaped wingless valves, having two round seeds without a raised margin in each loculus,—attached to long funiculi umbilicales, and containing the embryon without albuminous substance, its radicle being turned towards the opening of the cotyledons.

Diagnosis and Affinity.

This plant has some resemblance to *Draba verna,* which
blooms somewhat earlier however, and is much more common. Both·are of nearly similar stature, but *Draba verna*
is commonly the smaller; its individual flower-stalks are much
longer, the root leaves are undivided, and furnished with
three-pointed hairs. The petals are deeply indented, and the
fruit is a longish, pointed silicle, with many seeds. *Lepidium
nudicaule* L., which grows about Montpellier, and is figured
by Magnol (Bot. Monsp. p. 187.), is also distinguished from
it, by the smaller lobes of its lyre-shaped root-leaves. But
De Candolle has shewn (Flor. Franc. 4. 708.), that this plant
is only a subspecies of ours. *Thlaspi Bursa,* also, has sometimes an appearance, which might lead us to take it for *Teesdalia,* since it has lyre-shaped root leaves, and white flowers.
But its stem is always branched and furnished with leaves
the silicles, also, are inversely triangular, and contain many
seeds. *Lepidium alpinum* and *petræum* L. have also some
resemblance to it; but they have peculiarly pinnated leaves,
which grow on the stem, and lanceolate siliculæ, furnished
with pointed extremities.

This plant is commonly classed with *Iberis,* because its petals are somewhat unequal. But *Iberis* has the valves of its
siliculæ distinctly marginated, and has no appendages to the
filaments. These two circumstances constitute the diagnostic
character of the two genera. This plant cannot be classed
with *Lepidium,* because, in this latter genus, the petals are
uniform, the filaments are without appendages, and the radicle of the embryon is turned towards the ridge of the cotyledons. *Thlaspi* is still further distinguished by the winged
or marginated valves of the *siliculæ,* by the want of appendages to the filaments, by having many seeds in its loculi, and
by the direction of the radicle towards the ridge of the cotyledons.

Synonymes and Figures.

Pastoria bursa minor, *Dodon.* 103. *Park. Theatr.* 866. Minima, *Lob. Ic.* 221. *Ger. em.* 276.
Bursa pastoris parva, folio glabro, *J. Bauh. Hist.* 2. 937.
? Bursa pastoris media, *Moris.* sect. 3. t. 20. (*Nasturtium petræum, Ger. Park. Tabern. Moris.*, the plant more commonly identified with this is *Lepidium petræum.*)
Nasturtium minimum vernum, *Magnol. Bot. Monsp.* 186. *Tourn. Inst.* 214. Formerly, however, he classed *Nasturtium petræum* of his predecessors, with *Pastoria bursa minor*
Iberis foliis pinnatis, *Hall. Stirp. Helv.* n. 521.
Iberis nudicaulis, *Linn. Fl. Suec.* ed. 2. p. 228. *Willd. Sp. Pl.* 3. 458. *Fl. Dan.* 323. *Schk.* t. 179. *Engl. Bot.* 327. *Sturm. Fl.* 11.
Lepidium nudicaule, *Gouan, Ill.* p. 40. *Willd. Sp. Pl.* 3. 432.
Thlaspi nudicaule, *Desfont. Atl.* 2. 67. *De Cand. Fl. Franç.* 4. 708.
Guepinia Iberis, *Desvaux, Journ. de Bot.* 3. 167.
Teesdalia nudicaulis, *R. Brown,* in *Ait. Kew.*, ed. 2. tom. 4. p. 83.

Geographical Distribution.

This plant seems chiefly to inhabit the south-west regions of the Old World. Its eastern limits seem to be Grodno and Translyvania, (20° E. Lat.) Northward it extends to the 64°, for it does not grow in Lapland. Southward it extends, with some variety of form, (with smaller lobes of its root-leaves), as far as Peloponnesus, the south of Spain, and even Algiers, (35° N. Lat.)

ORDER II.

Siliquosæ, with long extended Siliquæ

22.

Erysimum cheiranthoides, *L.*

Leucoienartiger Hederich, Schotendotter.—French, *Velar giroflée.*—Engl. *Treacle hedge-mustard.*—Swed. *Aker-rym.*

This plant, the fibrous root of which lasts two years, is pretty common among the shrubs of our pastures, in meadows and moist fields. Its somewhat angular, branched, erect, leafy, herbaceous stem is furnished with numerous accumbent sharp hairs. The leaves of the stem stand alternate, are lanceolate, without stalks, or lengthened into the leaf-stalk, generally two inches long, and half an inch broad, furnished with indistinguishable, scattered teeth on the margin, which almost entirely disappear in the upper leaves. The surface appears smooth to the naked eye; but with a glass we observe small, contiguous, three-pointed hairs. The branches stand open, and carry rich bunches of flowers; the individual flower-stalks are a line and a half long; at first they are erect, afterwards, when the fruit has ripened, they become horizontal. The calyx consists of four lanceolate, erect, yellowish-green pieces, furnished with a membranaceous margin: the corolla consists of four spoon-shaped, more or less emarginated yellow petals, somewhat longer than the calyx. The filaments are six in number, four of which are longer than the other two; but they are all as long as the pistillum, and as the claws of the petals. The antheræ are oval, yellow, and bilocular. The nectaries stand at the base of the filaments, those on the shorter filaments being half ring-shaped, those on the longer crenated. After flowering, both the si-

liquæ stand erect, are quadrangular, having the remains of
the simple stigma at their points, and containing the seed in
two rows, the radicle of the embryon being turned towards
the ridge of the cotyledons.

Diagnosis and Affinity.

The plant which most resembles this is *Cheiranthus erysimoides* Linn., or *Erysimum hieracifolium* Linn. Ehrh. (not
Jacqu.), or *Er. lanceolatum* R. Br. But this latter plant is
distinguished by its habitat being confined to dry, sunny
places, along the wayside,—by its erect stem, distinctly dentated leaves, and by its white flowers being larger and whiter.
Er. hieracifolium Jacq., or *odoratum* Ehrh. Willd., is also
extremely similar, but it has commonly a simple stem, distinctly dentated leaves, golden-yellow flowers, which have a
pleasant smell, circular petals, and a two-lobed stigma at the
end of the siliqua. *Er. repandum* Linn. has an almost procumbent stem, the flower-branches of which stand opposite to
the leaves : it has also lanceolate, angular dentated leaves,
and horizontal siliquæ. *Er. diffusum* Ehrh. has linear,
greyish, slightly dentated leaves, the hairs of which have
merely a simple cleft, and the flowers are of a pale yellow colour, and large. *Er. angustifolium* Ehrh. has revolute, linear, sulcated, entire, greyish leaves, sulphur-yellow
flowers, and a long pistillum on the end of the siliqua.

Synonymes and Figures.

Viola lutea sylvestris, *Trag.* f. 212. b.
Myagrum alterum, Thlaspi effigie, *Lobel. Hist.* 112. *Ic.*
   225. *Dalech.* 1137. Amarum, *Park. Theatr.* 868.
Myagro affinis planta, siliquis longis, *J. Bauh. Hist.* 2. 894.
? Erysimon iii. *Tabern.* 840. (It may also, however, be
   Cheiranthus erysimoides).
Camelina, *Ger. em.* 273.
Eruca sylvestris, Thlaspios effigie, *Moris.* sect. 3. t. 5.
Turritis leucoii folio, *Tourn. Inst.* 224.
Erysimum cheiranthoides, *Linn. Willd. Sp. Pl.* 3. 511.

Jacq. Fl. Austr. t. 23. *Fl. Dan.* 731. 923. *Engl. Bot.*
942. *Schk.* t. 183.
Cheiranthus turritoides, *Lam. Enc.* 2. 716.

Geographical Distribution.

This plant is found from 68° to 44° N. Lat., throughout
the whole northern hemisphere. It grows in Lapland, Siberia, and North America ; but it does not grow farther south
than the South of France.

———

CLASS XVI.

23.

Geranium rotundifolium, *L.*

Rundblättrigger Storchschnabel.—French, *Geranium à feuilles rondes.*—Engl. *Dove's foot Cranesbill..*—Swed. *Gärd
Storknäf.*

This plant flowers about the middle of summer, in gardens and fruitful fields. From a soft, fibrous root arise several herbaceous, roundish stems, furnished with soft, white,
erect hairs, somewhat glutinous, much branched, of the thickness of a thread, and from one to two feet high. The branches
and leaf-stalks stand opposite to one another, and, where
they grow, the stem and branches are somewhat thickened
and reddish. Among these divisions stand reddish, membranaceous, lanceolate, pointed leafy appendages, which in
time become dry, and fall off. The leaf-stalks, which are
nearly horizontal, are almost an inch long, and furnished
with shaggy hairs. The lower stem-leaves are nearly circular, having a deep small indentation, where the leaf-stalk is
inserted : they are five-lobed, having the lobes standing thick
together, obtusely dentated, of a pale green, furnished on

both sides with the same soft shaggy hairs. The upper leaves
are more broad than long, so that they are almost reniform;
they have no indentation at the base, but sometimes they taper into a wide shape, sometimes they are directly truncated.
Their circumference is also divided into five obtuse lobes,
standing at some distance from each other, and furnished
with three obtuse teeth: their surfaces also are hairy. In
the neighbourhood of the flowers the leaves are frequently
three-lobed. Their size varies, the lowermost having a diameter
of an inch and a half, the uppermost of scarcely half an inch.

Opposite to the leaf-stalks, and higher up even between the
leaf-stalks, appear the flower-stalks, likewise an inch long.
They stand erect, or open, and are divided, half way their
length, into two distinct stalks, under which two fine, white,
pointed bracteæ, or stipulæ, stand. The calyx consists of five
oblong, pointed, streaked, strongly ciliated pieces. The
flower consists of five spoon-shaped, pale red, quite entire petals, the claws of which are ciliated at the base: they are rather longer than the calyx. In the bottom of the latter stand
ten filaments, enlarged at the base, and clinging to one another. On their outer side we perceive five nectaries, covered
by the ciliæ of the petals. The antheræ are reddish, and
contain a yellow, oval pollen, surrounded by three circles.
In the middle of the filaments rises the pistillum, inclosed by
the five, rostrate, connate, shaggy appendages of the germen.
At its summit, the pistillum is expanded into a five-lobed,
reddish stigma. After the fall of the flower, five ciliated, but
not bunchy or wrinkled utriculi, remain, which open laterally
from below upwards, and whose beaks are externally set
with the same parallel white hairs which cover the whole
plant. The spotted, or warty seed, contains the embryon
without albumen, with convoluted, membranaceous cotyledons, and the radicle turned upwards.

Diagnosis and Affinity.

This plant is most nearly related to *Geranium molle* and
pusillum. But the former of these two is distinguished by
its cleft, or deeply emargined petals, and by its wrinkled utri

culus having a smooth beak. *Geranium pusillum* has emarginated petals, only five antheræ, and its flowers are much
smaller, although also pale-red. The beak of the *utriculus*
is not furnished with distant, but with thick crowded hairs.
G. *pyrenaicum*, which is most nearly related to G. *molle*, has
a perennial root, is much larger in all its parts, has also large
flowers, the petals of which are deeply cleft, and it has not a
bunchy utriculus. G. *dissectum* is distinguished by its pal-
mate, cleft leaves, the lobes of which are linear, and stand at
regular distances from each other ; the emarginated petals
are as long as the awned calyx : the beak of the utriculus is
furnished with shaggy hairs : the flower-stalks are shorter
than the stem-leaves. G. *columbinum*, on the other hand,
which is very like G. *dissectum*, has very long, wavering,
flower-stalks, stem-leaves of the same kind, large flowers, and
a smooth beak of the utriculus.

The genus *Geranium* cannot be distinguished at first sight
from *Erodium* Ait. But the latter has among its ten filaments, five that are unfruitful and abortive : the beak of the
utriculus is turned into a spiral shape, and is internally furnished with hairs, which may be most easily seen in *E. cicu-
tarium*, the most common species. *Pelargonium* Ait. has a
similar beak of the *utriculus*, but the corolla is irregular and
two-lipped : the nectary is in the bottom of the tubular calyx.
The diagnostic character of *Geranium*, therefore, is as follows : the corolla regular, ten fertile filaments, five nectaries
at the base of the filaments, and a straight beak of the utriculus, internally without hairs. The Geraniæ constitute a
natural family, which stands between the Agrumæ and Malvaceæ, (Anleit. ii. vid. 793.)

Synonymes and Figures.

Geranium alterum, *Fuchs.* 205. *Matth.* 621.
Pes columbinus, *Dodon.* 61. *Lobel. Hist.* 376. *Ic.* 658
Ger. aliud fecundum, *Dalech.* 1277. *Tabern.* 123.
Ger. columbinum, *Ger. em.* 938. *Park.* 706.
Ger. folio rotundo, *J. Bauh.* 3. 473., is intended for it, but
the figure is quite a failure.

Ger. folio malvæ rotundo, *C. Bauh. Pin.* 318. *Tourn. Inst.* 268.

Ger. annuum, folio malvaceo rotundo, *Moris.* sect. 5. t. 15.

Ger. viscidum, caule decumbente, *Hall. Stirp. Helv.* n. 941.

Ger. rotundifolium, *Linn. Willd. Sp. Pl.* 3. 712. *Cav. Diss.* 4. t. 193. f. 2. *Engl. Bot.* 157.

Geographical Distribution.

This species delights in the temperate and warmer regions of the earth. In Sweden it does not grow beyond the 61°. On the other hand, it extends throughout the whole of Europe, as far as the islands of the Archipelagus, and it is even found in the northern coasts of Africa. Its diffusion eastward seems to be limited by the Wolga Heights ; for it is found in Lithuania, but not in the other parts of Russia, in Tauris, or in Asia.

———

CLASS XVII.

24.

Lathyrus tuberosus, *L.*

Erdnuss, Erdmandel, Grundeichell, Akereichell, Sand-brot. —French, *Anette, Marcusson.*

This beautiful plant is common enough in the corn fields of Germany. The stem rises from an irregular, yet, for the most part, round tuber, externally of a brown colour, internally white, and of an agreeable taste. The stem is herbaceous, erect, commonly quadrangular, without leafy or membranaceous appendages: its upper part is divided into branches, and it attaches itself, by cirrhi, to other plants and objects. It is an ell or arm in length, and, at its lower part, is of the thickness of a pack-thread. Where the leaf stalks and branches arise, there are linear, long, and fine pointed stipulæ, which are half arrow-shaped, by having the lower teeth reflex. The leaf-stalks stand open, are at least half an inch long, angular, and carry each two opposite, oblong, entire leaves, tapering at the base, somewhat rounded at the point, furnished with an herbaceous spine, penetrated by many nerves and veins, but in other respects smooth, and which are an inch long, and rather more than half an inch broad. The leaf-stalk passes above these leaves into divided crooked cirrhi.

At the extremity of the shoots grow the flower-stalks, about a finger in length, frequently still longer, without leaves, smooth, roundish, and erect. The beautiful red flowers stand in a six or eight blossomed bunch. The individual flower-stalks stand open, are from three to four lines in length, and have beneath them a linear stipula, which is about one half shorter than the flower-stalks.

The calyx is divided into five lanceolate teeth, two of which commonly lie on the vexillum of the corolla, and three stand beneath it. The corolla is papilionaceous. The vexillum is emarginated, reflex, white in the centre, of a beautiful red above, and marked with red streaks below. The alæ and the carina, likewise of a beautiful red, are inferior to it. The carina incloses a cylinder of filaments, one of which becomes separated from the rest towards the vexillum. The other nine are completely united : all of them carry round, yellowish antheræ. In the middle of them stands the longish, compressed germen, with the ciliated, broad pistillum, and a yellow, roundish stigma. The fruit, which is superior to the calyx, is an unilocular, two-lobed, rather compressed legume, which contains ten roundish seeds, fixed to one suture. The embryon fills the whole seed with its two roundish strong cotyledons.

Diagnosis and Affinity.

This species cannot easily be confounded with any other. It is true that in *Lathyrus pratensis* the cirrhus also springs from two opposite leaves ; but these leaves are lanceolate, tapering at the point, and the flowers are always yellow. In *L. sylvestris* and *latifólius*, also, the flowers are red ; but in the former they are more soiled, and the carina is green; both have leafy appendages on the stem and on the branches. The leaves of *L. sylvestris* are about three inches long, and sharply tapered at their point. *L. latifolius* has flowers of a rose-red colour; oblong, rounded leaves, with an herbaceous spine, and these leaves are also much larger than those of the other species, and almost coriaceous. The stipulæ are broad lanceolate, and rather dentated. With other species it has still less affinity.

The genus *Lathyrus* has a very distinct character in the flat pistillum, although in other respects it is nearly related to *Vicia*, which, however, is distinguished by the hairs of the roundish pistillum; and *Orobus*, which is related to both of them, is distinguished only by the want of cirrhi. These genera belong to the natural family of leguminous plants, which stand between the Polygaleæ and Capparideæ; (An. 2. vid. 740.)

Synonymes and Figures.

Apios, *Fuchs*, 131.   *Dalech.* 1596.
Pseudoapios, *Matth. ed. Bauh.* 876.
Terræ glandes, *Dodon.* 550.   *Lobel. Ic.* 2. 70.   *Ger. Emac.* 1237.
Chamæbalanus, *Tabern.* 891.   *J. Bauh. Hist.* 2. 328.
Arachydna Theophrasti, *Column. Ecphr.* 1. p. 304. t. 301.
Lathyrus arvensis, *Park. Theatr.* 1061, radice tuberosa, *Moris.* sect. 2. t. 2.   *Riv. Tetrapet. Irreg.*
L. tuberosus, *Linn. Willd. Sp. Pl.* 3. 1088.   *Fl. Dan.* 1463.

Geographical Distribution.

It is as yet completely unknown according to what laws this plant is distributed. We find it so dispersed from 30° to 56° N. Lat. in the Old World, that some countries have it, whilst others, lying in the same latitude, want it. Thus, it is very common on the north coast of Africa; on the other hand, it is wanting in Greece and Asia Minor: it is found in Tauris and Transylvania, in Germany, France, and Poland; but is wanting in Sweden and Great Britain. In Denmark it is only found near the fort at Copenhagen. In Siberia, again, it grows along the upper Jenisei as far Krasnojarsk.

Uses.

The tubers are edible. In Siberia they are much relished by the Tartars, under the name Tschina. The common people in Germany also use them. They contain three times more starch than potatoes.

———

CLASS XVIII.

25.

Hypericum montanum, *L.*

Berg-Johanniskraut, Grossblättriges Hartheu,—French, *Millepertuis de montagne.*—Engl. *Mountain St John's-wort.*

This graceful plant grows single, in woods and on calcareous soils. The stem, which is an ell long, round, simple, and smooth, springs from a brown, woody, fibrous root, and is about the thickness of a pack-thread. The stem-leaves are oblong, smooth, and stalkless: they stand above and opposite each other, at intervals, which are an inch and a half long, and they partially embrace the stem. The margin of the leaves is adorned with black points, and its lower surface is reddish. The leaves are rather tapering at the point, yet not pointed. On their lower surface we perceive some ribs proceeding from the base. The length of the leaves is an inch, the breadth half an inch. The leaves, when rubbed, before the flowering, give out a reddish juice. On the upper part of the stem they are less frequent, and much smaller. The flowers stand on the top of the stem in crowded panicles. Under each flower-stalk there is a lanceolate bractea, set round with black stalked glands. In the same manner are the five lanceolate leaves of the calyx inclosed. Of the several flowers in the panicle, only a pair is always quite open: they consist of five oblique, entire, citron-yellow leaflets. From the bottom of the calyx rise the yellow filaments, in indeterminate number, and in three bundles: they carry oval antheræ, likewise of a yellow colour, and are for the most part longer than the petals. The germen is superior, and carries three remote pistilla, with button-shaped stigmata. After flowering, the corolla withers, without falling off, and becomes somewhat twisted, for its æstivation is complex. The fruit is a three-lobed capsule, the valves of which form double dissepimenta with their inverted margins. The numerous fine seeds contain the embryon evolved, with albuminous substance.

Diagnosis and Affinity.

The species most related to this is the *H. elegans* Willd., (*Kohlianum Fl. Hal.*) But this is distinguished—by its shrubby stem, which with us is seldom longer than a small span,—by its leaves, which, being much smaller, are furnished on the margin, not with black, but with bright points, and have their margins for the most part reversed. This species is limited in Germany to a single calcareous hill. But it grows also in Volhynia, where it is an ell in length, and in Siberia. *H. perfoliatum* L. or *ciliatum* Willd. has a two-edged stem, and pellucid points in the leaves, which embrace the stalk. Of exactly the same nature is the structure of *H. Thomasii* from Calabria. It has cordate leaves, completely embracing the stem, furnished with bright points, and pellucid cartilaginous margins, an obtuse quadrangular stem, the bracteæ and the leaves of the calyx completely set round with glands having stalks. Very much resembling our plant is also *H. maculatum* Wall. Mich.; but here also the stem and the petals are furnished with dark points. The panicle is expanded, with branches distant from one another, and forming together an umbel. *H. corymbosum* Willd., and *punctatum* Lam., belong to this species. But *H. punctatum* Willd. seems to be a different species, although one that is related to *H. montanum*. The dark points shew themselves through the whole

3

leaves, and the round stem also is set with them; the leaves also are much smaller. *H. alpinum* Kit. (Hung. 3. t. 265.) is also very nearly related, but is distinguished by fine hairs on the petals, and by having the upper part of its stem two-edged. *H. dentatum* Loisel. (Fl. Gall. 2. p. 499. t. 17.) has completely the external appearance of *H. montanum*; but the leaves are not studded on the margin with black, but every where with pellucid points, and those farthest up are fringed and ciliated on the margin. We cannot, with Poiret, (Enc. Suppl. 3. 700.) suppose it to be a subspecies of our plant. *H. pulchrum*, to which *H. montanum* is also very similar, has quite short, ovate leaves, embracing the stem, and fine dark points. *H. hirsutum* is distinguished by the shaggy hair which covers the stem, and the lower surface of the leaves.

The genus *Hypericum*, along with some others, constitutes a natural tribe, which includes the Guttiferæ (*Marcgravieæ* and *Mesueæ*), and has, in common with them, the coloured peculiar sap, the indeterminate number of the almost united filaments, and the number five prevailing in the calyx and in the corolla. As we do not observe any nectaries in most of the species of *Hypericum*, we distinguish as a peculiar species, under the name Martia (*Elodea* Adans.) the *H. Virginicum* L. and *petiolatum* Wall., on account of three nectarious glands on the base of the petals, between the bundles of filaments. *Androsæmum* Tourn. also is distinguished by the berry-shaped fruit, and *Ascyrum* by its four-leaved calyx and corolla, and by its two-valved capsule. On the other hand, the diagnostic character of *Hypericum* is its five-leaved calyx; its corolla, consisting of five petals; its filaments in three or more bundles; no nectaries; three or five pistilla; a three or five valved capsule, with the margins of the valves involuted. To this genus also is now added *Sarothra* L., the filaments of which vary from five to ten.

Synonymes and Figures.

Hypericon II. *Trag.* 28.
Androsæmon, *Fuchs.* 76.

Hypericum elegantissimum non ramosum, *J. Bauh. Hist. 3.*
383. *Tourn. Inst.* 255.

Androsæmum campoclarense, *Column. Ecphr.* 1. 74.

Ascyron, s. *Hypericum bifolium, C. Bauh. Pin.* 280. (excl.
synon. Dodon. et Matth., quæ ad H. quadrangulare.)
Moris. sect. 5. t. 6.

Hypericum montanum, *Linn. Willd. Sp. Pl.* 3. p. 1463
Fl. Dan. 173. *Engl. Bot.* 371. *Trattin. Ostr. Flor. H.* 3.

Geographical Distribution.

Desfontaines places this species on the mountains of Algiers, and Sibthorp near Messenia in Peloponnesus. It is diffused from 30° N. Lat. throughout the whole of Europe.
How far it proceeds towards the north, is not yet established.
It grows at Bornholm, and in the south of Sweden; but not
in Lapland, nor in Norway. Its northern boundary in Sweden seems thus to be the 61° N. Lat. How far eastward it
extends, is not yet determined. It is found in *Lithuania,*
Gallicia, and *Transylvania,* but not in Tauris, the other
parts of Russia and Siberia. Its eastern limit may therefore
be the 30° E. Lat.

CLASS XIX

Order I.

Cichoreæ, or Semiflosculosa.

26.

Thrincia hirta, *Roth.*

This plant grows in July and August on some pastures and
fields in Germany. The root is perennial and fibrous. Here
and there appears a somewhat thickened, tuberous radix. From
this spring, in the first place, several lanceolate leaves, taper-

ing towards the base, having sparse teeth on the margin, and
sinuses between ; these leaves are set with scattered white
hairs, rather stiff, separated in the form of cirrhi at the extremity. The length of the leaves is at most that of a finger,
the breadth about three lines. At the side, not in the centre,
of this plot of leaves, there arise several round stalks, either
quite smooth, or having their under surface furnished with a
few scattered hairs, of a reddish colour below, a small or great
span long, of the thickness of a thread, and carrying single
flowers. The flowers, before blossoming, are pendant. The
calyx is smooth, or slightly ciliated, simple, divided into eight
or ten pieces, and furnished at the base with some small short
scales. The calyx has as many corners as parts.

The flower consists of an indeterminate number of uniform,
yellow, tongue-shaped florets, which are penetrated longitudinally by five fine parallel veins (Vid. 260. R. Brown in
meinen. Neuen Entdeckungen, i. 166.), and the upper end
of which terminates in five teeth, between which run those
veins or nerves. The ray-florets, or those on the margin,
are of a lead or copper colour on their lower surface. In
each of the florets is a cylinder of yellow antheræ, the filaments of which are ciliated, remote, and inserted into the
lower tubular part of the floret. Through the middle of the
cylinder of antheræ proceeds the pistillum, which terminates
above, in two linear stigmata, internally warty, outwardly
furnished with small hairs, (Collectores, vid. 273.) At first
the pistillum is included in the cylinder of antheræ; afterwards the former comes forth, and its stigmata become expanded, and revolute. The pollen is spherical, set round
with fine spines, (271.) The nectary is the upper part of
the germen, within the *pappus*. The receptacle is full of
very fine cavities, set with extremely delicate bristles. The
seed is longish, brown, sulcated, having cross ridges, and a
varying *pappus :* For the ray-florets bear seeds the *pappus*
of which consists altogether of short chaffy leaves ; the seeds
in the centre, on the contrary, have a pinnated *pappus*. The
seed is a caryopsis, in which the evolved embryon stands
erect, without albuminous matter.

D d

Diagnosis and Affinity.

This plant is commonly classed with *Hieracium pilosella* and *Apargia hispida*. With these two it might most readily be confounded. But *H. pilosella* is distinguished by trailing shoots,—by its entire, more strongly ciliated leaves, having their under-surface set with a white *tomentum*,— by its citron-yellow flowers, the lower parts of which are of a bright red, and which, upon the whole, are much larger than the flowers of *Thrincia*;—finally, by its simple, hairy pappus. *Apargia hispida* has runcinate leaves, the hairs of which are universally divided into a forked shape. In the centre, not at the sides of these leafy plots, arise, pretty perpendicularly, the flower-stalks having each one blossom, and more strongly ciliated. The calyx is covered by strong hairs, scaly, and the scales lie like tiles on one another. The flowers are larger than those of *Thrincia*, and of a darker yellow ; each floret has a bush of long yellow hair at the entrance of the tube, and the five teeth of the upper extremity have five brown glands on their lower surface. In other respects the lower surface of the florets is of the same colour as the upper. The pappus is ovate and pinnated. In the south of Europe there grows a species, which is uncommonly like this, namely, *Thrincia hispida* Roth. This is distinguished by its annual root,—by its runcinate leaves, every where furnished with forked hairs,—by its roughly ciliated calyx and long-stalked pappus. There grows also in the south of France and in Italy, a *Thrincia tuberosa* De Cand. Sav., which is distinguished by its beet-shaped tuberous roots, its runcinate, nearly smooth leaves, and by its calyx being but slightly ciliated. Its florets also are discoloured on the back. *Apargia tuberosa* Willd. does not altogether correspond with this, but the following more ancient synonymes are applicable to it : *Cichorium Constantinopolitanum, Matth. ed. Bauh.* 388. ; *Dens Leonis Monspeliensium,* Dodon. 636. ; *Monspeliensium dens Leonis,* asphodeli bulbulis, Lobel. Hist. 117. ic. 232. : *Chondrilla altera Dioscoridis,* Columb. Phytob. t. 4. These, therefore, are

the true *Thrincia tuberosa. Apargia tuberosa* grows smooth
and ciliated : the ciliæ are short and forked. The leaves are
shaped pretty much in the same way as those of *Thr. hirta.*
The flowers are large, the calyx distinctly scaly and ciliated,
the pappus uniformly pinnated. The plant which has the
greatest resemblance to this is *Apargia hastilis* Willd. particularly because here also the calyx has the same shape, and
the marginal florets have the same leaden colour on their
lower surface. I also think that the two species could only
be artificially distinguished. For *Ap. hastilis* has an ovate
pinnated *pappus,* and is quite smooth. Are there not transitions from the one form to the other ? I think there are.
Jacquin fifty years ago attempted in vain to discover these
transitions, (*Enumer. Plant. Vindob.* p. 270.) Lachenal,
(*Act. Helv.* 1. p. 272.), indeed, has remarked transitions to
Leontodon hispidum ; but can we believe that he had our
plant actually in his eye ?

Synonymes and Figures.

Hieracium χαμαιλασιοτραχύφυλλον, *Richard de Bellev. Ic.* t. 119.
Hier. hirsutum parvum, caule aphyllo *Magnol. Bot. Monsp.*
  131.
Hier. montanum latifolium non sinuatum, *Losel. Boruss.*
  p. 125. *
Dens leonis foliis hirsutis *Zannich Venet.* p. 86. t. 183. f. 1.
? Hieracium dentis leonis folio, *Buxb. Hal.* 157.
Rhagadiolus foliis semipinnatis asperrimis, *Hall. Stirp.* n. 7.
Leontodon hirtum, *Linn. Sp. Pl. ed. Reich.* 3. 634. (excl. sy-
  non. Bauh. et haud apta descriptione.) *Houtuyn.* 9. 60.
  Gouan Fl. Monspel. 349. *Wither. Arrang.* 3. 837. (excl.
  Syn. Fl. Dan. 901. quæ Apargia hispida.) *Roth. Fl.*
  Germ. 2. 2. 145. *Jacqu. Vindob.* 139. *Leers, Herborn.*
  168. * *Leyss. Hal. ed.* 2. p. 191.
Leontodon proteiforme, *Vill. Delph.* 3. p. 87. t. 24. (sed excl.
  plerisque synonymis.)
Apargia hirta, *Scop. Carn.* n. 984. (excl. syn. J. Bauh.)
  Hofm. Germ. 274. *Host. Syn. Austr.* 424. *Smith,*
  Prodr. Græc. 2. 131. *Schk.* t. 220.

Hyoseris taraxacoides, *Lam. Enc.* 3. 159. (Sed planta dicitur annua.)   *Sav. Pis.* 2. 230.   *Santi it. Tosc.* 3. p. 360.
t. 7.
Rhagadiolus taraxacoides, *Allion. Pedun.* p. 836.
Leontodon hispidum, *Pollich. Palat.* n. 737. * (Sed de pap-
pi diversitate nil dicit.)
Colobium hirtum, *Roth. in Rëm. Arch.* 1. 37.
Thrincia hirta, *Roth. Catal. Bot.* 1. 98, 2. 103.   *Willd. Sp.
Pl.* 3. 1554.   *Pers. Syn.* 2. 368.   *De Cand. Fl. Fran.* 4.
51.   *Sav. Bot. Etrusc.* 3. 122.   *Bertolon. Amœn. Acad.*
183.   *Hagen. Fl. Boruss.* 2. 153.   *Spreng. Hal.* 228.
Baumg. Transylv. 3. 15.

Geographical Distribution.

This plant seems to be limited to a few countries. Germany, from 50° N. Lat. going onwards to the south, France,
Italy, and Transylvania, are the countries in which it is found.
Whether it grows in the Island of Great Britain is still unde-
termined ; because *Hedypnois hirta* Huds., although called a
Thrincia, has many circumstances that distinguish it from
this, especially as *Hieracium pumilum saxatile* Rai Syn. 167.
along with which *Hier. montanum saxatile* C. Bauh. Prodr.
66. and Column. Ecphr. 1. 243. is classed, is by no means
our plant, but a subspecies of *Apargia hispida* or *A. Villar-
sii* Willd.   But the plant of Ray is figured in the Engl. Bot.
555.

Cynareæ ; (Anleit. ii. 532.)

27.

Cirsium Eriophorum, *Scop.*

Wolldistel.—French, *Chardon aux ânes.*—Engl. *Woolly-
headed thistle.*

This remarkable and beautiful thistle only grows in Germany
within very confined limits, on mountain meadows.   It has a

---

* Apargia hirta (distinct from A. hispida) is now ascertained to be a native
both of England and Scotland.

biennial, white root, of the thickness of a thumb, and an ell in length, from which, during the second year, a stem shoots up, about the height of a man, of the thickness of a finger or thumb, straight, green, angular, and entirely woolly. The leaves are deeply semi-pinnate, the lowermost being often two feet long, covered on their under-surface with thick, woolly, white *tomentum*, and having their upper surface green, set with stiff, somewhat crowded hairs. On the lower leaves the laciniæ are remote and upright, again divided into two other laciniæ, the larger of which is linear, the smaller spear-shaped ; both of them are entire, but furnished with thorns, and terminated in two strong yellowish thorns, which are placed alternately upwards and downwards. The middle rib is strong, projects downwards, and also terminates in a long, stiff thorn. The stem-leaves are not so long, embrace the stem without running downwards, and are not so regularly pinnated, but in other respects resemble the root-leaves. The flowers on the top of the shoot, together with the calyx, are about eight inches in circumference. The calyces, several of which often stand together, are furnished at their base with very small, semi-pinnate, thorny covering leaves, about the size of an ordinary apple : the small thorny scales of the calyx are set with thick white wool, which, however, in many instances passes into something resembling a mere cobweb. The florets are all uniform, tubular, of a purple red colour, and their margin is divided into five segments. The cylinder of antheræ is longer than the floret ; the pistillum is furnished with a divided stigma ; the receptacle is set with chaffy leaves, which are divided into bristles. The pappus, stalkless and pinnated, rests upon a ring, which seems to be in the act of disengaging itself from the oval seed. The seed is a caryopsis, in which the embryon, without albuminous matter, stands erect, with its cotyledons unfolded.

Diagnosis and Affinity.

The most nearly related to this species is *Cirsium lanceolatum* Scop. ; yet the calyx of the latter is not woolly, but merely covered by a fine web, and the leaves, which are not so deeply semi-pinnate, run down the stem. The stem itself is not so tall, and the flowers are not so large as in our species. *Cnicus laniflorus* M. B. is still more nearly related to it, and is distinguished by having the scales of its calyx broader;—these are of a reddish colour. As this corresponds with the figure in the *Engl. Bot.*, it becomes a question, whether the Tauric and British plants be not the same variety of *C. eriophorum*. The other species have a less perfect resemblance. *Cirsium*, long ago very correctly distinguished by Tournefort, has been named *Cnicus* by some later writers, who have not recollected, that Seb. Vaillant a century ago, had given the name to the well marked *Centaurea benedicta* as a peculiar genus. We appropriate the name *Cirsium*, therefore, to those thistles the pappi of which are pinnate and the scales of the calyx thorny. If the scales of the calyx are unarmed, it is the genus *Saussurea* De Cand. If the pappus is simply hairy, it is the genus *Carduus*.

Synonymes and Figures.

Carduus eriocephalus, *Dodon.* 723. *Clus. Pann.* 666. *Hist.* 2. 154. *Gerard, Emac.* 1152.

Carduus tomentosus, *Lobel. Hist.* 482. ic. 2. 9. C. capite tomentoso. *J. Bauh. Hist.* 3. 57. *Parkins,* 978.

Cirsium foliis pinnatis, *Hall. Helv.* n. 168.

Carduus eriophorus, *Linn. Hort. Upsal.* 249. *Mill. Ic.* 293. *Willd. Sp. Pl.* 3. 1669. *Jacq. Fl. Austr.* 171. ? *Eng. Bot.* 386.

Cirsium eriophorum, *Scop. Carn.* n. 1008.

Geographical Distribution.

In England, where this thistle is not rare, its farthest northern limit is Cumberland *, (between 54° and 55° N. Lat.) In Germany, again, it extends only a little beyond 51°; but southward from this, it is very common in all hilly regions, especially in the Palatinate, Austria, Hungary, Transylvania. If Gillibert's account (Jundzill, Fl. Lithuan. 244.) be correct, it grows again more eastward, as far as 55°, for it is found be-

* It grows also in Scotland, but sparingly ; as near the foot of Largo Law in Fifeshire.

tween Grodno and Wilna.   In France, Italy, and Greece, as
well as in Asia Minor, near Smyrna (39°), it grows pretty
frequently.

Order II.

Radiatæ, (Polygamia superflua.)

28.

Arnica montana, *L.*

Fallkraut, Wolverley.—French, *Tabac des Vosges, Tabac
Savoyards.*—Swed. *St Hansblomster.*

This remarkable plant grows in open woods and on mountain meadows.   Its root is perennial, brown, of the thickness
of a quill, almost horizontal : it shoots out several root leaves,
which are opposite to one another, oblong or elliptical, entire,
set on both sides, especially on the upper, with sparse, crooked hairs, ciliated, and furnished with five nerves.   They pass
downwards into a short sheath-shaped leaf-stalk, which is
composed of the two opposite leaves.   The lowermost leaves
are from two inches to a finger in length, from an inch to an
inch and a half broad, rounded at the point, and commonly
they stand in pairs, one above the other.   The simple, roundish, furrowed or angular stem, set with crooked glandular cilia, and of the thickness of a pack-thread, rises from about a
foot to an ell in height.   About its centre are two lanceolate,
small, pointed leaves, which embrace the stem on both sides.
Commonly the stem is divided at the upper end, but frequently
also it remains simple, and carries a single large copper-yellow
coloured flower, of a strong, peculiar smell.   The common
calyx consists of four ciliated, lanceolate leaflets, which stand
in two rows.   The ray-florets are lingulate, terminate in
three points, and are penetrated by nine parallel nerves.   At
the entrance of the tube of these ray-florets are four or five
free, short filaments, inserted in the floret, with the same
number of pointed, empty, evidently abortive antheræ.   The
florets of the disc are tubular, with a five-lobed margin.   The

yellow cylinder of antheræ surrounds the pistillum, which
here, as in the ray-florets, is divided. The receptacle is furnished with short cilia. The fruit is an angular caryopsis,
of which the inferior *umbilicus* is furnished with a persistent
funiculus umbilicalis. The caryopses are provided with sharp
cilia, and have a sharp-haired pappus.

Diagnosis and Affinity.

There is a distinct variety of this plant, with small
lanceolate leaves, which is figured in the Fl. Dan. 1524.
as the *Arnica angustifolia* Vahl. of Greenland. Linnæus
(Fl. Lappon. n. 305.) mentions the same plant under *Doronicum foliis lanceolatis,* and maintains that it is a peculiar species. But he cites, at the same time, *Doronicum IV. Clus.
Pannon.* 522. and *Alisma Matthioli* J. Bauh. Hist. 3. 20,
both of which figures perfectly correspond with our *A. montana. Chrysanthemum latifolium minus* Ger. Emac. 742.
might also be mentioned on the same occasion. In the later
editions of the *Species Plantarum* this plant is mentioned in
the Flora of Lapland, as a variety of *A. montana.* Nuttall (Amer. Plants, 2. 164.) also recognises *A. fulgens* and
plantagineæ Pursh, as varieties which grow in Labrador,
and on the Missouri. The former small-leaved species is
mentioned by Linnæus in a letter to J. G. Gmelin (Fl. Sib.
2. 153.) as a variety, produced by its situation on high mountains. That which bears the greatest resemblance to our
plant is *A. doronicum,* which is distinguished, however, by the
alternate position and dentated margin of its upper leaves.
The genus *Arnica* is in other respects difficult to be distinguished from *Doronicum,* for both have a double row of leaflets in
the calyx, both have a hairy receptacle, abortive filaments in
the ligulate florets, and a pappus with sharp-pointed hairs. The
only difference is, that in *Doronicum* the marginal seeds have
no pappus. *Doronicum plantagineum,* which is very like our
plant, is distinguished from it, partly by the generic character, partly by the alternate, imperfectly dentated teeth, and
by the pale yellow, almost inodorous flowers.

Synonymes and Figures,

Chrysanthemum latifolium, *Dodon.* 263. *Dalcch.* 1358,
    Gerard, Emac. 742.
Alisma, *Matth. ed. Bäuh.* 666. *J. Bauh. Hist.* 3. 20.
Doronicum IV. *Clus. Pannon.* 522. V. VI. *Clus. Hist.* 2.
    18.
Nardus Celtica altera, *Lobel. Ic.* 313.
Ptarmica montana, *Dalech.* 1169.
Damasonium s. Alisma Matth. *Dalcch.* 1057.   *Tabern.* 1116,
Caltha alpina, *Tabern.* 714.
Doronicum Germanicum, *Park. Theat.* 321.
Διουρητική *Renealm. Spec.* 119.
Doronicum plantaginis folio alterum, *C. Bauh. Pin.* 185.
    Tourn. Inst. 487.   *Linn. Fl. Lap.* 304, 305.
Arnica montana, *Linn. Sp. Pl.* 1245.   *Fl. Dan.* 63.   *Schk.*
    t. 248.
Doronicum oppositifolium, *Lam. Enc.* 2. 312.
Cineraria cernua, *Thon. Land.* 344,

Geographical Distribution.

If *A. angustifolia* Vahl., *fulgens* and *plantaginea* Pursh,
belong to our species, it is diffused northward as far as
Greenland and Labrador; which cannot be wondered at,
since Messerschmid and John George Gmelin found it on the
banks of the Tungusca and the Jenisei (beyond 60° N. Lat.),
and Steller found it on Behring's Island. It is remarkable
that Linnæus found it at Torneo (68° N. Lat.); but Wahlenberg found it not in Lapland. To the south it grows
throughout Sweden, Denmark, Germany, Prussia, Lithuania,
Galicia, Hungary, Transylvania, France, and Switzerland.
The farther south, the higher it ascends the mountains. It
is even found on the Pyrenees.

Uses.

It is one of the most important medicines, the stimulating
power of which is seen particularly in the vascular system, and
must be ascribed to the resinous ingredient which accompanies

the volatile oil. The root contains tanning matter also; the flowers are very rich in volatile oil and resin ; the leaves contain more soapy extractive matter. In warm solutions we employ it in palsy and typhus, for awakening susceptibility ; we employ it also for stopping the blood after external wounds, and as a diuretic and sneezing powder. From this last use its present name is derived, for *εϱϱινον* has passed into *Arnica*. Under the latter name it was used in the fifteenth century, and the first person who suspected its medicinal quality was unquestionably Lobelius (Adv. 133.), where he praises the plant on account of its diuretic powers. Tabernamontanus extols it for its power of stopping bleeding after wounds, (Krauterb. 417.) The name *Tabac de Savoyards et de Vosges*, is derived from its being smoked and snuffed by the Savoyards and inhabitants of the *Vogeses*.

ORDER III.

Centaureæ, (Polygamia frustranea.)

29.

Calcitrapa stellata, *Lam.*

Sterndistel, Wegedistel.—French, *Chausse-trape*, *Chardon étoilé.*—Eng. *Star-thistle.*—Ital. *La scardigliona.*

This plant grows abundantly in central Germany, by the way side, and on dry fields. It has a white, rather creeping root, which lasts only one summer. The woody, branchy stem is perhaps two feet high, divided from below upwards into branches which are squarrose and smooth, or furnished with a few soft hairs; they are round also, and of a yellowish white. The root-leaves are lyre-shaped, but soon fall off. The stem-leaves are alternate, embrace the stem, or are without stalks, lanceolate, pointed, an inch long or somewhat longer, semi-pinnated at the base, with sharp teeth on the margin,

smooth on both sides, or slightly ciliated. Above the divisions
of the branches spring the flowers, on very short stalks, surrounded by leaves similar to those on the stem. The calyces are ovate,
smooth, of a pale green colour, and about the size of a hazel-
nut. The scales of which they consist, pass into very strong
thorns, of a whitish yellow, half an inch or more in length, and
have subordinate thorns at their base. The florets are all of
a pale red colour and tubular, with a quinque-partite margin :
the marginal florets contain no sexual parts, and the seed
below them is therefore abortive. They are somewhat larger than the florets of the disc, which shew the pistillum with
a cleft stigma within the cylinder of antheræ. If the cylinder of antheræ be touched at a certain period during flowering, it contracts, and the pistillum comes more strongly forward, (816.) The germen is furnished with cilia. The fruit
is an oval caryopsis, without a pappus, with its umbilicus at
one side.

Diagnosis and Affinity.

Calcitrapa lanceolata Lam. (*Centaurea Calcitrapoides*
Linn.), is the most nearly related to this plant. But it is
distinguished by its taller growth, its linear-lanceolate leaves,
its woolly scales of the calyx, and its white pappus. *Centaurea myacanthus* De Candolle, also resembles our plant, and
the seeds are likewise without a pappus. But the leaves
are woolly, the scales of the calyx have appendages which
are surrounded with small thorns. *Centaurea solstilialis*, although the scales of the calyx are armed with similar thorns,
is yet sufficiently distinguished by its decurrent leaves, which
make the stem winged, and by its yellow flowers. The genus *Calcitrapa* is separated, by Vaillant and Jussieu, from
the Centaureæ : it is distinguished by its want of a pappus, and by compound or double thorns on the scales of the
calyx, (Anleit. ii. 540.)

Synonymes and Figures.

Eryngium, *Brunft.* 3. 59.
Carduus stellatus, *Dodon.* 733. *Matth. ed. Bauh.* 504.

428 29. CALCITRAPA STELLATA. [CL. XIX.

Lobel. Hist. 482. ic. 2. 11. *J. Bauh. Hist.* 3. 89. *Ger. emac.* 1166. *C. Bauh. Pin.* 387. *Zann. Ist.* t. 155.
Carduus muricatus, *Clus. Hist.* 2. 7.
? Myacanthus Theophr., *Dalech.* 1473.
Spina stella, *Tabern.* 1080.
Calcitrapa vulgaris, *Park. Theatr.* 989.
Hippophæstum, *Colum. Phytob.* t. 24.
Centaurea Calcitrapa, *Linn. Hort. Ups.* 273. *Willd. Sp. Pl.* 3. 2317. *Engl. Bot.* 125. *Sturm.* 4.
Rhaponticum Calcitrapa, *Scop. Carn.* n. 1018.
Calcitrapa stellata, *Lam. Fl. Franç.* 2. 34.
Calcitrapa Hippophæstum, *Gärtn. Fruct.* 2. 376. t. 163.

Geographical Distribution.

This plant is a proof of the principle laid down, p. 399. that, in the same latitude, the temperature diminishes towards the east; and that hence southern plants grow at a higher latitude in the west than in the east. This plant does not grow in Germany beyond 52° N. Lat. In England, again, it is found as far as Yorkshire, (54°). Eastward from Germany it seems to have a still more southern limit, since it has not once been found in Galicia, although it grows in Hungary and Transylvania. But towards the south it extends as far as Peloponnesus and Sicily.

Uses.

Formerly this plant was celebrated for its medicinal powers. The root was used in decoctions as a diuretic, of which use, in particular, Tournefort (*Hist. des Plantes aux Env. de Paris*, p. 12, 13.) has adduced proofs.

ORDER IV.

(*Polygamia necessaria.*)

30.

Calendula officinalis, *L.*

Ringelblume, Dotterblume.—French, *Souci des jardins.*—
Ital. *Fior rancio.*—Engl. *Marygold.*—Swed. *Ring-blom-
ma.*

This well known plant propagates itself by seed in our
gardens. It has a pretty strong, whitish root, and a branchy,
round, sulcated stem, furnished with short hairs, and having
open branches. All the leaves embrace the stem and
branches, are rather glutinous, and have a peculiar strong
smell: the lower ones are spathulate, quite entire, set with
short, soft hairs, which also make the margin ciliated. The
upper leaves are lanceolate, imperceptibly dentated, and furnished with a herbaceous spine at the extremity: they are
also more hairy than the lower. The flower-stalks, at the
points of the shoots, are woolly. The common calyx is
divided into several lanceolate, woolly laciniæ, with tapering points. The flowers are of a golden-yellow colour: the
ray florets are tongue-shaped, furnished with three pointed
extremities, and with several parallel nerves. The florets of
the disc, having a cylinder of anthers, are tubular, and have
commonly abortive seeds under them. The fertile are com-
.monly on the margin, are lanceolate, or boat-shaped, and internally have spines on their back.

Diagnosis and Affinity.

Calendula arvensis is very nearly related to our plant,
but it has no spathulate, but only cordate-lanceolate leaves:
it has smaller flowers, and the exterior seeds stand erect, and
are small lanceolate. *C. sancta* is distinguished by having

its calyx notched, or furnished with herbaceous spines. *C.
stellata* Cav. has flowers of a sulphur-yellow colour, and five
below the fruit on the margin are horned, and stand very remote, (Schk. t. 235.) The genus *Calendula* is related to few
others, principally to *Melampodium* and *Silphium*.

Synonymes and Figures.

Calendula, *Brunf.* 3. 77. *Dodon.* 254. i.—viii. *Tabern.*
711, 712. i.—vii. *Ger. em.* 739.
Ringelblumen, *Trag.* f. 55.
Caltha, *Fuchs.* 382. *Matth.* 894. *Dalech.* 811. *J. Bauh.
Hist.* 3. 101.
Chrysanthemum et Caltha poëtarum, *Lobel. Hist.* 298.
ic. 552.
Clymenum Colum. *Phytob.* t. 13.
Calendula officinalis *Linn. Willd. Sp. Pl.* 3. 2340. *Sturm.* 8.

Geographical Distribution.

Although this plant propagates itself in the gardens of
Germany, it is properly, however, a native of the south of
France, where it grows in the fields. It does not seem to
grow in the other southern countries of Europe.

Uses.

In the sixteenth century, this plant, from its strong smell,
was reckoned medicinal, and was employed in the case of
diseased female organs, especially in cancer, (*Matth. Valgris.* 628) This practice has been lately renewed by West-
ring, (Erfahrungen über die Heilung der Krebsgeschwure.
Aus. dem Schwed., Halle 1817, 8vo.) Exact chemical analyses have been given by Geiger (Diss. de *Calendula officinale,* Heidelb. 1818.), and by Stolze (Berlin. Jahrb. fur
die Pharm. 1820, s. 282. f.) According to the analysis of
the latter, its principal constituent parts are green vegetable
wax, albumen, lime united to malic acid, Myricin and Calen-
dulin, (a peculiar matter which might readily be mistaken
for jelly).

ORDER V.

Cynareæ. (Polygamia segregata.)

31.

Echinops sphærocephalus, *L.*

Kugeldistel.—French, *Echinope, Boulette.*—Engl. *Globe-thistle.*—Swed. *Bol-tistel.*

This plant is frequently observed in hedges and bushes, by the way side, and in rocky places. From a woody tap-root, the stem rises to the height of two ells, frequently to the height of a man. It is of the thickness of the finger, angular, completely covered with wool, and with a glutinous moisture, which may also be observed on the leaves. The leaves are alternate, short-stalked, the upper ones without stalks, a large span in length, frequently still longer, deeply semi-pinnate, having their upper surface green and hairy, their lower entirely white, covered with a woolly tomentum : the laciniæ of the leaves are angular, and terminate in thorns. The upper part of the stem is divided into several branches, on the tips of which grow the compound flowers, resembling bluish spheres of the size of a middling-sized apple. The spherical receptacle is properly naked, (*Meese Het.* xix. class, t. 3. f. 5.), and a calyx for each floret is formed of chaffy leaves, which are stiff and hairy. At the base of these calyces bristles appear, which are attached to them, and upon superficial observation seem to belong to the receptacle. The florets are all similar, tubular, with a quinque-partite margin, and of a whitish-blue colour. The cylinder of the antheræ is violet coloured : the stigma is divided. The æstivation is valvular, and the florets expand from the centre of the sphere towards the circumference. The roughly ciliated seed (a caryopse), is surmounted by a membrane ; (*Berkhey, Expos. Fl. Comp.* t. 3. f. 18.)

Diagnosis and Affinity.

The most nearly related to this species is *E. exaltatus* Schrad., which is distinguished, however, by wanting branches, by the absence of the glutinous integument, and by having the teeth of the laciniæ of the leaves more closely set; (Schrad. Hort. Gott. t. 9.) *E. paniculatus* Jacq., also is nearly related to it, and is distinguished by its very branchy stem, by having its flowers in panicles, and by its wrinkled leaves, the upper surface of which is without hairs, the under surface covered with a bluish-green tomentum, and the laciniæ of which are squarrose; (*Jacqu. Eclog.* t. 48.) *E. Ritro* L., has likewise its leaves smooth on the upper surface, and covered with a snow-white tomentum on their lower, and it has five flowers on one stem; (Schk. t. 269.) *E strigosus* L., has its leaves strigose on the upper surface, and tomentose beneath. The flower-tops are not spherical, but properly form tufts, the exterior calyx of which is protracted and abortive; (*Herm. Parad.* 224.) This species affords an opportunity of taking an important view of the inflorescence: it is a compound, compressed spike, the flowering of which is always from above downwards; (s. 74. Neue Entdeck. 1. 174.) The genus *Echinops* is related to the Boopideæ of Cassini, or the Calyceræ Br., to which belong *Boopis* Juss., *Calycera* Cav., and *Acicarpha* Juss. These also have a calyx for each floret, but they have a receptacle furnished with chaffy leaflets, and a common calyx. They are distinguished from all the other syngenesious plants by their simple pistil, which is united to the tube of the corolla, by the rich albuminous matter in the seed, and by the direction of the radicle upwards. In *Echinops*, on the contrary, the structure of the seed corresponds with the family character.

Synonymes and Figures.

Chamæleon verus, *Trag.* f. 322.   *Fuchs.* 883.
Carduus sphærocephalus, *Dodon.* 722.   *Tabern.* 1069. Acutus major, *Park.* 977.   *Besler. Syst. Æst.* xi. t. 7. f. 1. Latifolius major, *Moris.* sect. 7. t. 35.

Spina alba altera, *Matth.* 494.
Ritro, *Lobel. Hist.* 481. ic. 2. 8.
Chalceios, *Dalech.* 1483.
Spina Arabica, *Dalech.* 1467.
Echinops major, *J. Bauh. Hist.* 3. 69. *Tourn. Inst.* 463
Carduus globosus, *I. Ger. emac.* 1151.
Echinops sphærocephalus, *Linn. Sp. Pl.* 1314. *Willd. Sp.*
  Pl. 3. 2396. *Lam. Illustr.* 719.

Geographical Distribution.

This plant is diffused from the north of Africa (30° N.
Lat.), over all the temperate regions of Europe. But Barby, on the Elbe, (52° N. Lat.), seems to be its northern
limit. It grows, indeed, here and there, even in Sweden,
beyond Liljeblad ; but, as it is not mentioned by Linnæus, it
has probably been propagated from the gardens.

——◆——

CLASS XX.

Gynandria. (Orchideæ.)

32.

Ophrys myodes, *Jacqu.*

Insecten-Ragwurz.—French, *Ophrys mouche.*—Engl. *Fly-*
orchis.—Swed. *Flug-blomster.*

This beautiful and interesting plant grows sparingly in
June, in our mountain woods, on clay soils. From a tuber,
of a yellowish-brown colour, and of the size of a hazel-nut,
by the side of which a second commonly stands, and on the
top of which several fibrous roots are expanded, arise, in the
first place, convoluted, whitish sheaths, afterwards three or
four leaves, embracing the stem, oblong lanceolate, smooth on
both sides, quite entire, penetrated by parallel nerves, some-

E e

what pointed at the top, about the length and breadth of a
finger, in the centre of which the round, smooth stem, of the
thickness of a strong packing thread, rises straight, erect, to the
height of a foot. The blossoms are arranged on the top of the
stem in a spike, but distant from one another, seldom to the
number of six. Among them we observe erect, whitish-green,
small, nearly linear bracteæ, a little longer than the flowers.
These last consist of a three-leaved calyx, which externally
is green; internally, and at a later period, it is brown. The
leaflets are oblong, somewhat obtuse, and penetrated by three
nerves. The labellum is four-lobed or three-lobed, with the
central lobe emarginated or divided; the lateral lobes are
somewhat distant, but they are all obtuse. The whole labellum is reddish-brown, or of a rusty colour, hairy, ciliated,
and has a bluish spot in its centre, by which means the
whole flower resembles a fly. Above it, the reddish-brown
fruit-column rises, with two linear, remote, lateral horns, of
the same length. Two yellow pollenous masses, each of
which is divided into two parts, rest on stalks, which are attached to the fruit-column by small spheres; on the upper part
they are surrounded by two folds, from which they gently issue,
when they are fully ripe. The stigma, on its lower side, is
splendent with a glutinous moisture; (s. 90, 91.) The germen is cylindrical, a little twisted, stands below the fruit, and
opens with three valves, which last are connected by particular ribs, and carry on their sides the exceedingly fine, small
seeds surrounded by a spongy membrane.

Diagnosis and Affinity.

The most nearly related species are *O. aranifera* and *apifera*. But the latter, the flowers of which have a great resemblance to a bee, are distinguished by their broader and
shorter stem-leaves, by three large pale-red calyx leaflets, and
two others that are small, green, and fringed. The labellum
is brownish-red, with a yellow hairy setting, and the central
lobe is turned backwards with a small process. The fruit
column passes above the antheræ into a distinct rotellum. It
flowers in July. Three months earlier, in April, appears

O. aranifera, the flowers of which entirely resemble a spider. The leaflets of the calyx are all obtuse, and of a yellowish-green: the labellum is brownish-yellow, strongly set with hairs, three-lobed, with the margin bent inwards; the fruit column terminates in imperceptible rostella. *O. arachnites,* which bears the greatest resemblance to *O. apifera,* has also a brown, hairy labellum, but this part is three-lobed, the central lobe being again divided into three obtuse lobes. *O. tenthredinifera* Desfont., which also grows in Calabria and Sicily, has very long, rose-coloured bracteæ, three obtuse oblong calyx-leaflets, and two very short: the labellum is two-lobed, with a process between the two lobes.

The genus *Ophyrs* is most nearly related to *Epipactis* Sw.; but the latter is distinguished by the jointed structure of the middle lobe of the labellum, and by its four round pollenous masses. Richard distinguishes *Ophyrs Monorchis* as a peculiar genus, under the name *Herminium,* the character of which consists in the short, spur-shaped sack of the hastate labellum, and in the naked large retinaculum. Respecting the family of the Orchideæ, vid. Anleit. ii. 280. f. 880. f.

Synonymes and Figures.

Orchis Serapias tertius, *Dodon.* 238.
O. myodes, i. *Lobel. Hist.* 90. ic. 181. *Ger. em.* 213. *J. Bauh. Hist.* 2. 767, 768. *Park.* 1352.
Triorchis Serapias tertius, *Dalech.* 1555.
Testiculus muscarius, ii. *Tabern.* 1050.
Orchis muscæ corpus referens, *Bauh. Pin.* 83. (excl. synon., quæ hic non sunt.) *Rudb. Elys.* 2. 201. f. 11. *Vaill. Bot. Paris.* t. 31. f. 17, 18. (flores soli.) Minor, *Tourn. Inst.* 434.
Ophrys insectifera α. myodes, *Linn. Sp. Pl.* 1343. *Gunner. Fl. Norw.* 2. t. 5. (icon gigantea.)
Orchis n. 1265. *Hall. Stirp. Helv.* t. 24.
Ophrys myodes, *Jacqu. Misc.* 2. 373. *Ic. Rar.* t. 184. *Fl. Dan.* 1398.

O. muscifera, *Smith. Fl. Brit.* 3. 937. *Engl. Bot.* 64.
Withering, Arrang. 2. 43.

Geographical Distribution.

The northern limit of the growth of this plant is another
proof of the principle formerly laid down, that in the western
countries, plants are found at a higher latitude than in the
eastern, on account of the warmer temperature. In Sweden,
O. myodes is found only in Gothland and Oeland, (57° N.
Lat.) In Norway, on the contrary, it is found, according to
Gunnerus, at Snaasen, (64° N. Lat.) ; and, according to *Fl.
Dan.*, it is even found on the island Langoe, (69 N. Lat.)
It is diffused throughout England, France, Germany, Hun-
gary, Italy, and Transylvania. Its southern limit seems to
be Peloponnesus, (38°.) The four varieties mentioned by
Desfontaine (Fl. Atl. 2. 320.), under the name *Ophyrs insec-
tifera*, do not belong to this. I have also a doubt, whether
O. myodes of Hagen (Pruss. Flor. 2. 215.), be really our
plant. Its eastern limit northward would then be 20° W
Long., southward 25°.

CLASS XXI.

33.

Sparganium simplex, *Huds.*

Einfache Igelsknospe.—French, *Rubanier simple.*—Engl.
Simple Bur-reed.—Swed. *Rak-trägjan.*

This peculiar plant flowers in July and August, in our
standing waters and ditches, especially where the bottom is
gravelly.   From a creeping, perennial, fibrous root, arises
about perhaps a foot or a foot and a half high, a round,
green, smooth stem, which is wholly undivided, and about the
thickness of a quill.   All the leaves embrace half the stem with

a sheath, which is membranaceous on the sides; the lower leaves, when cut across, are triangular, with smooth interior surfaces: the upper are somewhat concave, and do not exhibit a triangular section. Besides, the leaves are almost all longer than the stem, frequently two feet high, and scarcely the breadth of a little finger, uniformly small throughout, smooth and entire, tapering at the point, and furnished with parallel soft nerves. Their cellular texture is spongy and compound; the cells full of air. There are slits on both sides of the leaves. In their centre rises a simple flower-stalk, which carries below two petiolated spherical flower-tufts, and above one stalkless female tuft, and several stalkless male spherical flower tufts. The individual female florets consist of three or four lanceolate scales or leaflets, in the centre of which rises, on an oval germen, the simple, green, sometimes cleft pistillum, having the stigma placed laterally at its summit. The male flowers contain, in the centre of the somewhat spoon-shaped scales, imperceptibly dentated at the top, commonly three filaments of a white colour, on the top of which stand the bilocular straw-yellow antheræ, containing an oval pollen. The fruit is a brown nut, or drupa, containing in the centre of the albuminous matter, the unevolved embryon in a reversed position.

Diagnosis and Affinity.

The most nearly related species is *Sp. ramosum*. But this species is much larger, its flower-stalk is branchy, the sides of the leaves are concave, not smooth. The scales of the calyx are also of a deeper brown colour. *Sparg. natans*, on the other hand, has leaves entirely of a grass shape, swimming on the surface of the water, rather concave, and very long; the flower-tuft is much smaller, and only the one that is uppermost is male. This genus evidently borders on *Typha*. I also find *Chrysithrix* related to it, the divided shaft of which pushes out laterally the flower-tuft. *Acorus* and *Orontium* form the transition to the Aroideæ, to which these plants belong; (Anleit. ii. 127.)

Synonymes and Figures.

Platanaria altera, *Dodon.* 601.

Sparganium alterum, *Lob. Hist.* 41. ic. 80. *J. Bauh. Hist.* 2.
541. *Dalech.* 1019. *Tabern.* 560.

Sp. latifolium, *Ger. emac.* 45.

Sp. non ramosum, *Park. Theatr.* 1206. *Moris.* sect. 8.
t. 13.

Sp. foliis natantibus planoconvexis, *Linn. Fl. Lapp.* ed. 2.
p. 280. *Fide Smith.*

Sp. erectum, *Linn. Var. β. Sp. Pl.* 2. 1378.

Sp. simplex, *Huds. Fl. Angl.* 401. *Engl. Bot.* 745. *Schk.*
t. 282.

Sp. americanum, *Nuttall,* 2. 203. Admunerandum juxta
descriptionem.

[Sp. maius S. ramosum virginianum, *Park. Theatr.* 1206.
repet. in *Moris.* sect. 8. t. 13. non Sparganii species, sed
forte Carex lupulina W ?]

Geographical Distribution.

Sp. simplex is a northern plant. According to Linnæus's
description and Smith's assertion, n. 345. in the *Fl. Lappon.*
can be nothing else but this plant. Yet Wahlenberg says
(*Fl. Lapp.* p. 822.), it is *Sp. natans;* and *Sp. erectum* (by
which he understands *Sp. ramosum* and *simplex*) does not
grow beyond Medelpadia 63° N. Lat. How shall we re-
concile such evident contradiction of two equally credible eye-
witnesses ? As Linnæus says respecting this plant, that it has
from ten to twelve male flower-tufts at its top, it cannot be
Sp. natans. Linnæus found this plant in the Great Calix-
elf, (67°), where, according to our opinion, Wahlenberg
never was. J. G. Gmelin found it in Siberia, in the Jenisei,
and in the province of *Isezk.* In North America it is found
every where as far as the river St. Lawrence, (50°). It does
not seem to extend far south, for it is not met with either in
Greece or Tauris; but it is found in Transylvania and the
south of France, (45°).

CLASS XXII.

34.

Salix Caprea, *L.*

Sohlweide.—French, *Saule marceau.*—Engl. *Great round-leaved sallow.*—Ital. *Salcio a grande foglie.*—Swed. *Sälg-pihl.*

As the willows of all known plants shew least constancy in their forms, so the Salix Caprea, is perhaps of all willows, that which is subject to the greatest variations of form. Situation has a powerful effect upon it; for commonly it grows on a somewhat dry soil, but frequently also upon a moist and boggy one, where it undergoes a remarkable change in the size of the stem, and in the shape of the leaves. But even on the same soil we observe such differences in its structure, and particularly in the form of the leaves, that those persons may easily be excused who have regarded such different forms, when they have seen them single, and in dried specimens, as peculiar species. We are most secure, when we hold to those characters which never vary. These are— 1. Early catkins, which appear before the leaves. 2. Female catkins, short and thick. 3. The germen tomentose, or covered with silky hairs, and also ventricose. 4. Broad, almost ovate or oblong leaves, quite entire when they are young, afterwards dentated and undulated, either green or hairy above, but always tomentose, and intersected by reticular veins below. 5. Lunulate dentate stipulæ, which are either persistent or fall off. 6. Smooth filaments.

The size varies uncommonly. The Salix Caprea is usually a moderate sized tree, from eight to ten feet high: frequently it is a branchy shrub, with a greyish-brown, pretty smooth bark. The shoots of the present year are commonly dark-brown and hairy, when they are viewed late in the summer. The leaf-stalks are alternate, strongly ciliated, and even tomentose, from two to four lines in length, with roundish or lunulate, dentated stipulæ at their base; and in the axillæ of the upper leaves are found the buds of the coming year, when they are examined in the middle of summer. The leaves are about three inches long, and half an inch broad; tapering slightly at their base, rapidly at their summit, dentated on the margin, the teeth being bent towards the point The upper surface is for the most part green, properly, however, not smooth, but partly wrinkled, partly furnished along the nerves and veins with white, soft, short hairs. Hairs also are found on the interstices, when we avail ourselves of the aid of the microscope. The lower surface is more or less strongly ciliated; often it is covered with silky hairs, often also it is shaggy, and even tomentose.

The male and female flowers appear, in April, in catkins on different trees, on the smooth shoots of the preceding year. Both of these spring from buds, the splendent silky scales of which are persistent, and give a pleasant appearance to the flowering, but still leafless tree: both of them, during the time of flowering, are scarcely an inch long, obtuse, and oblong; the female flowers afterwards increase to three inches in length. The scales of the male catkins are brown, oblong, and ornamented with long soft hairs. Two filaments, which are longer than the scales, carry quadrilocular yellow antheræ, and have at their base a longish, nearly cylindrical honey-gland. The female flowers have the same scales and glands, and also a petiolated, strongly ciliated germen, the lower part of which is swollen, and crowned with three or four short stigmata. The fruit is a bivalved capsule, having several seeds attached to the inner surface of the valves, and surrounded from the base upwards by long soft hairs.

Varieties.

1. This species is not unfrequently found almost quite entire, and without perceptible teeth, whence we might be

tempted to consider it as a peculiar species. But, when transplanted, it shews the transition by distinct teeth on the margin of the leaves. As the tip of the leaf is often withered and discoloured, Smith has named it *S. sphacelata.*

2. On the young shoots, which spring from the cut root-stems, the leaves are uncommonly long and broad. They are sometimes seen more than six inches long, with very long tapering points, and the stipulæ very large, cordate, and strongly dentated ; (*Sal. tomentosa*, J. *macrophylla*, Ser. p. 17.)

3. This species grows often with small, almost lanceolate leaves. In this case, the germen also is usually drawn out to a greater length. This is S acuminata, Mill., Hofm., Smith., Willd. But this variety has many evident transitions into the usual form of *S. caprea.*

4. With completely round, and even with cordate leaves, which scarcely taper at the point. This is *S. tomentosa*, H., *rotundifolia*, Serv. p. 17.

5. With oblong leaves, the lower surface of which is greyish, and furnished with a few hairs, (S. aquatica, Smith). The moist situation seems to produce this variety.

6. With androgynous catkins, (S. Tinimii, Schk. 3. s. 457. *S. tomentosa* D., *androgyna* Ser. p. 16.), Vorgl. s. 322. Numerous malformations also are produced by the puncture of insects, and by parasitical plants.

Diagnosis and Affinity.

This species is most nearly related to *S. aurita* L., (*S. rugosa* Ser.), and as both of them are alike various in their aspects, they approach each other in their forms. But *S. aurita* is principally distinguished by its wrinkled leaves, which taper at the base, and have their tips drawn obliquely (*mucrone adunco*), by its great cordate stipulæ, by its hairy filaments, united at their lower part, by its long tapering germen, and by its low growth, seldom exceeding six feet, commonly only from two to four feet in height. To *S. aurita* L., belong, as varieties, *S. ambigua* Ehrh., *spathulata* W., and *uliginosa* W.

The second species, with which *S. caprea* might easily be confounded, is *S. grandifolia* Ser. In particular, it has a great resemblance to the third variety. But the principal character of *S. grandifolia* consists in this, that the catkins appear at the same time with the leaves, whilst, in *S. acuminata*, they appear earlier: that the leaves are much larger, often six inches long, and properly lanceolate, nearly quite entire,—the stipulæ large, pointed, and semi-cordate,—and the honey-gland drawn out to a great length. *S. stipularis* Sm., does not belong to this species, because its catkins appear much earlier.

S. patula Ser., *oleæfolia* Vill., has also some resemblance to the small-leaved variety of *S. caprea*, but the catkins appear late, first in May, when the leaves come forth. The other species are less liable to be confounded.

The natural affinity of Willows to Poplars is obvious : the distinction of genera lies in the flowers, which, in the case of Willows, consist only of simple scales. But in Poplars, beside scales, there is also a funnel-shaped corolla, with eight or more filaments. Both belong to the catkin-bearing trees, or the Amentaceæ ; (Anl. ii. 344.)

Synonymes and Figures.

Seilweiden, *Trag.* f. 406. a.

Salix aquatica, *Lobel.* ic. 2. 137. (var. 3.)

S. platyphyllos leucophloeos, *Dalech.* 276.

S. caprea latifolia, *Tabern.* 1452. *Ger. em.* 1390.

S. latifolia inferne hirsuta, *J. Bauh. Hist.* 1. p. 2. 215. Rotunda et oblongior, *Park. Theatr.* 1432. *Bauh. Pin.* 474. *Rai Syn.* 449. *Tourn. Inst.* 591.

Salix foliis obscure crenatis, *Linn. Fl. Lapp.* n. 365. t. 8. f. 1.

S. caprea, *Linn. Sp. Pl.* 1448. *Fl. Dan.* 245. *Engl. Bot.* 1488. *Hofm. Sal.* t. 3. f. 1, 2. t. 5. f. 3, 4. t. 21. f. a, b, c, d. *Schk.* t. 317. c. n. 15.

S. foliis ovatis rugosis, *Hall. Stirp. Helv.* n. 1653.

S. acuminata, *Hofm. Sal.* t. 6. f. 1, 2. t. 22. f. 2. *Engl.*

Bot. 1434. (var. 1.) *Ser. Sal.* p. 12. *Schk.* t. 317. c. n. 12.

S. sphacelata, *Engl. Bot.* 2333. (var. 3.)
S. lanata, *Lightf. Scot.* 602. (var. 3.)
S. aquatica, *Smith, Fl. Brit.* 3. 1065. *Engl. Bot.* 1437. (var. 5.)
S. tomentosa, *Ser. Sal.* p. 14.

Geographical Distribution.

Few trees have so extensive a distribution. It grows throughout the whole of Europe, from the forests of Arcadia (37°), to the woody heights of Kautokeimo, ont he Alten-Elf, in Lapland, (69°). It passes also into Siberia, where it is found abundantly in all its varieties. But it is not a native either of Japan or of North America.

Uses.

The bark contains tanning matter, and is hence used in the preparation of leather; and in *Smaland* it is employed in the manufacture of the gloves called *Klipping.* The bark is also used in the making of Danish gloves, and of Russian leather. The bark of the young shoots is used, instead of Peruvian bark, in medicines for the poor. It operates simply as an astringent and tonic, and, as it wants cinchonin, and the spicy matter of Peruvian bark, it may be used in common intermittent fevers, with ammonia, and to prevent inflammation. In other diseases it lies too heavy on the stomach, and must be taken in too great quantity, if it is expected to operate. The bark also contains colouring matter, which may be applied to woollen cloth, that has previously been treated with bismuth, and produces a beautiful apricot-yellow colour. Linen yarn is dyed black by it, when it has previously been mixed with alder bark.

The wood is very tough, and is easily cleft. Hence in Thuringia, sieves, and other plaited utensils, are constructed of it. Ray maintains, that very good straps are made of it, on which knives may be sharpened. It is also used for making handles to knives, and other instruments. Although it

3

properly affords no good burning wood, because it consists
almost entirely of alburnum, and the heat soon passes away,
it yet affords a light charcoal, which readily takes fire, and can
hence be used, in particular, for gun-powder. The tree is
also well adapted for forming quick hedges, because it grows
rapidly, because its branches can readily be twisted, and in
early spring the flowers are anxiously sought for by bees.

The leaves are eagerly eaten by all sorts of cattle. In
Sweden, calves are foddered with them.

The wool of the seeds is often used as native cotton; (Her-
zer's Gesch. der hierlandischen Baumwollenarten. Munchen,
1788. Rafn's Danmarks Flor. 1. s. 417. f.)

CLASS XXIII.

35.

Atriplex patula, *L.*

Sparrige Melde.—French. *Arroche étalée.*—Engl. *Spreading
orache.*—Ital. *Atrepice spalancante.*—Swed. *Gull-frö, Aker-
molla.*

This plant grows in August and September, commonly
on salt soils, or on fatty soils, on heaps of rubbish and dung.
From a short, thin stake-shaped root, rises directly upwards a stem, which for the most part is thin and angular. This pushes out its branches from below upwards nearly in a horizontal, and therefore a squarrose direction on
all sides, and is about an ell high. The leaf-stalks are
half an inch long, somewhat concave and squarrose. The
leaves are triangular, hastate, tapering at the base, furnished with three nerves, and set on both sides, but especially on the lower side, with whitish scales, which give the leaf
a brilliant appearance. The margin of the leaf, for the most
part red, terminates below on both sides in two wing-shaped
distant points, which produce the hastate form. Towards the

point are projecting teeth, and sinuses between them. The latter disappear in the uppermost leaves, which are hence simply
hastate, and very small. The frequently red coloured flowers are in fasciculi or glomeruli, which form spikes at intervals, and are separated by leaves. The flowers are partly
male, partly hermaphrodite, partly female. The two former
consist merely of a quinque-partite calyx, with five filaments,
and the same number of yellow bilocular antheræ. The hermaphrodite and the female flowers have two linear pistils.
The calyx of the female flowers consists of two thick leaflets,
which are triangular, have long projecting points, dentated
margins, and herbaceous warty spines on the surface. This
calyx also is covered with whitish scales. Between these
elastic leaflets of the calyx, lies the fruit, a caryopse, which
contains the evolved, bent embryon around its circumference,
and in its centre the remainder of the albuminous matter.

Diagnosis and Affinity.

This species has the nearest affinity to *A. angustifolia* Smith,
with which it has also been frequently confounded. But the
angustifolia is much more common in central Germany, grows
every where by the road side, is much taller, has not the
reddish colour nor the crowded stem, but has its branches at
a distance from each other. Fewer scales are observable, and
the lowermost leaves only are hastate, those farther up being
lanceolate and quite entire. The leaflets of the calyx are
nearly smooth, at least they have only some small bunches on
the margin.

A. hastata L. is also frequently confounded with our species.
This indeed has similar leaves, but the leaflets of the calyx
are intersected by reticular veins, and have bristly teeth.
A. laciniata L. is less nearly related : it grows only by the
sea-shore, and is entirely covered with white and red spots;
the stem lies low ; the leaves are more oblong, sinuated, and
dentated; the leaflets of the calyx are enlarged, and have
strong warty bunches. *A. nitens* Schk. has similar leaves,
but it grows much taller, and has quite smooth, and entire calyx leaflets.

Atriplex is essentially distinguished from *Chenopodium* by its polygamous flowers, and by the two-leaved calyx of the female flowers ; (Anleit. ii. 308.)

Synonymes and Figures.

? Atriplex sylvestris, *Dodon.* 615.

Atriplicis marinæ species Valerando, *J. Bauh. Hist.* 2. 974.

Atriplex folio deltoide triangulari sinuato, *Moris.* sect. 5. t. 32.

A folio hastato s. deltoide, *Tourn. Inst.* 585. *Hist. des Plantes aux Envir. de Paris*, p. 10. *Magnol. Bot.* 34.

A foliis sagittato-lanceolatis, *Linn. Lapp.* n. 377.

A foliis triangularibus basi productis, *Hall. Stirp. Helv.* n. 1617.   A caule herbaceo, valvulis fœmineis magnis deltoidibus, *Ger. Prov.* 329.

A. hastata, *Linn. Fl. Suec.* n. 921. (Manifesto nostra, cum Morisonii synonymon citet.   Hinc confusio posteriorum, quæ tandem herbario Linnæano soluta est.)   *Scop. Carn.* n. 1245.   *Pollich. Pallat.* n. 942.   *Scholl. Barb.* n. 809. *Leyss Hall.* n. 1015.   *Schk.* t. 348.   *Huds. Angl.* p. 443. *Lightf. Scot.* p. 636.   *Gouan, Fl. Monsp.* 433.   *Host, Austr.* 545.   *Rafn. Dan.* 2. 238.   *Vill. Delph.* 2. 566.   *Lam. Enc.* 1. 275.   *Fl. Dan.* 1286.   *De Cand. Fl. Franc.* 3. 386. (Omnes hi aliique auctores, si A. patulam enumerant, A. angustifoliam intelligunt.)

A. patula, *Linn. Herb. Smith, Fl. Brit.* 3. 1091.   *Eng. Bot.* 936.   *Fl. Dan.* 1285.   *Marsch. Bieb. Taur. Cauc.* 2. 443.   *Spreng. Hal.* n. 293.

A. laciniata, *Besser Gallic.* 1. 194.

Geographical Distribution.

This species is distributed from 44° to 64° throughout the whole of Europe ; for Tauris, Bologna, and Montpellier, seem to be its southern limits ; Umea and Angermanland its northern.   How far it stretches eastward is not clear, because the Siberian plants which pass under this name are still doubtful.

CLASS XXIV

I. *True Ferns.*

36.

Blechnum boreale, *Sw.*

Nordlicher Rippenfarrn.

This handsome fern grows in our woods. The root is about the thickness of a little finger, of a brownish red colour, covered with scales, chaffy leaves, and the remains of old stalks. It pushes downwards fibrous roots, which are expanded on all sides. From the root arise the fronds, at first of a snail-shape, and every where set with hairs and chaffy leaves. Afterwards these last-mentioned substances disappear in a great measure, so that the mature stalk is only furnished at its lower part with a few scattered chaffy leaflets. The fronds are partly fertile, partly barren. These last are from a large span to a foot in length, lanceolate, deeply half-pinnated. The laciniæ alternate in such a manner, that one is always in the centre of two others that are opposite to it, and, when seen from above, it seems to unite with them. The laciniæ are from half an inch to eight lines in length, quite smooth and entire, somewhat falcated, pointed or somewhat obtuse, and penetrated by a principal nerve and parallel veins. The lower laciniæ are always more obtuse, rounder and shorter, the wider they stand; the uppermost unite into one entire point. The stalks of these barren shoots are sharp, angular, yellowish brown below, and become whitish where the frond begins; but sometimes also the yellowish brown colour stretches a little higher. The barren fronds are always green and lie in a circle upon the ground. The stalks of the fertile fronds rise in the centre of those that are barren : they are

dark brown to the point, compressed, and an ell in length.
The fronds are also lanceolate, pinnated below, deeply half-
pinnated above. The leaflets are linear, opposite, and distant or alternating, pointed, half an inch or more in length.
On both sides of the middle rib, on the back, lie the sori in continuous lines, and are covered by a membranaceous indusium, which opens towards the middle rib. When
young, the capsules are petiolated, and have sap-tubes between them. Afterwards they become brown, consist of a reticular membrane, and are surrounded by an elastic, jointed
ring, by the springing of which, rough, angular seeds are
thrown out. In a state of greater maturity, the whole back
is so covered with sori, that the more early structure is no
longer distinguishable : hence the plant can now be readily
classed with the genus *Achrostichum.*

Diagnosis and Affinity.

The older fronds can scarcely be distinguished from *Acrostichum,* but by attending to the presence of the indusium, which
is always persistent at the margin of the leaflets ; but in *Acrostichum* it is wanting. The genera *Struthiopteris* and *Lomaria* Willd. might also be confounded with this plant, were it
not that in the former the indusium is formed by the margin
of the frond, and is laid in the form of scales over the sori.
In *Lomaria* the two margins of the leaf unite continuously
over the sori, (*Billard. Nov. Hall.* t. 246.) Compare Anleit.
2. 101.

Synonymes and Figures.

Walt-asplenon, *Trag.* f. 208. b.
Lonchitis aspera minor, *Matth.* 661. *Dodon.* 469. *Dalech.*
    1221. *Ger. Emac.* 1140. *Park.* 1042.
Lonchitis altera Neotericorum, *Clus. Hist.* 2. 213. *Lobel.
    Hist.* 475. *Ic.* 815. Fœmina, *Tabern.* 1190. *J. Bauh.
    Hist.* 3. 737.
Asplenium sylvestre, *Dalech.* 1217.
Lonchitis minor, *C. Bauh. Pin.* 359. Altera foliis Polypo-
    dii, *Moris.* sect. 14. t. 2.

Polypodium angustifolium folio vario, *Tourn. Hist. des Plantes aux Envir. de Paris*, p. 519. *Inst.* 540.

Spicant Tragi et Germanorum, *Rupp. Ien. ed. Hall.* 346.

Acrostichum Osmunda, *Linn. Sp. Pl.* 1522.

Osmuda spicant, *Linn. Fl. Suec.* n. 936. *Fl. Dan.* 99. *Bolt. Fil.* t. 6. *Gouan Fl. Monsp.* 439. *Lightf. Scot.* 2. 654.

Struthiopteris, *Hall. Hist.* n. 1687. *Böhm. Fl. Lips.*, 296.

Str. Spicant, *Scop. Carn.* n. 1258. *Weis, Crypt. Gott.* 287.

Acrostichum Spicant, *Vill. Delph.* 4. 838.

Acr. nemorale, *Lam. Enc.* 1. 35.

Blechnum Spicant, *Roth. Germ.* 3. 44. *De Cand. Fl. Franc.* 2. 551.

Onoclea Spicant, *Hofm. Germ.* 2. 11. *Liljibl. Fl. Suec.* 385.

Blechnum boreale, *Sw. Syn. Fil.* 115. *Eng. Bot.* 1159. *Schk Fil.* t. 110.

Geographical Distribution.

This fern grows in Europe from 43° to 60°. The south of France and Genoa (Bertol. Amœn. Ital. 212.) seem to be its most southern stations; the south of Norway and Scotland its most northern. Whether it grows in Siberia, I know not; but Lewis found it on the north-west coast of America, (*Pursh, Amer. Sept.* 669.)

Uses.

The power of healing wounds has been ascribed to this plant. It is also mixed, in some places near Jena, with beer, to increase its salutary qualities, (*Rupp. Fl. Ien.* p. 346.)

II. *Pteroida.*

37.

Botrychium lunaria, *Sw.*

Mondkraut.—French, *Lunaire.*—Engl. *Moonwort.*—Swed *Läsgräs.*

This plant grows with us in June, on stony, almost on
barren places, and woody heights. The root consists of thick
brown fibres, of the thickness of a pack-thread. These send
out first two sheath-shaped leaflets, and from them a smooth,
round, herbaceous stem, a large span high at most, and of
the thickness of a pigeon's quill. This stem stands quite
erect, and is divided at about two inches high, pushing out
laterally a pinnated frond, about a small finger long. This
frond consists of nine or ten dull green, flabelliform (29.) leaflets, which stand alternate, are irregularly crenated on their
convex margin, and penetrated by radiated fine nerves. They
are about the size of the nail on the little finger. Sometimes
the capsules shew themselves on the margin of these leaflets;
not unfrequently the frond sends out, besides the principal
fruit-stalk, one or two subordinate stalks, which are formed
in the same manner. The principal fruit-stalk forms a branched, compound spike, with angular, somewhat open branches,
on which the yellowish brown, spherical, smooth fruit is placed,
without stalks, for the most part on one side. These spherules, of the size of mustard seed, split transversely when
they are ripe, and scatter their fine seeds. The plant springs
from these, like a green, lobed, cellular texture.

Diagnosis and Affinity.

This species is related to two others of more rare occurrence,—B. rutaceum Sw. and matricarioides Willd. The former has a bi-pinnated frond, the extreme lobes of which are
obtusely dentated. In this species the frond comes along with
the stem out of the root-sheath; the frond is tri-partite, bi-
pinnate, and the extreme lobes are oblong and obtuse. The
latter species is very rare: I have it from Courland. The ge-
Botrychium borders on Ophioglossum, which is distinguished
by its simple spike: together they form the tribe of Stachy-
opteridæ, under the Pteroidæ, (Anl. ii. 107.)

Synonymes and Figures.

Lunaria minor, *Fuchs.* 482. *Dodon.* 139. *Matth.* 647. *Dalech.* 1313. *Ger. Em.* 405. *Park.* 507. *Moris.* s. 14. t. 5.

L. racemosa, *Lob. Hist.* 470.   *Ic.* 807.   Botrytis minor, *Clus.*
Hist. 2. 118.   *J. Bauh. Hist.* 3. 710, 711.
Ruta lunaria, *Tabern.* 413.
Osmunda foliis lunatis, *Tourn. Inst.* 547.
Osmunda lunaria, *Linn. Sp. Pl.* 1519.   *Fl. Lapp.* n. 389.
Fl. Dan. t. 18. f. 1.   *Sturm, Fl.* 1.   *Eng. Bot.* 318.
Botrychium lunaria, *Sw. Syn. Fil.* 171.   *Willd. Sp. Pl.* 4.
61.   *Schk. Fil.* 154.

Geographical Distribution.

In the north this plant extends not only to Iceland and
Lapland, where it is found in those crevices of rocks which
are turned to the sun, but in Western Finnmark, as far as
the Island of Masöe, (70° N. Lat.)   How far south it extends, I know not exactly, but it grows at Montpellier and in
Calabria, but not in Greece.

As the plant grows and fades so quickly, without leaving a
trace behind it, it was formerly believed, that during the increase of the moon it sprung up, and that with the waning of
of the same it died.   Hence the alchemists of the middle ages
made use of it in their researches ; (*C. Gesner de herbis, quæ
Lunariæ nominantur. Tiguri* 1555. 4.)

III. *Musci Frondosi.*

38.

Cinclidotus fontinalioides, *Pal. Beauv.*

This beautiful moss grows on stones and wood in our
streams and flowing waters.   It sends out brown fibres into
the mud with which the stones or wood are covered ; and
from thence rise several branched stems, from four to six
inches long, of the size of a linen thread, and of a green colour, and covered from below upwards with leaves.   These
last-mentioned parts grow close together, without lying like
tiles on one another : they half embrace the stem and branches,
are oblong-lanceolate, quite entire, a little tapering at the

point, of an olive-green colour, which, by drying the part, becomes dark green. The texture of the leaves is very thick
granular, almost opake, indistinctly cellular. A strong green
nerve runs to the point : this also is persistent, when the parenchyma of the leaves is consumed by water, and these remains of the leaves appear as merely short fibres on the lower part of the stem.

In the axillæ of the leaves we find sometimes male, sometimes female buds, in considerable numbers. The covers
of these buds are of a yellowish red colour. In the male
buds the covering leaflets have no nerves, and are ovate;
in the female they are lanceolate. In the male buds we find
green knobs, with intermingled sap-tubes ; in the female we
find tender red pistilla, with fine sap-tubes standing between
them. From these buds rises the fruit, but on a very short
stalk, scarcely half a line in length, whence it is commonly
overtopped by the covering leaves. The capsule is perfectly
elliptical, smooth, olive-green, and afterwards of a brownish
red colour. Where the operculum rests on the capsule, the latter is of a red colour. The former is conical, with the point
standing rather obliquely, and the twisted peristome is im-
pressed upon it. The calyptre is even, mitre-shaped, and
splits transversely at the base. After the operculum has fallen off, the peristome appears, which is beautifully red and
simple. It consists of thirty-two pretty long, hair-shaped,
divided, and in the dried state, twisted teeth, which at the
base are partly united with, and partly penetrate each other.
The seeds are small dirty-green spheres.

Diagnosis and Affinity.

The moss which most nearly resembles this is *Anœctangi-
um aquaticum* Hedw., which likewise grows in running water,
and has a dark green colour. But the leaves of the latter
are much longer and smaller, always falcated, and bent to one
side : the fruit-stalks are rather longer, and hence the capsules are more prominent than in our species. But especially the peristome is wanting in An. aquaticum : it likewise grows only in waters south of Austria. Our moss can

less readily be confounded with *Fontinalis antipyretica*, which grows almost in all running waters, but is distinguished by its triple direction, by its want of nerves, and by the somewhat keel-shaped nature of its leaves, which are also much larger, and more remote. The capsules are enveloped by short round scales : the peristome consists in a trellis-shaped net. The other aquatic mosses are still less similar. *Hypnum fluitans* L. is distinguished by very small leaves, separate from one another and alternating ; by long fruit-stalks, and by a double peristome ; *H. ruscifolium* Neck. by its ovate, serrated leaves, the nerves of which extend almost to the point, by its long petiolated nodding capsules ; *H. riparium*, by the fibres which every where proceed from the axillæ of the leaves ; *H. fluviatile* Hedw. almost solely by its petiolated capsules with a double peristome.

The genus *Cinclidotus*, first estabished by Palisot-Beauvois, and admitted by Hooker, stands between *Trichostomum* and *Barbula*. From the former, with which this moss used to be classed, it is distinguished by having the teeth of the peristome united below, and twisted above. *Barbula* is distinguished by this circumstance, namely, that the twisted cilia of the peristome are tender, and must be considered merely as prolongations of the internal membrane of the capsule. The calyptre, too, is laterally divided.

Synonymes and Figures.

Fontinalis minor lucens, *J. Bauh. Hist.* 3. 770. Foliis triangularibus minus complicatis, *Rai Syn.* p. 79. Triangularis minor carinata, *Dill. Hist. Musc.* p. 257. t. 33. f. 2.
Muscus aquaticus frutescens pinnatus, *Moris.* 3. p. 626. s. 15. t. 6. f. 32.
Muscus squamosus, foliis acutissimis, in aquis nascens, *Tourn. Inst.* 554.
Fontinalis minor, *Linn. Sp. Pl.* 1571. *Ed. Reich.* 4. 452. *Gunner, Fl. Norv.* n. 969. t. 3. f. 2. *Eng. Bot.* 557.
Hypnum, *Hall. Stirp. Helv.* n. 1795.
Hypnum nigricans, *Vill. Delph.* 3. 905.
Trichostomum fontinalioides, *Hedw. Stirp. Crypt.* 3. t. 14.

Roth. Germ. 3. 195. *Smith, Fl. Brit.* 3. 1248. *Turn. Musc. Hibern.* 41. *Hedw. Sp. Pl.* 114. *Bridel. Muscol.* 2. 133. *Schwägrich. Suppl.* 1. 160.
Hypnum fontinalioides, *Hofm. Germ.* 2. 79.
Fontinalis alpina, *Dicks. Fasc.* 3. p. 2. t. 4. f. 1.
Cinclidotus fontinalioides, *Palis. Beauv. Aëthelog.* p. 28. 52.
Hook. Mus. Brit. p. 29. t. 11.

Geographical Distribution.

Norway and the north of Scotland seem to be the most northern regions where this moss is found. It extends, so far as I know, only into the south of Germany. But it is not found either in Hungary or in the more southern countries.

IV. *Musci hepatici.*

39.

Jungermannia trilobata, *L.*

This beautiful moss grows, early in spring, on stiff clay soils, in our woods. It climbs on the roots and stems of our forest-trees. The stem, which is about five inches high, is dichotomous, branched, every where set with leaves, attaches itself by its tendrils, which spring at considerable distances, and are covered by small scales, to all objects. The leaves are green, but are generally inclined to yellow, stand thick together in two lines, nearly opposite to one another, and horizontal. They are oblong, almost quadrangular, embrace the stem or branches with their base, have a thick, granular, cellular texture, no nerves, and are provided at their broad extremity with three, more rarely with four distinct teeth ; in other respects they are entire. On the upper surface they are somewhat convex, on the lower a little concave. On the lower surface of the stem are found very small *amphigastriæ*, which are also nearly square, with three larger doubly dentated teeth, and with their margin somewhat bent. The

membranaceous, cylindrical, deeply notched calyx springs
from the axillæ of the lower part of the stem, (by no means
from the top, as Roth maintains.) The fruit-stalk is tender,
white, pellucid, erect, smooth, about the size of a fine linen
thread. It carries a brownish-red capsule, which at first is
closed like a splendent sphere; afterwards it opens with four
valves, which stand cruciform from one another, and have
the seeds hanging by a chain-shaped appendage.

Diagnosis and Affinity.

The species most nearly related to this are *J. Flörkii* Web.
Mohr. and *Naumanni* Nees. But the former is distinguished by sending out tendrils only below the stem, by
having its leaves not so much lengthened, and rather verti-
cal, and chiefly by the size and multifarious division of the
amphigastriæ, (*Mart. Fl. Crypt. Erlang.* p. 144. t. 4. f. 17.)
J. Naumanni has still shorter leaves, placed still more remote from each other, and deeply divided, ciliated amphigastriæ, (*Mart.* 143. 4. f. 16.) *J. quinque-dentata* L. is very
like our species, but it wants the amphigastriæ, and the leaves
are much shorter. *J. nitida* (*convexa* Thunb.) is distinguishes by its uniform, quadri-partite amphigastriæ, the laciniæ of which are subulate. There is a smaller variety,
which passes into that of Roth, and the amphigastriæ of which
are merely crenated: This is *J. tricrenata* Wahlenb. *Car-
path.* p. 364.

Synonymes and Figures.

Muscoides terrestre repens, ex obscuro virescens, *Michel.
Nov. Gen.* p. 10. t. 6. f. 2.
Lichenastrum pinnulis obtuse trifidis, *Dillen. Hist. Musc.*
p. 493, t. 71. f. 22. A. B. (*Encl. Syn. Michel.*)
? L. multifidum majus, ab extremitate florens, *Dill. Hist.
Musc.* p. 494. t. 71. f. 23.
Jungermannia alpina nigricans major, *Rupp. Ien.* 404.
Jungermannia, *Hall. Stirp.* n. 1866.
J. trilobata, *Linn. Sp. Pl.* ed. *Reich.* 4. p. 507. *Roth. Germ.*

3. 396. *Engl. Bot.* 2232. *Web. Prodr. Hepat.* p. 42.
Mart. Fl. Crypt. Erlang p. 141. t. 3. f. 14.
J. radicans, *Hofm. Germ.* 2. 87.
J. stolonifera, *Sw. Fl. Ind. Occid.* 3. 1862.

Geographical Distribution.

This plant is dispersed throughout the temperate zone of
the whole northern hemisphere. It is found in the south of
Sweden, in Norway, and Scotland, as well as in France,
Italy, and Germany. I have also received it from New
York. As *J. stolonifera* Swartz, and *J. tricrenata* Bridol,
brought by Bory S. Vincent, are the same with this plant,
it grows also between the tropics, at least in Jamaica, and
Mascaren's Island.

V. *Lichens.*

40.

Lecanora saxicola, *Ach.*

This lichen grows abundantly on field-stones, on porphyry
and sandstone rocks. Its thallus is thick accumbent, scaly,
wrinkled, dissimilar, often as it were broken, and of a dirty
palegreen, or yellowish-green colour, whitish, and smooth beneath, but formed into radiated lobes on the margin. The
whole lichen has often a diameter of several inches, and in
its perfect state, on level stones, it is circular. The pale gem-
miferous dust exudes in spring, on the surface. The apparent fruit is flat, round, without stalks, and pressed together,
of a dirty, reddish-yellow colour, surrounded by a pale margin resembling a thallus, and of the size of lentil or mustard
seed. It stands in closely crowded heaps, commonly in the
centre of the thallus, is frequently irregular, without the
margin resembling a thallus, and contains opake grains in
fine tubes.

Diagnosis and Affinity.

This species has a great resemblance to *Lecanora stra-minea* Ach., (Tab. II. Fig. 3). But, in this last named specics, the thallus is straw-coloured. The lobes are quite linear on the margin, roundish when cut through: the central lobes are also as it were inflated. The colour of the apparent fruit is darker, and the margin, which resembles a thallus, is as it were swollen. *Lec. versicolor* Ach. also resembles it : but this species grows almost exclusively on calcareous rocks. The thallus is whitish on the margin, of a a dirty-green in the centre. The apparent fruit, when young, is of a flesh colour, and has a white margin : afterwards it is of a dark dirty red, and the margin vanishes. *Lec. crassa* Ach. corresponds, indeed, in the colour of the thallus, and of the apparent fruit ; but the structure is quite different, for it consists of single, cartilaginous, crenated, undulated, bent lobes, which lie like scales on one another : the lower surface, too, is brown, and the apparent fruit stands dispersed.

Synonymes and Figures.

? Lichen pulmonarius saxatilis farinaceus glauco-virescens, *Michel. Nov. Gen.* p. 34. t. 51. f. 4.

Lichen saxicola, *Pollich. Palat.* n. 1098. *Engl. Bot.* 1695. *Achar. Lichenogr. Suec. Prodr.* p. 104. *Wahlenb. Fl. Lappon.* p. 415.

L. muralis, *Schreb. Spec. Fl. Germ.* p. 130. *Wither. Arrang.* 4. 31.

L. ochroleucus, *Wulff.* in *Jacqu. Coll.* 2. t. 13. f. 4. a.

Psora muralis, *Hofm. Pl. Lichen.* t. 16. f. 1. (malè.)

Parmelia saxicola, *Ach. Method.* p. 191. *Mart. Fl. Crypt. Erlang.* p. 215.

Lecanora saxicola, *Ach. Lichenogr.* p. 431. *Syn.* p. 180.

Geographical Distribution.

Although this lichen is dispersed through Europe, from Lapland to Italy, and from the Pyrenees to Tauris, no traces of it are to be found in more distant countries.

VI. *Algæ.*

41.

Scythosiphon intestinalis, *Lyngb.*

In summer, there appear upon our running and stagnant waters, yellowish-green, membranaceous, spongy tubes, which frequently cover a large surface of the water. When young they are filiform, and of a bright green. These threads are firmly attached to stones with a shield-shaped expansion, and ascend as tubes, which gradually become wider, but are contracted in particular places, without exhibiting any partitions, and swim on the water, where they are then filled with water, and throw up air-bubbles. When in their perfect state, they resemble the intestines, and are about the thickness of a finger : towards harvest they always become discoloured, more yellow, and finally dissolve into slimy matter. When we examine the fine sides of the tubes with a microscope, we observe fine grains collected together in fours.

Diagnosis and Affinity.

The species which stands nearest to this is *Sc. compressus* Lyngb., but this latter plant grows only in salt water, is branched and compressed, and is not so thick as *Sc. intestinalis.* Formerly both of these were arranged under *Ulva*; but Roth remarked (Catal. 1. 158.), that the definition of *Ulva* includes the level, expanded, but not tubular nature of the frond. Hence he classed the *Ulva intestinalis* with the Confervæ; but the latter has always jointed tubes, and these furnished with partitions. Hence Lyngbye very properly formed the genus *Scytosiphon*, the character of which consists in its membranaceous, uninterrupted tubes, the seeds of which contain the granular germs.

Synonymes and Figures.

Cava, *Imperat. Hist. Nat.* p. 858.
Fucus cavus, *C. Bauh. Pin. J. Bauh. Hist.* 3. 791.
Lactuca marina tubulosa, *Rai Hist.* 1. p. 77.
Folliculus marinus, *Lösel, Fl. Pruss.* p. 75.
Conferva marina tubulosa, *Dillen. Giess. App.* p. 16.
Conferva latifolia flavescens et tubulosa major, *Buxb. Hal.* p. 79.
Fucus tubulosus intestinorum forma, *Tourn. Inst.* p. 568. *Buxb. Cent.* 5. t. 13. f. 1. (malè).
Ulva marina tubulosa, *Rai Syn.* p. 62.
Ulva tubulosa simplex, *Linn. Lapp.* p. 348.
Tremella marina tubulosa, *Dill. Hist. Musc.* p. 47. t. 9. f. 7.
Ulva tubulis cylindricis, *Hall. Stirp.* n. 2128.
Ulva intestinalis, *Linn. Fl. Suec.* n. 1154. *Sp. Pl. ed. Reich.* 4. 583. *Lightf. Scot.* p. 368. *Huds. Angl.* 568. *Wither. Arrang.* 4. 141. *De Cand. Fl. Franç.* 2. p. 8. *Lamour. Thalass.* p. 65. *Agardh. Syn.* p. 45.
Conferva intestinalis, *Roth, Catal.* 1. 154. *Wulff. Crypt. Aqu.* p. 13. *Schumach. Suelland.* 2. p. 103.
Scytosiphon intestinalis, *Lyngb. Hydroph.* p. 67.

Geographical Distribution.

This alga is found in all waters, from the polar circle to the tropic, in the northern hemisphere: whether it also grows in the southern hemisphere I know not.

VII. *Myeolomyci.*

42.

Ceratostoma fimbriatum, *Fries.*

On the leaves of beeches and hazel bushes, we observe dark spots during summer,—results of the decaying vegetation of the leaf, and of another kind of vegetation which has begun, but which is often interrupted in the germ. In this last

3

case it remains a *Xyloma*, (Anl. ii. 17.) Otherwise small
spherules, of a black colour, and scarcely distinguishable
by the naked eye, arise during harvest on these spots. It
sometimes happens, especially with regard to hazel leaves,
that single spherules project above the common layer. These
pass into stiff, straight rostella, thick above, about a line in
length, and of the fineness of a hair, between which and the
spherule, a white, membranaceous rim, resembling a ruffle,
stands in a circular form. This rostellum is not the peristome,
which in the Sphæria supplies the place of the operculum of
the Mosses, but it is a continuation of the perithecium ;
(*Fries. Obs. Myc.* 2. 318, 319.) The spherule contains the
germ-sacks, which are club-shaped, nearly pellucid bodies,
filled with eight fine grains. But what purpose does the
white rim of the rostellum serve ? Batsch thought that it
was the reflex interior cellular membrane, which opinion I do
not assent to. But Rebentisch's idea is more probable, namely, that it is the remainder of the epidermis of the leaf, and
disappears in spring.

This species cannot be confounded with others, because
the above-mentioned rim is not found in any other species.
In other respects it is nearly related to *C. pulchellum* and
cornutum.

Synonymes and Figures.

Sphæria spiculosa, *Batsch, El. Fung.* 1. p. 273. t. 30. f. 182.
Sph. Coryli, *Ib.* 2. p. 131. t. 32. f. 231.
Sph. Carpini, *Hofm. Veg. Crypt.* 1. t. 1. f. 1.   *Timm. Fl.
Megap.* p. 279.
Sph. fimbriata Pers.,. *Obs. Myc.* 1. p. 70.   *Syn.* p. 36.   *Alb.
et Schwein. Fung. Niesc.* p. 17.   *Rebentish, Neom.* p. 329.
Schultz, Fl. Starg. p. 425.   *Mart. Fl. Crypt. Erlang.*
p. 479.
Ceratostoma fimbriatum, *Fries, Obs. Myc.* 2. p. 340.

VIII. *Proper Fungi.*

43.

Spathularia flavida, *Pers.*

Leistenschwamm, Nees.

In our fir woods, seldom among hard woods, this fungus is observed, in the end of summer, and in harvest. It grows on the foliage of the fir, and on other fallen leaves, commonly surrounded by moss. The whole fungus is commonly only two inches, at most a little finger in length. The stem has its base thick, like a tuber : it is of a straw-yellow colour, smooth, about the size of a writing quill. Its length is somewhat more than an inch. Internally it is sometimes hollow when it is old : commonly it consists of a double substance, an exterior fibrous part, and an interior cellular. It carries a compressed pileus, of a yellow or reddish-yellow colour, which runs downwards on the side of the stem, and has the shape of a spade ; at most an inch long and broad, it is smooth all round, but frequently crenated or notched on the margin. It consists of two united hymenia, which internally contain a fine, white cellular texture. At a more advanced age these hymenia separate from each other : the pileus appears then to be inflated, and internally full of soft fibres. From the stem, branchy wrinkles are extended through the cap, which have been regarded by many as veins. In the hymenium we observe, by the microscope, fine pellucid, club-shaped sporidia, with intervening sap-tubes. In the former lie five double-ringed sporæ, exactly in the same manner as in *Geoglossum viride* (Tab. I. Fig. 34.), with which this fungus is found in company. During warm sunshine these sporæ give out dust like a fine shining cloud, and the plant becomes, by this means, like *Peziza*, a bladder fungus; (*Fungus Utrinus, Nees*, p. 243.)

Diagnosis and Affinity.

This genus is most nearly related to the genus *Helvella.* The two hymenia, which are here united, are separated in *Helvella,* and hang down in the form of a mitre. The colour of the pileus is also commonly darker. *Geoglossum* forms a peculiar club, which is distinctly separated from the stem.

Synonymcs and Figures.

Elvella clavata, *Schäft. Fung.* t. 149.
Clavaria Spathula, *Flor. Dan.* 658.
Cl. spathulata, *Schmid.* ic. p. 196. t. 50. f. 1. (alter).
Helvella spathulata, *Afzel. Stockh. Handl.* 1783. p. 302. *Sowerb. Fung.* t. 35.
Helvella farctoria, *Bolt. Fung. ed. Willd.* 3. p. 10. t. 97.
Spathularia flavida, *Pers. Dispos. Meth. Fung.* p. 36. *Comment. de Fung.* clavæform. p. 34—36. *Syn.* p. 601. *Nees, Syst.* p. 174. t. 17. f. 156. (Sp. rufa eadem, icones malæ).

IX.   *Gastromyci.*

44.

Craterium pyriforme, *Ditmar.*

(Tab. I.   Fig. 25—28.)

In harvest, this fungus is observed on the rind of birch trees in moist places. It is scarcely a line in size, and of a brownish-yellow colour. On a short stalk stands the pear or flask shaped peridium, contracted below the mouth. The mouth is covered by a whitish round operculum, which opens and shews the sporæ, with intermingled tufts of hairs as the contents of the peridium.

Ditmar, in den Pilzen Deutschl. b. 1. n. 10.

X. *Nematomyci.*

45.

Botrytis polyspora, *Link.*

(Tab. I. Fig. 31.)

This fungus appears in harvest, a line in height, in thick plots, on dry twigs. The branchy flocci are furnished below with dissepimenta, and are of a greyish-green colour. The spherical, olive-green sporæ are placed in heaps on the branches, especially towards their extremities.

Link, in Berl. Mag. 3. s. 14.
Ditmar, in den Pilzen Deutschl. n. 35.

XI. *Coniomyci.*

46.

Fusidium gryseum, *Link.*

(Tab. I. Fig. 32.)

On dry beech leaves we observe this fungus, during harvest, in the form of heaps of whitish-grey spots, which, being examined by the microscope, are seen to consist of fusiform, very fine sporæ.

Link, in Berl. Mag. 3. s. 8.
Ditmar, in den Pilzen Dentschl.

INDICES.

INDICES.

INDEX

LATIN TERMS.

The Numbers indicate, not the Page, but the Section.

(479)

INDEX

TO THE

PRINCIPAL DEFINITIONS AND NAMES.

The Figures indicate the Sections.

A

ABORTION of organs, 60, 177
Acharius, 467
Acids of plants, 363,—in the soil, 374
Acotyledonous, 386
Adanson, 164, 456
Aerial smoke, what it is, 417
Agamia, 139
Agardh, 467
Agrumæ, 115
Aiton, 468
Albertini, 467
Albumen in seeds, 121, 383,—in plants, 353
Alburnum of trees, 67,—its structure, 298,—debility of, 420
Alexandria, botany studied in its schools, 432
Alkalies, 358
Allioni, 465
Alpini, 442
Alterations of organs, 184
Ammonia in plants, 358
Analogy, its use, 179
Andrews, 467
Anguillara, 441
Anomalœcia, 139

B

Anthers, 107,—their structure, 334
Aphides, hurtful to plants, 426
Appendage of parts, 28
Apple, 115
Arabians, 436
Aristotle, 431
Armour of plants, 80
Arsenic, its effects on plants, 375
Ashes, in what way they improve the soil, 347
Aublet, 458
Auriculæ of plants, 78
Axilla, 76
Axis of parts, 28
Azote, whence derived to plants, 349
Azotic gas, inhaled from flowers, 329

B

Balbis, 465
Barrelier, 450
Base, 28
Bastard plants, 381
Batsch, 463, 467
C. *Bauhin*, 211, 443.
J. *Bauhin*, 443
Baumgarter, 465

FINIS.

PLATE I.

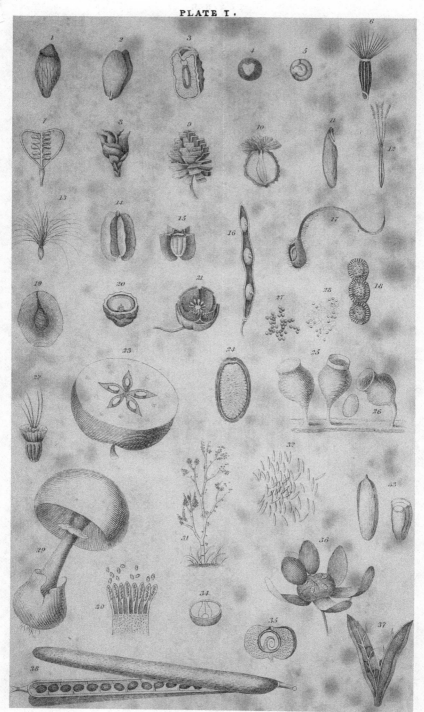

W H Lizars Sculpt

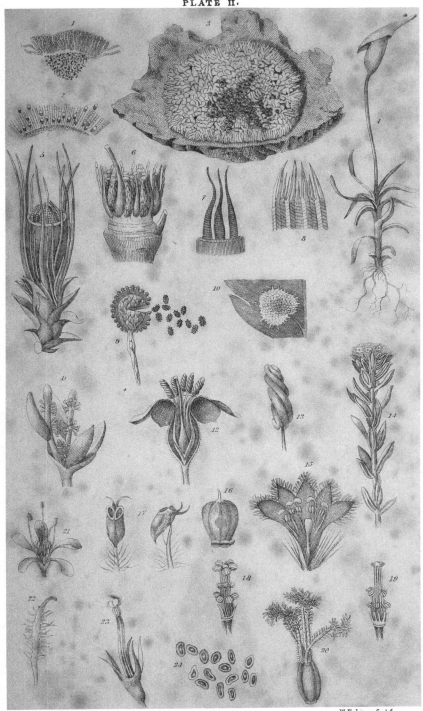

PLATE II.

PLATE III.

PLATE IV.

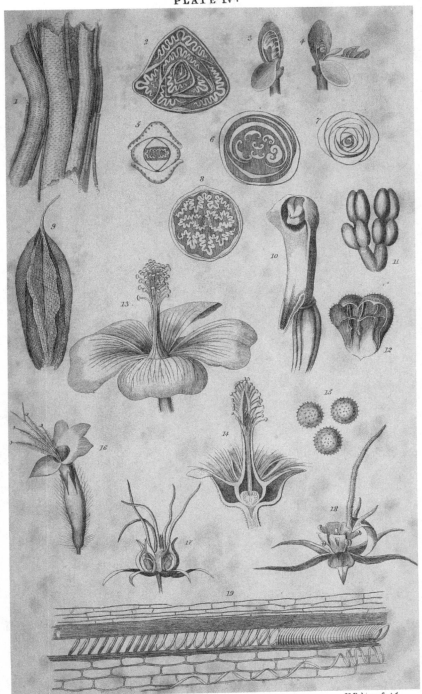

W H Lizars Sculpt

PLATE V.

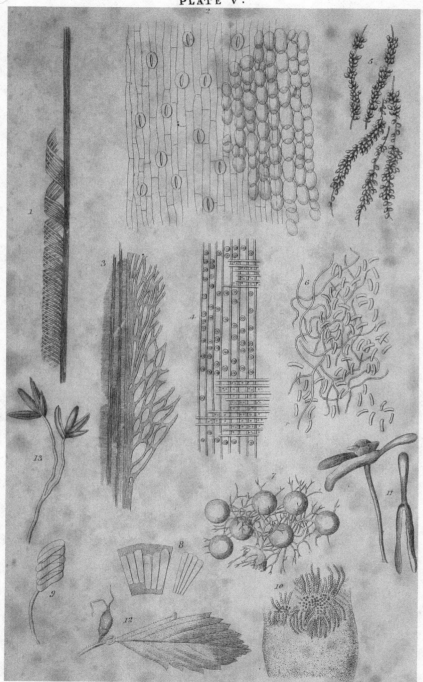

PLATE VI.

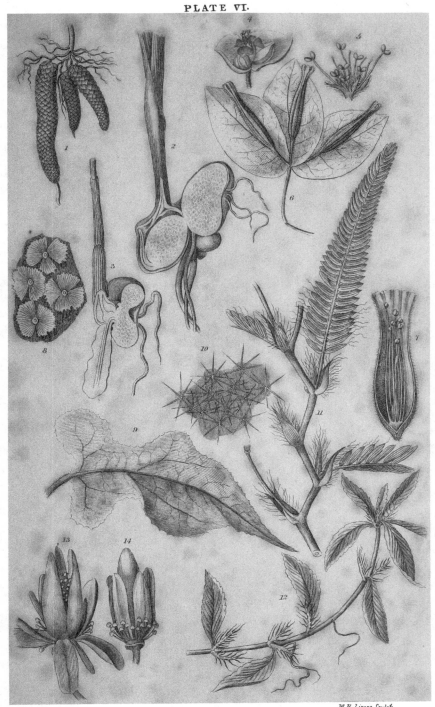

PLATE VII

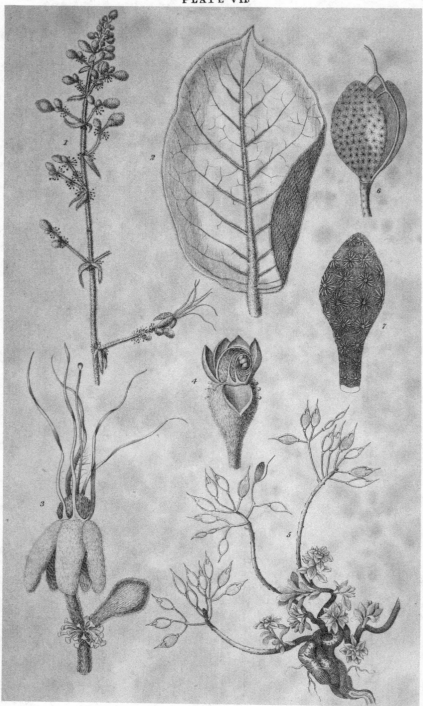

W H Lizars Sculp.t

PLATE VIII.

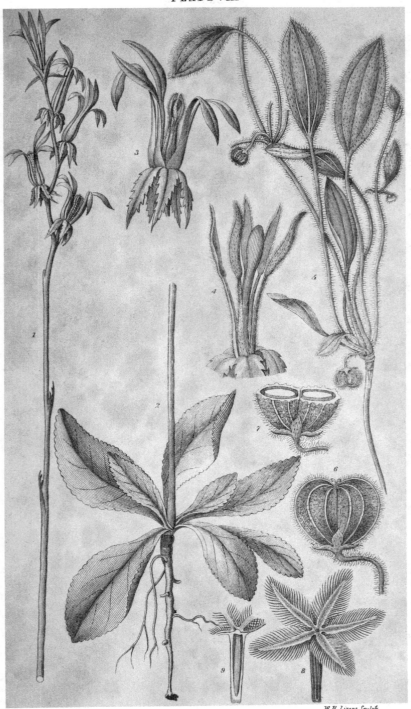

Printed in the United States
By Bookmasters